Advances in
Nuclear Dynamics 3

Advances in
Nuclear Dynamics 3

Wolfgang Bauer

Michigan State University
East Lansing, Michigan

and

Alice Mignerey

University of Maryland
College Park, Maryland

Springer Science+Business Media, LLC

Library of Congress Cataloging-in-Publication Data

Advances in nuclear dynamics 3 / edited by Wolfgang Bauer and Alice
Mignerey.
        p.   cm.
    "Proceedings of the 13th Winter Workshop on Nuclear Dynamics, held
February 1-8, 1997, in Marathon, Florida Keys"--T.p. verso.
    Includes bibliographical references and index.
    ISBN 978-1-4613-7224-0   ISBN 978-1-4615-4905-5 (eBook)
    DOI 10.1007/978-1-4615-4905-5
    1. Heavy ion collisions--Congresses.   I. Bauer, W. (Wolfgang),
1959-   . II. Mignerey, A. (Alice) III. Winter Workshop on Nuclear
Dynamics (13th : 1997 : Marathon, Fla.)
QC794.8.H4A43 1997
539.7'232--dc21
                                                            97-40605
                                                            CIP

Proceedings of the 13th Winter Workshop on Nuclear Dynamics,
held February 1 – 8, 1997, in Marathon, Florida Keys

Preface

The 13th Winter Workshop on Nuclear Dynamics is the latest installment in a series of workshops that was started in 1978. This series has grown into a tradition, bringing together experimental and theoretical expertise from all areas of the study of nuclear dynamics.

As always, the organizers had placed emphasis on the important aspect of cross-fertilization between the different energy regimes and ways of viewing a collision of energetic nuclei. This emphasis is reflected in the broad range of topics covered in these proceedings.

Phase transitions in nuclear collisions received most of the attention during this workshop, as indicated by the number of contributions on this subject. Many of the questions in connection with these topics remain not settled and will have a huge impact on the physics that can be extracted from experiments at the future Relativistic Heavy Ion Collider.

While the experimental program at the AGS is winding down, the NSCL upgrade and RHIC promise a bright future for our field in the USA. In Europe and Asia, major new facilities are under construction as well. The excitement and anticipation in connection with these new opportunities are reflected in these proceedings as well.

Wolfgang Bauer
Michigan State University

Alice Mignerey
University of Maryland

PREVIOUS WORKSHOPS

The following table contains a list of the dates and locations of the previous Winter Workshops on Nuclear Dynamics as well as the members of the organizing committees. The chairpersons of the conferences are underlined.

1. Granlibakken, California, 17-21 March 1980
   W.D. Myers, J. Randrup, G.D. Westfall

2. Granlibakken, California, 22-26 April 1982
   W.D. Myers, J.J. Griffin, J.R. Huizenga, J.R. Nix, F. Plasil, V.E. Viola

3. Copper Mountain, Colorado, 5-9 March 1984
   W.D. Myers, C.K. Gelbke, J.J. Griffin, J.R. Huizenga, J.R. Nix, F. Plasil, V.E. Viola

4. Copper Mountain, Colorado, 24-28 February 1986
   J.J. Griffin, J.R. Huizenga, J.R. Nix, F. Plasil, J. Randrup, V.E. Viola

5. Sun Valley, Idaho, 22-26 February 1988
   J.R. Huizenga, J.I. Kapusta, J.R. Nix, J. Randrup, V.E. Viola, G.D. Westfall

6. Jackson Hole, Wyoming, 17-24 February 1990
   B.B. Back, J.R. Huizenga, J.I. Kapusta, J.R. Nix, J. Randrup, V.E. Viola, G.D. Westfall

7. Key West, Florida, 26 January - 2 February 1991
   B.B. Back, W. Bauer, J.R. Huizenga, J.I. Kapusta, J.R. Nix, J. Randrup

8. Jackson Hole, Wyoming, 18-25 January 1992
   B.B. Back, W. Bauer, J.R. Huizenga, J.I. Kapusta, J.R. Nix, J. Randrup

9. Key West, Florida, 30 January - 6 February 1993
   B.B. Back, W. Bauer, J. Harris, J.I. Kapusta, A. Mignerey, J.R. Nix, G.D. Westfall

10. Snowbird, Utah, 16-22 January 1994
    B.B. Back, W. Bauer, J. Harris, A. Mignerey, J.R. Nix, G.D. Westfall

11. Key West, Florida, 11-18 February 1995
    W. Bauer, J. Harris, A. Mignerey, S. Steadman, G.D. Westfall

12. Snowbird, Utah, 3-10 February 1996
    W. Bauer, J. Harris, A. Mignerey, S. Steadman, G.D. Westfall

13. Marathon, Florida, 1997
    W. Bauer, J. Harris, A. Mignerey, H.G. Ritter, E. Shuryak, S. Steadman, G.D. Westfall

14. 1998 Committee
    W. Bauer, J. Harris, A. Mignerey, H.G. Ritter, E. Shuryak, G.D. Westfall

CONTENTS

x

ELLIPTICAL FLOW: A PROBE OF THE PRESSURE IN ULTRARELATIVISTIC NUCLEUS-NUCLEUS COLLISIONS

Heinz Sorge [1]

[1] Physics Department
State University of New York at Stony Brook
New York, NY 11794-3800

INTRODUCTION

Much experimental and theoretical effort has been devoted in the past to study the transition between quark matter and hadronic gas in ultrarelativistic nucleus-nucleus collisions (≥ 10 AGeV). How does the equation of state (EoS) reflect the transition from the hadronic matter to the quark-gluon-plasma (QGP)? A first-order phase transition is generically associated with the presence of a 'softest point' in the EoS. The pressure increases less than the energy density close to the critical temperature, because only the energy density 'jumps'. Quantum chromodynamics (QCD) calculations on lattices show indeed a softening of the EoS at temperatures in the transition region between pion gas and QGP. It is not yet settled whether QCD at finite temperature exhibits a true phase transition or just a rapid cross-over between the two phases. On the hadronic side, the softening is expected from the many resonance degrees of freedom which are getting populated at temperatures ≥ 140 MeV (resonance matter). The density of hadronic resonance states increases nearly exponentially with mass – a feature of QCD which is linked to its *confinement* property. Amuzingly, already at temperatures around 160 MeV hadronic matter becomes so dense due to the resonance excitations that hadrons tend to overlap and quarks may become deconfined.

Ultrarelativistic heavy ion collisions are a sensitive probe of the EoS in the transition region. The expansion dynamics of equilibrating matter is governed by the EoS, because pressure gradients drive collective flows. Flow analysis can thus be utilized to study the transient pressure in nuclear collisions. However, the dynamics of these collisions reflects other factors than the EoS at finite temperature as well. Foremost, these are

- preequilibrium stage:
  it is already clear from AA collisions in the 1AGeV region that the assumption of ideal fluid dynamics – instantaneous equilibration of the interpenetrating two

nuclei – breaks down. This is more so the case with higher beam momenta. Furthermore, even less time is available for the nuclei at higher beam energies to stop each other due to Lorentz contraction of the in-going nuclei.

- finite baryon density:
it was predicted and subsequently measured that a considerable amount of nucleons are stopped in today's fixed target experiments. The influence of the resulting large baryon density on bulk properties like the EoS must be taken into account.

- postequilibrium stage:
In the final stages matter becomes too dilute that equilibrium can be maintained. At first, chemical equilibrium is lost (if achieved at all), lateron kinetic equilibrium [1].

The strong role which non-equilibrium effects play in the dynamical evolution calls for transport theory as the appropriate theoretical tool to understand the physics of AA collisions quantitatively. On the other side, hydrodynamic studies have proven very useful to gain qualitative insight how matter expansion depends on the EoS.

What have we learnt so far about the EoS from ultrarelativistic AA collisions? Experiments in the 'stopping region' (including the beam energies utilized at the BNL-AGS from 10-15 AGeV) have revealed that baryon-rich matter behaves rather 'stiff'. For instance, the directed flow or dipole component of the transverse flow has been measured by the experimental group E877. They found maximum p_x values above 100 MeV in non-central Au(11.6AGeV) on Au collisions [2], pretty much the same values as in the SIS-BEVALAC experiments at lower energies. One can test the presence of pre-equilibrium dynamics by doing calculations which assume hydrodynamics from the beginning, i.e. the touching point of the two impinging nuclei. Only poor agreement with experimental data for longitudinal momentum spectra can be found at AGS energies, independent on the EoS [3]. It is inherent in this kind of hydrodynamic studies that the instantaneous creation of transverse pressure is compensated by choosing a too soft EoS to fit transverse spectra and directed flow. Interestingly, even these hydrodynamical calculations give nontrivial lower bounds on the softness of the EoS. The recent directed flow measurements done by E877 and the EOS collaboration seem to rule out ultrasoft EoSs of the type employed in [4]. Alternatively, transport calculations can be employed to extract information about the transient pressure in nuclear collisions. Since ideal-gas type cascading does not produce sufficient pressure at energies 1-15 AGeV, transport calculations generally invoke additional *repulsion* between the baryons to account for the magnitude of the directed flow. This repulsion may result from a complicated interplay between momentum and density dependence of nuclear forces. In particular, the directed flow which is produced very early receives a strong contribution from the pre-equilibrium stage during which target and projectile are 'running' into each other. The situation is simplified if we suppress the role of baryons by going to higher beam energies. Currently, Pb(160AGeV) on Pb collisions are being experimentally studied at the CERN-SPS. Analysis of experimental data reveals that the transverse momenta of hadrons level off in heavy ion collisions between beam energies of 10 AGeV and 200 AGeV [5]. The observation means in turn that the underlying collective motion does not increase sizably. On the other side, the total energy which is dumped into the central rapidity region is approximately 70 percent larger at the higher beam energy. It looks as if the EoS driving the expansion at the high energies is much softer, possibly because the repulsion between the baryons becomes less important.

In this contribution we will extract the hydrodynamic variables like pressure and energy densities from transport calculations (based on the RQMD model [6]) and link their time evolution to final observables. In particular, I will discuss how the measurement of elliptical flow can further constrain the EoS in the quark-hadron transition region.

PRESSURE IN BARYON-RICH MATTER

One way to get further information about the transient pressure is to combine information from the radial and elliptical flow (monopole, respectively quadrupole component of the transverse flow) in AA collisions [7]. What is the physics of elliptical flow in the stopping region? At low beam energy, matter at central rapidity escapes preferentially orthogonal to the reaction plane. The spectator nucleons block the path of participant hadrons which try to escape from the collision zone. This is the experimentally observed squeeze-out effect. At ultrahigh energies, particles produced in the central region do not interact with spectators, since the crossing time of projectile and target shrinks with the Lorentz factor gamma. If collective motion in transverse directions develops the almond-shaped geometry of the collision region clearly favors the directions (anti-) parallel to the impact parameter vector [8]. An interesting situation emerges for collision energies between the low and the ultrahigh energies. Taking collisions of equal mass nuclei moving nearly with the speed of light, the passage time of projectile and target spectators is approximately given by $2R_A/(\gamma c)$ (around 5 fm/c for Au(11.6AGeV) on Au). Such time scale neither covers the whole reaction time nor becomes irrelevant at these intermediate energies. As a consequence, the centrally produced matter is initially squeezed-out orthogonal to the reaction plane. Afterwards, however, the geometry of the central region favors in-plane flow. The orientation of the final azimuthal asymmetry in particle, momentum and energy flow is chiefly determined by the relative strength of the pressures in the initial compression and later expansion stage. The virtue of elliptical flow is that it helps to distinguish scenarios which lead to the same radial flow. For instance, a scenario in which softening from a strong 1st order phase transition is compensated by repulsion at a later stage favors in-plane emission. In contrast, if the pressure develops already very early squeeze-out should be dominating up to larger beam energies.

In the following, the question will be addressed how sensitive the azimuthal asymmetry of energy flow is to the pressure in the compression and expansion stages. For this purpose, I have analysed the evolution of the pressure in a particular reaction, Au(11.7AGeV) on Au at an impact parameter of 6 fm (calculated in the 'mean-field' mode of RQMD). Applying the virial theorem, the pressure from the potentials has been added to the kinetic pressure. The time evolution of (transverse) pressure, local energy and baryon density are displayed in Fig. 1. It becomes apparent from this figure that the time of maximum compression coincides with the time at which the spectators are about to leave the collision zone. The pressure in the expansion stage is somewhat larger than in the compression stage. This reflects the presence of a strong preequilibrium component. In Fig. 1 the contribution from the kinetic part of the pressure is also shown (open symbols). Roughly, the kinetic and the potential part contribute equally at the time of maximum compression.

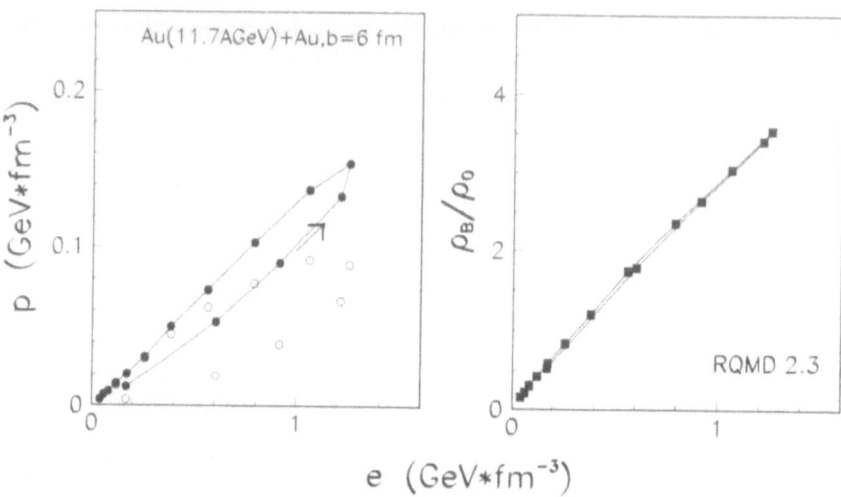

Figure 1. Time evolution of transverse pressure p, local maximum energy and baryon density (e, respectively ρ_B/ρ_0) in the collision center for the reaction Au(11.7AGeV) on Au at impact parameter 6 fm. The results were obtained using the RQMD model (version 2.3) in mean-field mode. Time direction is indicated by an arrow. The contribution from the kinetic part of the pressure is also displayed (open circles).

The azimuthal asymmetry in the energy flow can be quantified by defining the following variable:

$$E_{dir} = \sum_M E(M) \cdot \text{sgn}(\phi) \tag{1}$$

The summation over M includes hadrons only which are close to center-of-mass rapidity. ϕ is defined as the angle of a hadron's momentum with respect to the impact parameter vector. $\text{sgn}(\phi)$ is defined to be $+1$ in the cones with opening angle of $45°$ around $\phi= 0$ and $180°$, -1 elsewhere. Fig. 2 displays the time evolution of E_{dir}. In the time span up to maximum compression, E_{dir} acquires negative values (squeeze-out), because the pressure mostly from the repulsive potentials pushes the hadrons against the confining spectator material in the reaction plane and into the vacuum orthogonal to it. After the spectators are gone (t>5 fm) E_{dir} gets positive contributions with increasing time. Finally, the in-plane flow dominates for this reaction. Therefore the major energy flow axis is parallel to $\vec{b}$. For comparison, Fig. 2 shows the evolution of the same quantity, but calculated in the RQMD cascade mode. Due to the preequilibrium effect, the effective transverse equation of state is ultrasoft. There is no visible squeeze-out present at the early times. The pressure at later times is smaller than in the mean-field mode. However, the final azimuthal asymmetry expressed by its E_{dir} value is approximately 60 percent larger, because the initial squeeze-out is absent. In addition, Fig. 2 shows the evolution of the average nucleon transverse momentum (in the same rapidity window as for E_{dir}). The mean fields push the nucleons to larger p_{tr} values which is favored in comparison to measurements. However, experimental data for transverse spectra do not provide a clue whether the observed larger stiffness in comparison to the pure cascade result results from additional pressure in the early or in the late stage. Analysis of the azimuthal asymmetries may tell the difference.

4

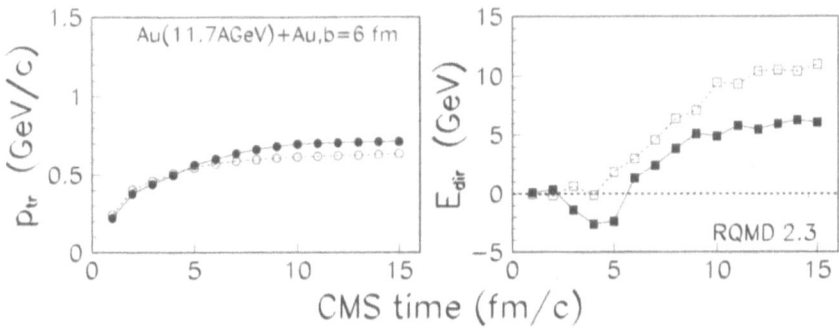

Figure 2. Time evolution of the average transverse momentum of nucleons p_{tr} (left) and E_{dir} (right). Closed (open) symbols refer to the calculation in RQMD mean-field (cascade) mode. A rapidity cut ± 0.7 around center-of-mass rapidity was imposed.

E877 has analysed the transverse energy and charged particle multiplicity distributions using the technique of Fourier decomposition [2]. A nonvanishing quadrupole moment was found by E877 with preferential emission (anti-)parallel to the impact parameter. We take the *qualitative* agreement with preliminary E877 data which show the major flow (anti-) parallel to the impact parameter as an encouraging sign that the model can be useful to extract the early produced pressure from experimental data.

The power of elliptical flow analysis becomes apparent if it is combined with measurement of the average flow in transverse directions, the radial flow. Early and late pressure contribute with opposite sign to the elliptic deformation but with equal sign to the average transverse flow. This beautiful feature makes it possible to gain separate information on the early and late pressure. The sensitivity of elliptic flow to the early pressure opens up possibilities for a rich research program. For instance, the beam energy at which final squeeze-out disappears and is replaced by domination of in-plane flow will strongly depend on the early pressure.

SOFTNESS: PRE-EQUILIBRIUM OR 1ST ORDER PHASE TRANSITION?

Matter in AA collisions at 160-200 AGeV seems to expand rather softly. The formation of a mixed phase seems a natural candidate for an explanation. The preequilibrium stage of nucleus-nucleus collisions is characterized by strong damping of transverse motion as well. How can one distinguish mechanisms which soften the pressure and therefore the expansion in the central rapidity region of ultrarelativistic nucleus-nucleus collisions? As suggested in [9] the time orderings of when the expansion of matter becomes soft differ in the two cases. Preequilibrium softening is bound to happen initially. In contrast, the softness of the mixed phase evolution may be preceded by a fast evolution in the quark-gluon stage. A very hot QGP expands presumably according to a hard equation of state close to $p = e/3$.

Using RQMD the softening of transverse expansion caused by preequilibrium effects is estimated in the following. The RQMD calculation provides an example of an expansion dynamics *without* a first order phase transition. I will demonstrate that elliptical flow distinguishes the RQMD-type preequilibrium scenario from hydrodynamical evolution based on a strong first order phase transition. How do the ingredients of the RQMD model [6] affect the transverse expansion? RQMD is based on string and resonance excitations in the primary collisions of nucleons from target and projectile. Overlapping color strings fuse into ropes, flux-tubes with sources of larger than fundamental color charges. Color strings and ropes model the prehadronic stage in 1+1 dimensions. By construction, they do not exert any transverse pressure on their environment. Transverse expansion in the central region starts only after hadronization, because the initially generated transverse momenta from string and rope fragmentation are oriented randomly. The hadronic expansion stage in RQMD starts off far from kinetic equilibrium. One reason is that the nuclear thickness sets a minimum time for nonequilibrium due to the finite crossing time of projectile and target. The crossing time of the two Pb nuclei in an observer frame with CMS rapidity is 1.4 fm/c at 160 AGeV and therefore even larger than the hadronization time from string and rope fragmentation. Hadrons in the same space-time area may have very different rapidity initially – just because they are produced in elementary collisions with different locations along the beam axis. The spreading of longitudinal rapidities from the dispersion of collision points can be easily estimated in the Bjorken scenario, roughly 1.1 units for Pb(159AGeV)+Pb collisions. This dispersion comes on top of rapidity fluctuations from hadronization which have the same order of magnitude. The total dispersion of local hadron rapidities is clearly much larger than in thermal equilibrium in which the width of rapidity distributions is restricted not to exceed 0.7 units. Initially, the local momentum distributions in RQMD are therefore elongated along the beam axis. This diminishes expansion in transverse direction in comparison to the kinetic equilibrium case. The softening of the transverse expansion from preequilibrium anisotropies shows up as a reduction of the transverse pressure.

In the hadronic stage of RQMD, the fragmentation products from rope, string and resonance decays interact with each other and the original nucleons, mostly via binary collisions. These interactions drive the system towards equilibrium [1]. The equilibrium state in RQMD is an ideal gas of hadrons and resonances, up to small corrections from strings and neglecting contributions from mean-field type potentials between baryons (which have not been employed in the following calculation, anyway). The relevant quantity for hydrodynamic expansion is the pressure as a function of energy density. At relevant temperatures around 150-180 MeV the ratio p/e stays approximately around 1:6 (for μ_B=0) if the experimentally well confirmed hadronic states with masses below 2 GeV/c^2 are included.

Fig. 3 displays the time evolution of the transverse pressure and energy density in the center of Pb(160AGeV) on Pb collisions. Initially, the energy density is very large, close to 3 GeV/fm^3. However, a rather large fraction of this energy resides in the 'hidden' collective motion along the beam axis. As a consequence, the transverse pressure is considerably softened for a time interval of about 4 fm/c. Taking the prehadronic and hadronic evolution together the preequilibrium stage lasts for approximately 5 fm/c. Such large values make the preequilibrium evolution utterly relevant for the transverse expansion dynamics.

After equilibration, the ratio p/e approaches values around 1:6 which are expected for the hadronic system in kinetic and chemical equilibrium. It provides indirect evidence that chemical equilibrium must be close at this later stage. Soon afterwards, the p/e ratio drops again signaling the break-down of near-equilibrium dynamics. The

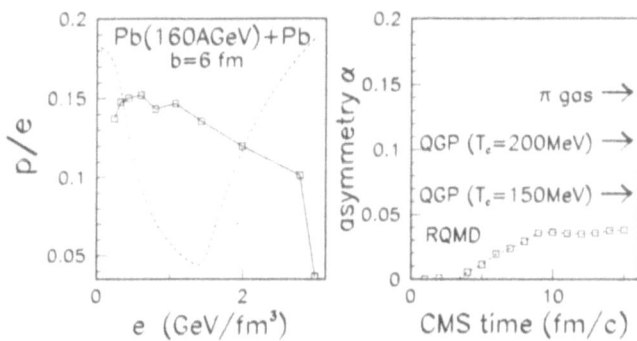

Figure 3. Time evolution of transverse pressure p and energy density e in the collision center of the system Pb(159AGeV) on Pb with impact parameter 6 fm (left side). Evolution starts 1 fm/c after initial touching of the two Pb nuclei in time intervals of 1 fm/c each (in the CMS of the two nuclei). The dashed line represents an equilibrium equation of state with 1st order phase transition at T_c=160 MeV. Time evolution of transverse momentum anisotropy parameter α calculated with RQMD (right side). Only hadrons with rapidities in the window $y_{CMS} \pm 0.7$ are included. The squares represent the RQMD values, the arrows the final α values from hydrodynamic calculations for the same system and three different EoS which have been published in the first of Refs. [8].

pressure goes down faster than the energy density, because the dilute gas in the central region is characterized by more massive constituents. One reason is complete loss of chemical equilibrium.

In order to assess the importance of preequilibrium dynamics on the transverse expansion the 'effective EoS' $p(e)$ from RQMD is compared to an equilibrium EoS with a 1st order phase transition (PT) in Fig. 3. The equilibrium EoS with critical temperature at 160 MeV was employed for the hydrodynamic studies in Ref. [10]. The comparison reveals that preequilibrium in RQMD softens its effective EoS as much as the latent heat does for the equilibrium EoS.

How can one distinguish the equilibrium scenario with strong 1st order transition from the RQMD-type preequilibrium softening of transverse expansion? The different time ordering of hard and soft expansion stage in the two cases which is visible from Fig. 3 is quite suggestive for a solution. One may look for observables other than radial flow which are also sensitive to the pressure, however, with different relative weight of early and late stage. Here the anisotropies of the transverse flow in non-central collisions are a promising candidate, because they are arguably more sensitive to early pressure. They evolve only as far as the system retains some memory of the initial anisotropy, because the anisotropic overlap zone of projectile and target nucleus is responsible for the asymmetries. Furthermore, the developing stronger in-plane flow acts against its cause, the asymmetry of the collision zone. In contrast, the late stage is weighted more heavily in the evolution of the average flow. The weight is essentially proportional to the system size (due to the pdV term in the thermodynamic approximation). Fig. 3 displays how the azimuthal asymmetry of transverse flows develops with time according to the RQMD calculation. The azimuthal asymmetry of transverse hadron momenta can be quantified by defining the dimensionless variable α via

$$\alpha = \frac{\langle p_x^2 \rangle - \langle p_y^2 \rangle}{\langle p_x^2 \rangle + \langle p_y^2 \rangle} \qquad (2)$$

where

p_x (p_y) denotes the transverse momentum component of hadrons parallel (orthogonal) to the impact parameter vector. The time evolution of α in Pb(159AGeV) on Pb collisions with b=6 fm generated from RQMD has been calculated and is plotted in Fig. 3. $\alpha(t)$ stays close to zero in the preequilibrium stage which reflects the small transverse pressure during this stage. Only after the system approaches kinetic equilibrium the flow develops a stronger asymmetry. The final α value is reached after approximately 10 fm/c. Fig. 3 exhibits also the final values of α for the same system but calculated with boost-invariant hydrodynamics [8]. Since thermal pressure which drives the expansion in these hydrodynamic calculations starts already at very early times (1 fm/c), the anisotropies become much stronger than in RQMD. The asymmetry parameter has been calculated by Ollitrault for different equations of state, of a pure π gas and with 1st order phase transition into a QGP (and critical temperature T_c either 150 or 200 MeV). Comparing the hydrodynamic and non-equilibrium RQMD values for α one finds that the π gas value is larger by a factor 3.84 and the QGP values by factors 1.74 (150 MeV) or 2.84 (200 MeV). It should be noted that the same hydrodynamic calculation with EoS as in the equilibrium limit of RQMD would give an α value between π gas and QGP value. We find indeed that the anisotropies are more sensitive to an early softening from preequilibrium than to a later phase transition.

Fortunately, important progress has also been achieved recently on the experimental side. NA49 has reported preliminary data on elliptical transverse energy flow patterns for non-central Pb(158AGeV) on Pb reactions [11]. Energy flows in neighboring pseudo-rapidity windows are clearly correlated. First comparisons to models indicate that the experimental signals may be strong enough to distinguish between different scenarios [11].

REFERENCES

1.   H. Sorge: Phys. Lett. **B 373** (1996) 16.
2.   T. Hemmick (for the E877 collaboration): Proc. of Quark Matter 96, Nucl. Phys. **A 610** (1996) 63c.
3.   N. Arbex et al.: Phys. Rev. **C55** (1997) 860.
4.   D. Rischke et al.: nucl–th/9505014 (to be published in Heavy Ion Phys.).
5.   N. Xu (for the NA44 collaboration): Proc. of Quark Matter 96, Nucl. Phys. **A 610** (1996) 175c.
6.   H. Sorge: Phys. Rev. **C52** (1995) 3291.
7.   H. Sorge: Phys. Rev. Lett. **78** (1997) 2309.
8.   J.Y. Ollitrault: Phys. Rev. **D46** (1992) 229; Phys. Rev. **D48** (1993) 1132.
9.   H. Sorge: Phys. Lett. **B** (1997) in print.
10.   C.M. Hung and E.V. Shuryak: Phys. Rev. Lett. **75** (1995) 4003.
11.   T. Wienold (for the NA49 collaboration): Proc. of Quark Matter 96, Nucl. Phys. **A 610** (1996) 76c.

A STUDY OF LOW-MASS DILEPTONS AT THE CERN SPS

Joelle Murray,[1] Wolfgang Bauer,[1] and Kevin Haglin [2]

[1]National Superconducting Cyclotron Laboratory
Michigan State University
East Lansing, MI, 48823
jmurray@nscl.msu.edu
[2]Department of Physics
Lawrence University
Appleton, WI, 54912

INTRODUCTION

Measuring and analyzing electromagnetic radiation from heavy-ion collisions represents a significant experimental challenge compared to hadronic signals owing to the relatively small cross sections. The additional information they provide certainly justifies the undertaking. Hadrons produced in the initial stages of the collision interact on average several times before leaving the reaction zone. Consequently, any information embedded in hadronic dynamics is completely masked by multiple scatterings. Dileptons are not disturbed by the hadronic environment even though they are produced at all stages of the collisions as they have long mean free paths. They are dubbed "clean probes" of the collision dynamics.

Recent results from CERN [1] have brought about a surge of activity in search of quantitative interpretation. The proton-induced reactions (p+Be and p+Au at 450 GeV) are consistent with predictions from primary particle production and subsequent radiative and/or Dalitz decays suggesting that the e^+e^- yields are fairly well understood. Yet, the heavy-ion data (S+Au at 200 GeV/n) show a significant excess as compared to the same model for meson production and electromagnetic decays. When integrated over pair invariant mass up to 1.5 GeV, the number of electron pairs exceeded the "cocktail" prediction by a factor of 5±2. It is clear that two-pion annihilation contributes in the heavy-ion reactions as fireball-like features emerge and support copious pion production [2]. Vector dominance arguments would naturally lead to extra production around the rho mass. Yet, the excess is most pronounced between the two-pion threshold and the rho mass.

The nature of the enhancement suggests several possibilities. Medium modifications resulting in a shifted rho mass could be responsible [3]. Along these lines, consequences arising from a modified pion dispersion relation have been investigated considering finite temperature effects [4] and collisions with nucleons and Δ resonances [5]. Secondary scattering of pions and other resonances has also been studied[6] focusing on the role of the a_1 through $\pi\rho \rightarrow a_1 \rightarrow \pi\ e^+e^-$. The contribution was shown to be relevant but not sufficient for interpreting the data. We extend the secondary scattering investigation in the present calculation by including non-resonance dilepton-producing $\pi\rho \rightarrow \pi\ e^+e^-$ reactions[7].

PRIMARY SCATTERING

Future collider energies, several thousand GeV per nucleon in the center of mass, probe distances much smaller than the nucleon. Models must of course incorporate QCD to describe the subnucleonic features. Quarks and gluons then comprise the appropriate degrees of freedom for a QCD transport theory. They are propagated through spacetime approximating the dynamics of collisions to be explored at RHIC and LHC [8, 9, 10, 11]. Evolution continues until soft processes dominate and hadronization occurs.

Whether or not a quark-gluon plasma can be experimentally detected depends largely on the characteristics of the collision in its absence, something we shall call background. In order to better quantify this background, simulations without "built-in" plasma formation that still assume a QCD description of nucleon scattering must be explored. HIJING, developed by Wang and Gyulassy[12], is precisely this type of model and has been used to look at multiple minijet production, shadowing and jet quenching in pA and AA collisions.

The simulation we develop is similar to HIJING. It is based on a simple prescription that uses QCD to characterize the individual nucleon-nucleon collisions and uses Glauber-type geometry to determine the scaling. The kinematics of the nucleon-nucleon collisions are handled by PYTHIA and JETSET [13], high energy event-generators using QCD matrix elements as well as the Lund fragmentation scheme. A somewhat detailed description of the model is outlined below.

- *Initialization of nucleons inside nuclei:* The nucleons are positioned randomly inside each nucleus according to the size of the nucleus and are given random Fermi momentum in the x-y plane and are given z-momentum proportional to the lab energy (or center of mass energy).

- *Number of collisions:* The number of collisions is determined geometrically [14]. For a proton-nucleus collision,

$$n(b) = \sigma_{NN} \int dx\, dy\, dz\, \rho(\sqrt{b^2 + z^2}) \tag{1}$$

For nucleus-nucleus collisions,

$$N(b) = \sigma_{NN} \int dx\, dy\, dz_1\, dz_2\, \rho_A(\sqrt{x^2 + y^2 + z_1{}^2})\, \rho_B(\sqrt{x^2 + (y-b)^2 + z_2{}^2}) \tag{2}$$

- *Picking scattering partners:* Two nucleons are chosen at random from each nuclei and are allowed to scatter when, and if, they meet several criteria: First, the two nucleons cannot have scattered previously. Second, the nucleons must be within one cross-sectional radius of one another in the transverse beam direction,

$$\sqrt{(x_t - x_p)^2 + (y_t - y_p)^2} \leq \sqrt{\sigma_{NN}/\pi} \qquad (3)$$

Thirdly, the pair must be moving toward one another in the transverse plane. And lastly, the center of mass energy, $\sqrt{s}$, must be above 6 GeV. This limit is chosen because it is on the order of the energy where perturbative QCD is no longer applicable. If the two nucleons meet these criteria, they are allowed to scatter.

- *Scattering and Rescattering:* PYTHIA chooses partons to participate in the hard scattering from each nucleon. The partons that are chosen, as well as the momentum fraction they carry, are based on known parton distributions [15]. After the individual partons have had a hard scattering and are color-connected with the diquarks from the remaining nucleon, strings are formed. The kinematics of the fragments from the string are determined by JETSET and are stored in a temporary array. Any partonic radiation that is not color-connected to either string goes directly into the nucleus-nucleus final state. This string is then put back into the nuclei and allowed to rescatter as a "wounded" nucleon. The wounded nucleon has the string's momentum while its position is updated to halfway between the original nucleons' positions through,

$$(x_1, y_1)(x_2, y_2) \rightarrow (\frac{x_1 + x_2}{2}, \frac{y_1 + y_2}{2}). \qquad (4)$$

- *Final State:* After all nucleons have experienced their geometrically determined number of collisions or they have center of mass energies below the cutoff, particles present in each nucleon's temporary array constitute the final state.

Hadronic observables from the model have been compared against data for several systems. Although the model is based on very simple premises, it reproduces the main features characterizing hadronic final states.

SECONDARY SCATTERING

Dileptons from pseudoscalars (π^0, η, η') and vectors (ω, ρ^0, ϕ) produced in the primary scattering phase are not enough to account for the S+Au data. Our model also incorporates secondary scattering of hadronic resonances. All π s and ρ s formed during the primary collisions of nucleons will have a chance to scatter amongst themselves before decaying. The reactions we consider are of two types, one which produces a resonance that decays to dileptons and the other which goes to dileptons directly.

Of the first type, $\pi^+ \pi^- \rightarrow \rho^0 \rightarrow e^+ e^-$, $\pi^0 \rho^\pm \rightarrow a_1^\pm \rightarrow \pi^\pm e^+ e^-$, $\pi^\pm \rho^0 \rightarrow a_1^\pm \rightarrow \pi^\pm e^+ e^-$ have been included. To accomplish these types of scattering, pions and rhos must of course appear in the final state of the model described in the previous section. As the default, JETSET automatically decays all hadronic resonances, but it also contains provisions to prohibit them. We thus allow neutral pions to scatter from charged rhos when conditions are favorable. Technically, the steps involved in secondary scattering are similar to those for primary scattering. The main component that is changed is the cross section used to determine whether or not the particles are allowed to scatter.

The cross section for creating a resonance is taken to be

$$\sigma(\sqrt{s}) = \frac{\pi}{\mathbf{k}^2} \frac{\Gamma_{\text{partial}}^2}{(\sqrt{s} - m_{\text{res}})^2 + \Gamma_{\text{full}}^2/4} \tag{5}$$

with $\mathbf{k}$ being the center-of-mass momentum. The full and partial decay widths for $\rho^0 \to \pi^+\pi^-$ are set to 152 MeV. The full a_1 decay width is 400 MeV and the partial width for $a_1^\pm \to \pi\rho$ are energy dependent[17]:

$$\Gamma_{a_1 \to \pi\rho} = \frac{g_{a_1\pi\rho}^2}{24\pi m_{a_1}^2} |\mathbf{k}|[2(p_\pi \cdot p_\rho)^2 + m_\rho^2(m_\pi^2 + \mathbf{k}^2)] \tag{6}$$

The kinematics of the resonances are determined from the pair of hadrons while JETSET decays the resonance into dileptons using appropriate functions for $d\Gamma/dM^2$ and $|\mathcal{M}|^2$ resulting from analysis of the same Lagrangian used to derive Eq. (6).

Dileptons from secondary scatterings of the resonance type increase the number significantly in the region around the ρ^0 mass, but not in the region with the largest gap, $0.2 \text{ GeV} \leq M^2 \leq 0.5 \text{ GeV}$.

The nonresonant component is estimated here by computing the sole process $\pi^0\rho^\pm \to \pi^\pm e^+ e^-$ and then assuming the other isospin channels contribute equally. The other channels involve Feynman graphs that result in a singularity and must be regulated in a full T-matrix or some other effective approach [18]. To this level of estimate, isospin averaging and ignoring interference effects between these and the resonant a_1 contributions is not worrisome.

The cross section for $\pi^0\rho^\pm \to \pi^\pm\rho^0$ determines how many and how often the charged rhos scatter with pions. Using an effective Lagrangian approach to be described momentarily, we generate a set of graphs. The resultant cross section can be nicely parameterized for $\sqrt{s} \leq 2.5$ GeV by

$$\sigma(\sqrt{s}) = 1.8\,\text{mb} + \frac{8}{s^2}\,\text{mb GeV}^4 + 0.5\,s\left(\frac{\text{mb}}{\text{GeV}^2}\right) \tag{7}$$

Since this is a non-resonant process, the Monte Carlo directly determines the kinematics of the final state. Necessary ingredients for such procedures include an interaction Lagrangian and a resulting squared matrix element. The Lagrangian employed is[19]

$$\mathcal{L} = |D_\mu\Phi|^2 - m_\pi^2|\Phi|^2 - \frac{1}{4}\rho_{\mu\nu}\rho^{\mu\nu} + \frac{1}{2}m_\rho^2\rho_\mu\rho^\mu - \frac{1}{4}F_{\mu\nu}F^{\mu\nu} \tag{8}$$

where $D_\mu = \partial_\mu - ieA_\mu - ig_\rho\rho_\mu$ is the covariant derivative, Φ is the complex charged pion field, $\rho_{\mu\nu}$ is the rho field-strength tensor and $F_{\mu\nu}$ is the photon field strength tensor. From this Lagrangian, the matrix elements can be determined. In the calculation, the graphs involving the a_1 are neglected as the contribution from a_1 has already taken into account in the resonance portion of the model. There are three graphs, Fig. 1, whose matrix elements are listed in Eq. 9 -11.

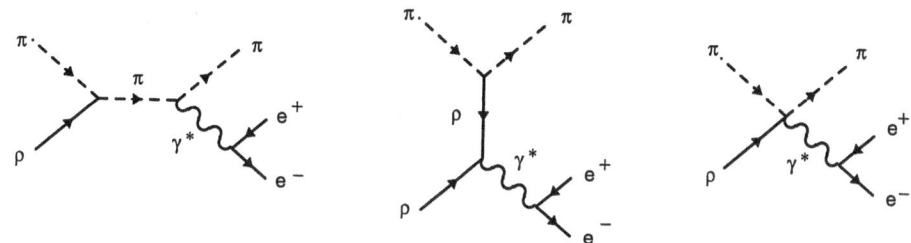

Figure 1. Feynman diagrams that contribute to $\pi\rho \to \pi e^+ e^-$.

$$\mathcal{M}_1 = \frac{g_\rho e^2}{M^2(s - m_\pi{}^2)}\epsilon^\mu(p_a)(2p_b + p_a)_\mu(2p_1 + q)_\nu \bar{u}(p_-)\gamma^\nu \bar{v}(p_+) \qquad (9)$$

$$\mathcal{M}_2 = \frac{-h_+(t)g_\rho e^2}{M^2(t - m_\rho{}^2)}\epsilon^\mu(p_a)(2p_a - q)_\nu(p_b + p_1)_\mu \bar{u}(p_-)\gamma^\nu \bar{v}(p_+) \qquad (10)$$

$$\mathcal{M}_3 = \frac{g_\rho e^2}{M^2}\epsilon^\mu(p_a)[X_{\mu\nu}]\bar{u}(p_-)\gamma^\nu \bar{v}(p_+) \qquad (11)$$

where $X_{\mu\nu} = a g_{\mu\nu} + b(p_{1\mu}p_{b\nu} + p_{b\mu}p_{1\nu}) + c(p_{b\mu}p_{b\nu} + p_{1\mu}p_{1\nu})$.

In the t-channel matrix element, a form factor, $h_+(t) = (\Lambda^2 - m^2)/(\Lambda^2 - t)$, appears to account for the finite size of the mesons. Its presence breaks gauge invariance. In order to completely restore gauge invariance, other terms must be added to the four point diagram $\mathcal{M}_3$: $a = -1$ $b = c = (h_+(t) - 1)/(p_b \cdot q + p_1 \cdot q)$. The parameters Λ and m are set to 1 GeV and m_ρ, respectively.

The absolute square $|\mathcal{M}_1 + \mathcal{M}_2 + \mathcal{M}_3|^2$ and $d\sigma/dM^2$ were used to Monte Carlo the three-body $\pi^\pm e^- e^+$ final state.

The cross section resulting from the above matrix elements is accurately parameterized by the following function in the energy region of interest.

$$\sigma(\sqrt{s}) = 0.018\,\text{mb} + \frac{0.9(\sqrt{s} - 1.05)^2}{s^3}\,\text{mb}\,\text{GeV}^4 \qquad (12)$$

These lepton pairs are of nonresonant origin and are now added to the pairs from resonance decays. The total dilepton yield from our model is the sum of lepton pairs from primary plus secondary scattering. The invariant mass distributions of the dileptons from all contributions will be discussed in the last section.

RESULTS

The invariant mass spectra of dileptons from the primary scattering part of our model for the S+Au and p+Au systems are displayed in Fig. 2. The simulation agrees with the proton-induced data and it is reassuring that our S+Au model-results are consistent with the cocktail from the CERES collaboration[1]. Plotting against the actual S+Au data reveals a significant enhancement over predictions in the invariant mass region between 200 and 500 MeV. There is also a modest enhancement for masses above this range.

Dileptons from secondary scattering for the S+Au system in our model are added to those from primary scattering. The contribution from pion annihilation increases

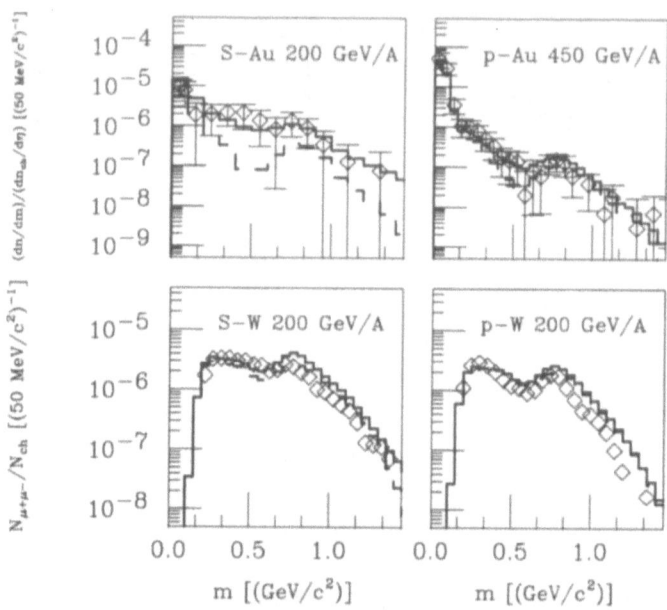

Figure 2. Dilepton invariant mass spectrum from primary scattering alone (dashed-histogram) and primary plus secondary scattering (solid-histogram). The top panels shows the comparison to CERES data for S+Au (left top panel) and p+Au (right top panel). The bottom panels show the comparison to HELIOS/3 data for S+W (left bottom panel) and p+W (right bottom panel).

the distribution significantly in the rho mass region, but still leaves an excess below the rho mass. We should stress that we have taken vacuum rho properties throughout. Radiative a_1 decay contributes a minimal amount in the excess (or deficit) region, but the contribution from non-resonance $\pi\rho$ scattering provides the most significant increase.

The secondary scattering previously described, has been included in the invariant mass distributions of dileptons. The dilepton spectra from the proton-induced interactions are not significantly changed, Fig. 2. This result is as expected—dileptons from the smaller systems are quantitatively described by primary hadronic decays. The proton-nucleus collisions do not create a heated nuclear medium large enough or dense enough to bring about significant collective effects. Conversely, the S+Au collision has a marked increase in lepton-pair production between an invariant mass of 200 and 500 MeV as well as a noticeable increase in the higher mass region, Fig. 2. It is not surprising that secondary scattering becomes important in the S+Au system, as a dense nuclear medium is created during the collision.

As a further check to our calculations, the simulation was modified so that we could study muon pair production and compare to HELIOS/3[25] results, Fig. 2. The experiment measured the invariant mass distribution of muon pairs from S+W and p+W at 200 GeV. The $\pi\rho$ cross section used in our model was recalculated with muons in the final state instead of electrons and all relevant branching ratios were changed to the correct values for resonances decaying into muons. In the invariant mass region between 0.2 and 0.5 GeV, there is an excess in the S+W data over the proton-induced data, although less pronounced than in the S+Au reaction. The resulting spectra are consistent with the study of electron pairs.

Allowing for the short-comings of our model, results still suggest that secondary scattering is a viable explanation for the excess found in dilepton data. Inclusion of secondary scattering, 1) preserves the consistency the primary scattering in our model has with proton-induced data and, 2) enhances the number of dileptons within the region of excess discovered in S+Au and S+W data. Although this agreement cannot rule out other possible explanations for the excess electrons, our model's simplicity is attractive.

REFERENCES

1. G. Agakichiev *et al.*, 1995, Enhanced production of low-mass electron pairs in 200 GeV/nucleon S-Au collisions at the CERN super proton synchrotron, *Phys. Rev. Lett.* 75:1272.
2. W. Cassing, W. Ehehalt, and C.M. Ko, 1995, Dilepton production at SPS energies, *Phys. Lett. B* 363:35.
3. G. Q. Li, C. M. Ko, and G. E. Brown, 1995, Enhancement of low-mass dileptons in heavy ion collisions, *Phys. Rev. Lett.* 75:4007.
4. C. Song, V. Koch, S. H. Lee, and C. M. Ko, 1996, Thermal effects on dilepton production, *Phys. Lett. B* 366:379.
5. R. Rapp, G Chanfray, and J. Wambach, 1996, Medium modifications of the rho meson at CERN/SPS energies, *Phys. Rev. Lett.* 76:368.
6. K. L. Haglin, 1996, Excess electron pairs from heavy-ion collisions at CERN and a more complete picture of thermal production, *Phys. Rev C* 53:R2606.
7. For first suggestions and preliminary results of the importance of this mechanisms, see, K. Haglin, proceedings of INT/RHIC Workshop *Electromagnetic Probes of Quark Gluon Plasma* 24–27, 1996; and proceedings of *International Workshop on Hadrons in Dense Matter*, GSI, Darmstadt, 3–5 July 1996.
8. K. Geiger and B Müller, 1992, Dynamics of parton cascades in highly relativistic nuclear collisions, *Nucl. Phys. A* 544:467c.
9. K. Geiger and B. Müller, 1992, *Nucl. Phys. B* 369:600.
10. G. Kortemeyer, J. Murray, S. Pratt, K. Haglin, and W. Bauer, 1995, Causality violation in cascade models of nuclear collisions, *Phys. Rev. C* 52:2714.
11. G. Kortemeyer, J. Murray, S. Pratt, K. Haglin, and W. Bauer, 1994, Parton cascade for RHIC, *NSCL Annual Report*, 63.
12. X. Wang, M. Gyulassy, 1991, HIJING: A monte carlo model for multiple jet production in pp, pA, and AA collisions, *Phys. Rev. D* 44:3501.
13. T. Sjöstrand, 1994, *Computer Physics Commun.* 82:74.
14. A. D. Jackson and H. Boggild, 1987, A model of trasverse energy production in high energy nucleus-nucleus collisions, *Nucl. Phys. A* 470:669.
15. H.L. Lai, J. Botts, J. Huston, J.G. Morfin, J.F. Owens, J. Qiu, W.K. Tung and H.Weerts, 1995, Global QCD analysis and the CTEQ parton distributions, *Phys. Rev. D* 51:4763.
16. R. Santo *et al.*, 1994, Single photon and neutral meson data from WA80, *Nucl. Phys A* 566:61c.
17. L. Xiong, E. Shuryak and G.E. Brown, Photon production through a_1 resonance in high energy heavy ion collisions, 1992, *Phys. Rev. D* 46:3798.
18. K.L. Haglin and C. Gale, to be published.
19. J. Kapusta, P. Lichard, and D. Seibert, 1991, High-energy photons from quark-gluon plasma versus hot hadronic gas, *Phys. Rev. D* 44:2774.
20. B. Holstein, 1990, Pion polarizability and chiral symmetry, *Comm. Nucl. Part. Phys.*, 19:221.
21. L. D. Landau and I. Ya. Pomeranchuk, *1953 Dokl. Akad. Nauk SSSR* 92:535.
22. K. Haglin and S. Pratt, 1994, Coherent pion pairs in heavy-ion collisions, *Phys. Lett. B* 328:255.
23. W. Bauer, G.F. Bertsch, W. Cassing, and U. Mosel, 1986, Energetic photons from intermediate energy proton-and heavy-ion-induced reactions, *Phys. Rev. C* 34:2127.
24. Particle Data Group, Phys. Rev. D **50**, Part I (1994).
25. M. Masera for HELIOS/3 collaboration, 1995, Dimuon production below mass 3.1 GeV/c^2 in p-W and S-W interactions at 200 GeV/c/A, *Nucl. Phys. A* 590:93.

EXCLUSIVE STUDY OF HEAVY ION COLLISIONS USING 2-8 A GeV Au BEAMS: STATUS OF AGS EXPERIMENT E895

Morton Kaplan

Carnegie Mellon University
4400 Fifth Avenue
Pittsburgh, Pennsylvania 15213

for the E895 Collaboration

ABSTRACT

The Time-Projection-Chamber (TPC) from the Equation-of-State (EOS) experiment at the Bevalac has been successfully installed and operated at the BNL Alternating Gradient Synchrotron (AGS). Using Au beams accelerated to energies in the 2-8 A GeV range, we are investigating the global characteristics of nuclear matter at high energy and baryon density, with specific emphasis on the energy dependence of collective-phenomena observables. In two runs completed to date, high statistics data were recorded at four Au beam energies using Au targets, and with less statistics on Be, Cu, and Ag targets. The E895 experiment will be described and some preliminary results presented.

INTRODUCTION

The objective of experiment E895 at the Brookhaven Alternating Gradient Synchrotron is to carry out a systematic study of Au + Au reactions at several energies using the EOS TPC as a primary tracking detector. To accomplish this task, a new collaboration has been formed consisting of the institutions listed in Table 1. The operating characteristics of the TPC device permit exclusive measurements of particle production, correlations, collective flow effects, and strangeness production in Au induced reactions, and their dependence on incident beam energy. The mass dependence of these observables will also be studied by taking data with several lighter-mass targets at each Au beam energy. Recent experiments and theoretical studies suggest that maximum baryon densities, as high as eight times normal nuclear matter density, may be attainable in central Au + Au collisions between 2 and 10 A GeV beam energies.

Table 1. The EOS Experiment at the AGS -
Collaborating Institutions in E895

Lawrence Berkeley National Laboratory
Brookhaven National Laboratory
University of California, Davis
Kent State University
Purdue University
SUNY, Stony Brook
Columbia University
Carnegie Mellon University
Ohio State University
Harbin Institute of Technology, China
University of Auckland, New Zealand

Under such conditions E895 is a logical extension of the physics program carried out at the Bevalac to understand the collision dynamics and obtain information about the nuclear matter equation-of-state. In addition, E895 is expected to be sensitive to the change of medium effects, since the baryon densities are high enough to alter the masses and widths of hadrons, which in turn, could signal the occurence of chiral symmetry restoration or the formation of baryon-rich Quark Gluon Plasma.

THE EXPERIMENT

A schematic diagram of the experimental setup is shown in Fig. 1. The heart of the experiment is the EOS Time-Projection Chamber, installed in the gap of a large 1 Tesla dipole magnet in the MPS area of the AGS facility. The TPC provides continuous tracking, nearly 4π acceptance, and dE/dx particle identification for light-mass particles over a wide rigidity range. Downstream of the TPC is a **MU**ltiple **S**ampling **I**onization Chamber (MUSIC) for heavy-mass fragment measurements.

Within the TPC, momenta can be measured for $\pi^{\pm}$, $K^{\pm}$, K^0_s, $p^{\pm}$, Λ^0, (Ξ^-, and Ω^-), d, ^3H, ^3He, ^4He, ^6He, and the isotopes of Li and Be, while projectile fragments

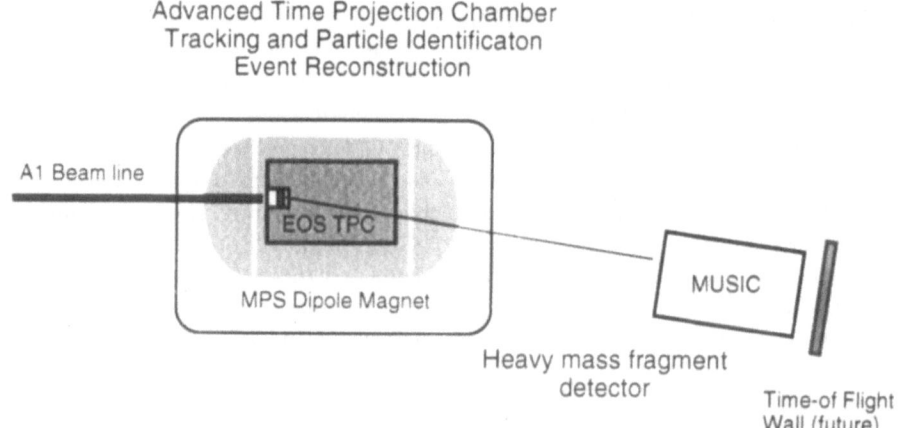

Figure 1. Schematic diagram of the E895 experimental setup showing the TPC in the MPS dipole magnet and the MUSIC heavy fragment detector.

from Z=6 to Z=79 are identified and recorded in the MUSIC detector. As the data are collected on an event-by-event basis, full event reconstruction is possible allowing simultaneous observations of many experimental quantities. Thus the E895 experiment has the capability of providing correlated information for individual reaction events. In addition to the TPC and MUSIC detectors indicated in Fig. 1, the experiment employs a variety of beam-defining and triggering counters upstream of the TPC.

What are the research goals of experiment E895? Basically, the goal is to produce high-density nuclear matter by colliding very heavy nuclei and to investigate its properties. Recent theoretical and experimental results suggest that the highest density baryon matter is most likely to be created in the energy range ~2-10 A GeV, or perhaps even somewhat higher. The availability of Au ion beams in this energy regime at the AGS provides the opportunity for E895 to achieve its goal, since the detectors indicated in Fig. 1 were specifically designed to find and identify highly compressed and excited nuclear matter. By studying the behavior of hot, dense nuclear matter, e.g., by determining the equation of state and reaction dynamics, as a function of incident Au beam energy, it should be possible to map out estimates of the densities attained in Au+Au collisions at the various energies. The most direct signatures of nuclear compression (or density increase) are in the collective effects known as nuclear flow. The determination of flow, however, requires exclusive measurements of specific particles (e.g., protons, pions, kaons), a task for which the TPC is well suited. At the same time, the exclusive data can be used for examining fine details of individual collision events, for identifying exotic events, and for studying multiparticle correlations and strangeness yields. These capabilities take on particular significance because very little exclusive data currently exists for heavy ion collisions in this energy range. Consequently, theoretical models suffer in their predictive power by the lack of experimental constraints.

We show in Fig. 2 an offline computer display of charged particle tracks in the TPC for a specific Au+Au central collision, at an incident beam energy of 6 A GeV. The analysis software required to identify such particle tracks is largely in hand, deriving extensively from the earlier coding used in the EOS experiment.

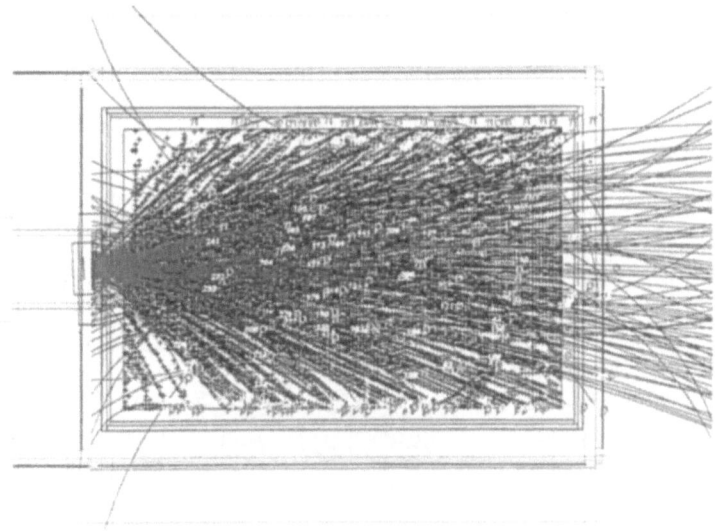

Figure 2. Computer generated display resulting from charged particle track data recorded in the TPC for a specific Au+Au central collision, at an incident beam energy of 6 A GeV.

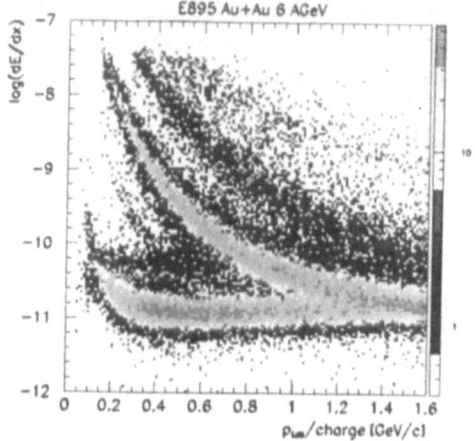

Figure 3. Particle identification via dE/dx in the TPC for 6 A GeV Au+Au reactions.

The quality of the light particle identification in the TPC (via dE/dx) can be seen from the illustration in Fig. 3. This semilog plot of energy loss vs. particle momentum indicates the relatively clean separation of the pion, kaon, proton, and deuteron bands.

The excellent performance of the TPC permitted the extraction of some preliminary representative results, even prior to applying calibrations to the data, in order to gain an overview of the global features of the reactions. Several examples are presented in the next section.

INITIAL PRELIMINARY RESULTS

We present in Figs. 4-8 the results obtained from preliminary analyses of samples of the TPC data. As the immediate objective was to provide a quick overview, these results are derived from uncalibrated data and will undergo additional refinement.

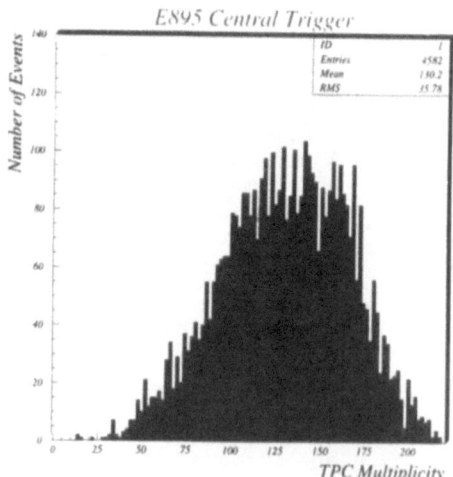

Figure 4. Measured TPC charged-particle-multiplicity distribution for central collision events in the reactions 4 A GeV Au+Au.

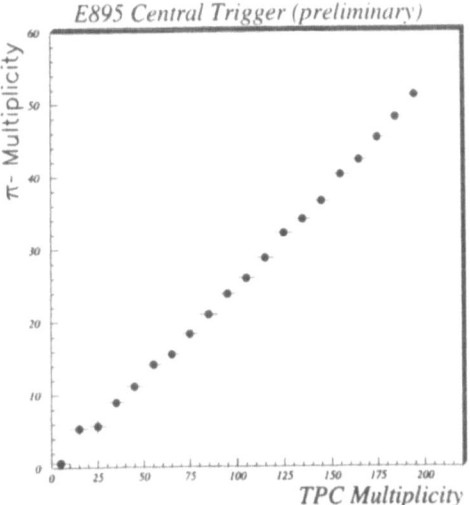

Figure 5. Correlation between the observed π^- multiplicity and the total TPC multiplicity for Au+Au centrally triggered reactions at 4 A GeV.

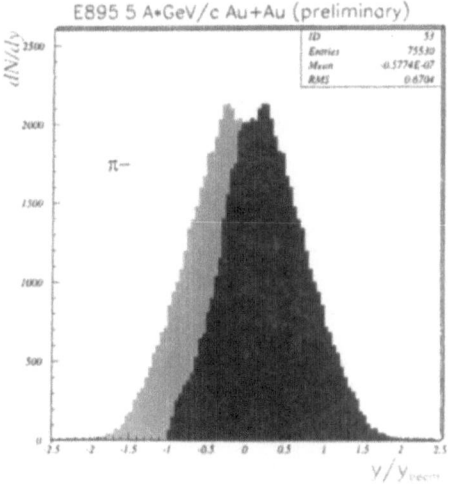

Figure 6. Rapidity distribution of negative pions in 4 A GeV Au+Au reactions. The black area indicates the measured π^- distribution, and the gray area is an estimate of missing particles (see text).

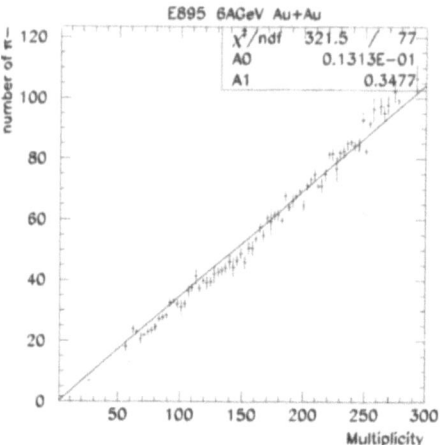

Figure 7. Correlation between measured π^- multiplicity and the total TPC multiplicity for Au+Au collisions at a beam energy of 6 A GeV.

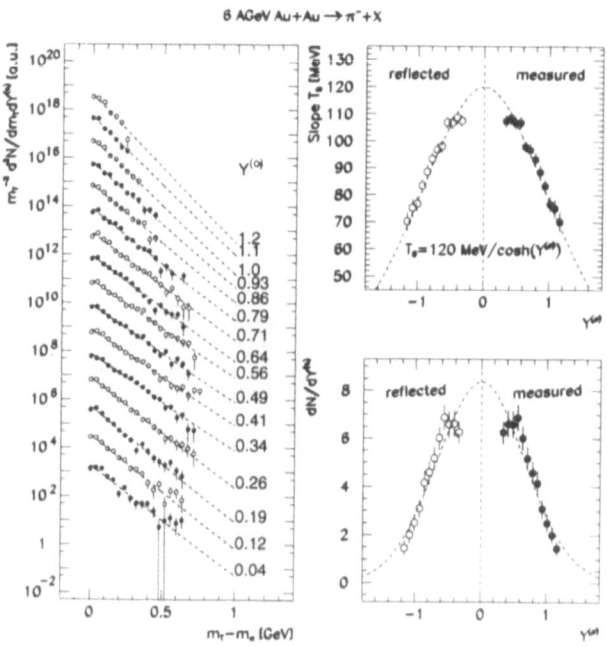

Figure 8. Transverse energy spectra in various rapidity intervals, the slope parameters derived from them, and the rapidity distribution, all for negative pions produced in 6 A GeV Au+Au reactions.

Table 2. TPC Data Summary for Experiment E895

Energy	Target	Trigger	Magnet	Events
8 A GeV	Au [3%]	Central	1.0 Tesla	364 K
	Au [3%]	Min. Bias	1.0 Tesla	59 K
	Ag [3%]	Central	1.0 Tesla	12 K
	Ag [3%]	Min. Bias	1.0 Tesla	41 K
6 A GeV	Au [3%]	Central	1.0 Tesla	645 K
	Au [1%]	Central	1.0 Tesla	54 K
	Au [3%]	Min. Bias	1.0 Tesla	100 K
	Ag [3%]	Central	1.0 Tesla	60 K
	Ag [3%]	Min. Bias	1.0 Tesla	45 K
	Cu [3%]	Central	1.0 Tesla	60 K
	Cu [3%]	Min. Bias	1.0 Tesla	17 K
	Be [3%]	Central	1.0 Tesla	9 K
	Be [3%]	Min. Bias	1.0 Tesla	40 K
4 A GeV	Au [3%]	Semi	0.75 Tesla	284 K
	Au [3%]	Semi	1.0 Tesla	186 K
	Au [3%]	Central	1.0 Tesla	295 K
	Au [1%]	Central	1.0 Tesla	70 K
	Au [2%]	Central	1.0 Tesla	192 K
	Au [1%]	Min. Bias	1.0 Tesla	43 K
	Cu [3%]	Central	1.0 Tesla	23 K
	Cu [3%]	Min. Bias	1.0 Tesla	20 K
	Be [3%]	Central	1.0 Tesla	10 K
	Be [3%]	Min. Bias	1.0 Tesla	25 K
2 A GeV	Au [2%]	Central	0.75 Tesla	300 K
	Au [2%]	Semi	0.75 Tesla	58 K
	Ag [3%]	Central	0.75 Tesla	56 K
	Cu [3%]	Semi	0.75 Tesla	31 K
	Cu [3%]	Min. Bias	0.75 Tesla	30 K
	Be [3%]	Semi	0.75 Tesla	14 K
	Be [3%]	Min. Bias	0.75 Tesla	61 K

In Fig. 4 the charged-particle-multiplicity distribution is shown for central Au+Au collisions at 4 A GeV. The mean TPC multiplicity is ~130 under these triggering conditions, and the distribution is quite broad.

We give in Fig. 5 a plot of the measured π^- multiplicity vs. the total TPC charged-particle multiplicity, for Au+Au central collisions at 4 A GeV. As can be seen, there is a strong correlation between the multiplicity of identified π^- particles and the total multiplicity. From this correlation we would infer that approximately 26% of the observed TPC multiplicity consists of negative pions.

The rapidity distribution, dN/dY, of π^- particles observed for the 4 A GeV Au+Au reactions is shown as the black area in Fig. 6. Since there is a symmetry requirement about Y=0, it is apparent that significant numbers of pions are absent from the data at the larger negative rapidities. By reflection symmetry we can estimate this particle loss, as indicated by the gray area in Fig. 6. We think this discrepancy is due to an acceptance problem in the analysis software, and expect to recover much of the missing data with improved software. It is interesting that the strong correlation between π^- yield and total charged-particle multiplicity (see Fig. 5) remains, even with the appreciable number of missing π^- particles.

Fig. 7 presents the correlation between the π^- multiplicity and the total charged-particle multiplicity (in the TPC) for Au+Au collisions at 6 A GeV. These data are analogous to those in Fig. 5, but at a higher beam energy. Note that at 6 A GeV the total multiplicity and the π^- multiplicity have both increased, compared to the 4 A GeV data, but the measured correlation is just as strong here as at the lower energy. Furthermore, at 6 A GeV the contribution of negative pions to the total multiplicity has increased to 35% (from 26% at 4 A GeV).

A sampling of the π^- results for 6 A GeV Au+Au reactions is given in Fig. 8.

On the left side of Fig. 8 are plots of π^- transverse energy spectra per unit rapidity interval, for a series of rapidity bins. As can be seen, the data in each bin are consistent with an exponential fall-off, and hence can be characterized by a slope parameter. The derived slope parameters are given as a function of rapidity in the upper-right part of Fig. 8. In the lower-right part of Fig. 8, we show the rapidity distribution for produced π^- particles. Although these data are still preliminary, Fig. 8 is representative of the quality of the pion data we have obtained, and highlights the very low p_t acceptance of the experiment.

CURRENT STATUS OF E895

A summary of the TPC data collection to date is outlined in Table 2.

The E895 experiment has completed two data collection runs at the AGS, the first using Au beams at 2 and 4 A GeV and the second with 6 and 8 A GeV Au beams. In both runs, high statistics were obtained with Au targets, and with lower statistics measurements were made on Ag, Cu, and Be targets. Major offline data analysis has begun using the Parallel Distributed Scientific Farm (PDSF) computing facility at NERSC, as well as the computational facilities at E895 collaborating institutions.

ACKNOWLEDGEMENTS

M.K. would like to thank the Division of Nuclear Physics, U.S. Department of Energy, for support of this work at Carnegie Mellon University.

ANALYSIS OF THE d/p RATIOS IN Au + Au COLLISIONS AT 11.1 GeV/c

E.J. Garcia-Solis[1] for the E866 Collaboration:
BNL-Columbia-INS Tokyo-Kyoto-LLNL-MIT-UC Berkeley-
UC Riverside-Maryland-Tokyo-Tsukbua-Yonsei

[1]Department of Chemistry and Biochemistry
University of Maryland
College Park, MD 20742

INTRODUCTION

The most interesting aspect of high energy heavy-ion collisions is that they provide the opportunity to produce matter at high net baryon density. In particular, under these conditions a signature of the quark-gluon plasma (QGP) formation may be found[1]. Beam energies of around 10 GeV per nucleon seem to achieve the stopping of two colliding nuclei creating a high density matter region [2, 3]. It is important to look for experimental probes that allow the study of the properties of this region, such as lifetime, volume, entropy, etc. Deuteron production by phase space coalescence is interesting because it can be used to study the space-time structure of the participant zone [4]. Once the participant region is formed in a heavy-ion collision, the baryon density increases up to a maximum, followed by a radial expansion of nuclear matter. At some point in this expansion (freeze-out time), composite particles are formed by coalescence. Since a proton and neutron form a deuteron when their relative momentum in phase space is small [6], the deuteron production is a measure of the source baryon density and/or the extension of the freeze-out zone [7]. Furthermore, a strong first-order QGP with a large latent heat may expand before hadronization so that the baryons would be formed at a lower density. Hence, suppression in the deuteron formation may be an indication of plasma formation [8].

In this paper, the proton and deuteron $m_t - m_o$ and dn/dy distributions from 11.1 GeV/c Au + Au collisions measured by the E866 collaboration in 1993 are presented. The centrality and rapidity dependence of the d/p yield ratio is discussed in terms of the number of participants. The d/p ratio decreases as a function of the number of participants for cuts on rapidities closer to the target rapidity, and the d/p ratio remains constant for bins closer to mid rapidities. In the following section a brief discussion of the experimental set-up and data analysis is presented. Then the transverse mass and

dn/dy distributions and the d/p yield ratios are presented as a function of rapidity and participant number, followed by a summary argumentation and final notes.

Experiment and Data analysis

The E866 experimental set-up used in this analysis consisted of three sets of detectors, one for triggering (beam counters), one for the identification and tracking of particles, and another for the global characterization of the events.

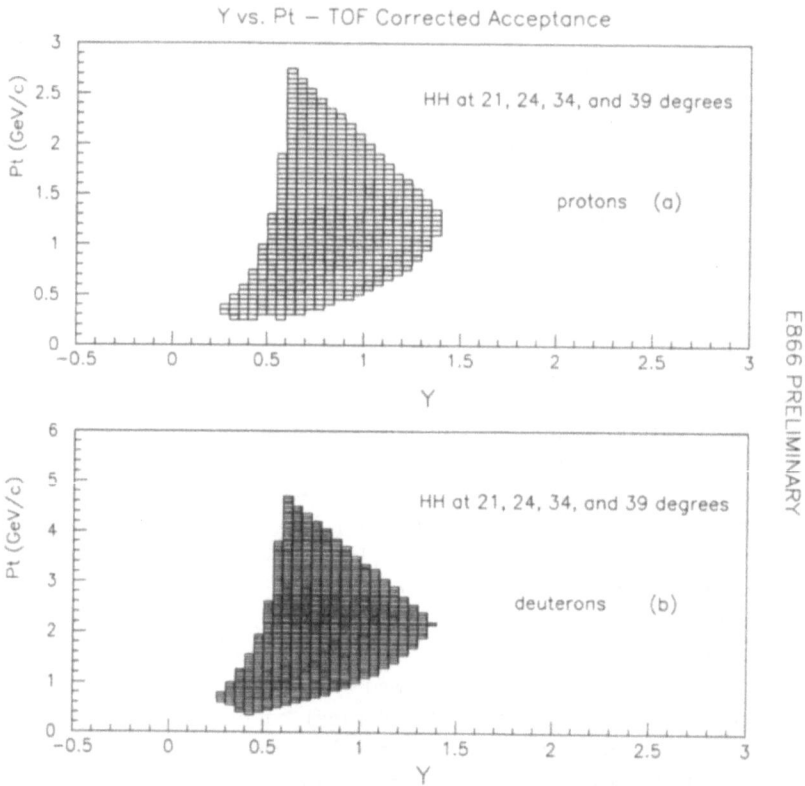

Figure 1. Summed acceptance for spectrometer settings of 21° ,24°, 34°, and 39°, for protons (a) and deuterons (b).

The tracking and particle identification (PID) were achieved using the Henry Higgins (HH) magnetic spectrometer [9], together with tracking detectors, and a time-of-flight wall (TOF). The acceptance of the Henry Higgins arm for protons and deuterons is shown in Fig. 1 (a) and (b), respectively. The figure shows the sum of the acceptances for spectrometer settings of 21°, 24°, 34°, and 39°; note in these figures that the acceptance for protons and deuterons is similar. The data were taken with a minimum bias trigger, which required a signal from the beam, a charge loss of a few units from the full charge of the beam, and a hit in the time-of-flight wall.

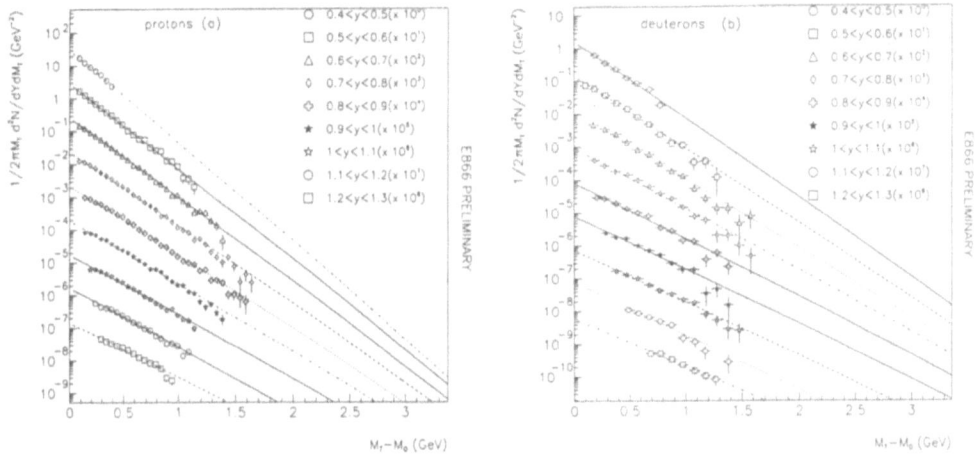

Figure 2. Measured invariant cross section for protons (a) and deuterons (b) as a function of the transverse kinetic energy $m_t - m_o$, and for a the centrality cut $N_{pp} > 75$.

The centrality selection for the spectra is made using the ZCAL detector [10], which measures the energy remaining in the projectile fragments traveling forward, subtending an angle of 1.5° from the beam axis. Since the energy in the ZCAL is predominantly from projectile spectators, the number of projectile participants, N_{pp}, can be estimated by

$$< N_{pp} >= 200.3 \; (1 - \frac{< E >}{2000}), \tag{1}$$

where $< E >$ is the average energy measured in GeV in the ZCAL and 2000 GeV is the approximate energy of the beam. According to this formula, for example, $75 > N_{pp} > 0$ corresponds to a 40% centrality cut of the interaction cross section ($\sigma_{int} \sim 5.3$ barns). It has been estimated that the systematic errors in measuring $< N_{pp} >$ are less than 8%.

The overall systematic error is estimated to be in the order of 10% in the measurement of the cross section. However, in the figures shown, only statistical errors are displayed.

Preliminary Results

Figure 2 (a) and (b) show the measured invariant cross sections over different rapidity intervals for protons and deuterons, respectively. These cross sections were calculated for 40% of σ_{int} or $N_{pp} > 75$. The normalization factor of σ_{int} needed to give the absolute yield is not included in the ordinate because it is canceled out in the yield ratios. For this reaction, target rapidity is at $y = 0$ and mid-rapidity is about $y = 1.6$.

The abscissa is $m_t - m_o$, the differences between the transverse mass, $m_t = \sqrt{p_t{}^2 + m_o{}^2}$, and the rest mass m_o. The curves in the figures are fits to the distributions using the function

$$\frac{1}{2\pi m_t} \frac{d^2 N}{dy\, dm_t} = A\, m_t\, e^{\frac{-(m_t - m_o)}{B}}. \tag{2}$$

The slope parameters for the fits shown in Fig. 2 are given as a function of rapidity in Fig. 3 (c) and (d) for protons and deuterons, respectively. For the protons, the slope parameters increase continuously as a function of the rapidity, from 0.15 GeV at $y \simeq 0.5$ to 0.22 GeV at $y \simeq 1.2$. For deuterons the same trend is observed, however the slope values are larger than the ones observed for the protons.

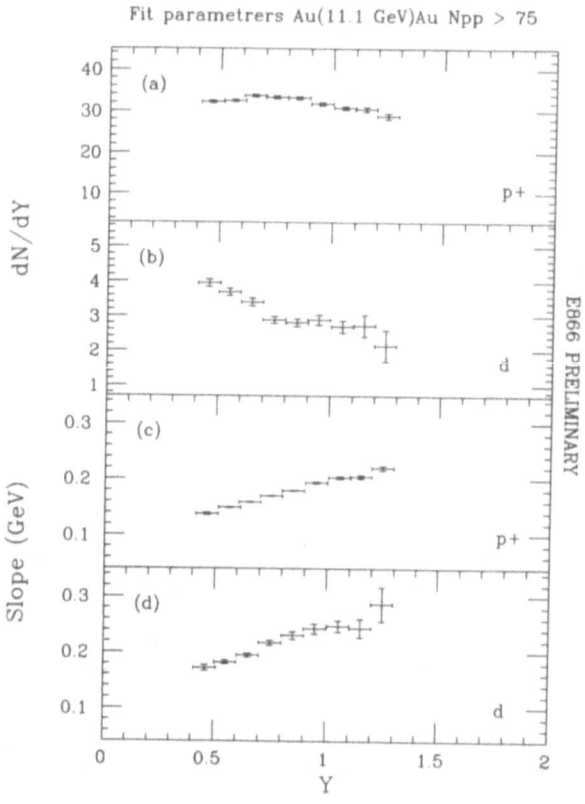

Figure 3. Resultant particle yield for protons (a) and deuterons (b) and slope parameter for protons (c) and deuterons (d), as a function of the rapidity, for the centrality cut $N_{pp} > 75$.

Integrating Eq. 2 over the interval $m_t = m_o$ to $m_t = \infty$, gives the total yield dn/dy as a function of rapidity. Note that for $y > 1$ the m_t distributions are more biased towards high m_t, and a larger extrapolation is needed in those rapidity bins to calculate the total yield. The total yield dn/dy as a function of rapidity is shown in Fig. 3 (a) and (b) for the protons and deuterons, respectively. The proton yield is approximately constant with respect to rapidity, while the deuteron yield decreases with increasing rapidity. Given the wide cut in the centrality used to generate these plots, a significant contribution in the yields is expected to come from spectators. To investigate if the decrease in deuteron production observed in Fig. 3 (b) is due simply to the disappearance of spectator deuterons, it is necessary to examine the yields as a function of centrality as measured by N_{pp}.

The slope parameter as a function of rapidity for protons and deuterons is shown in Fig. 4, for different cuts in centrality. For the less central bin, corresponding to Fig. 4 (a) and (b) the values of the proton slopes seem to be comparable to the deuteron values. On the other hand, for the more central bin (Fig. 4 (c) and (d)), the values of the deuteron slopes are larger for regions closer to mid rapidity. This is consistent with radial expansion following the formation of the participant zone. In this picture, particles with larger masses would gain larger transverse momentum, reflected in larger slope parameters [5].

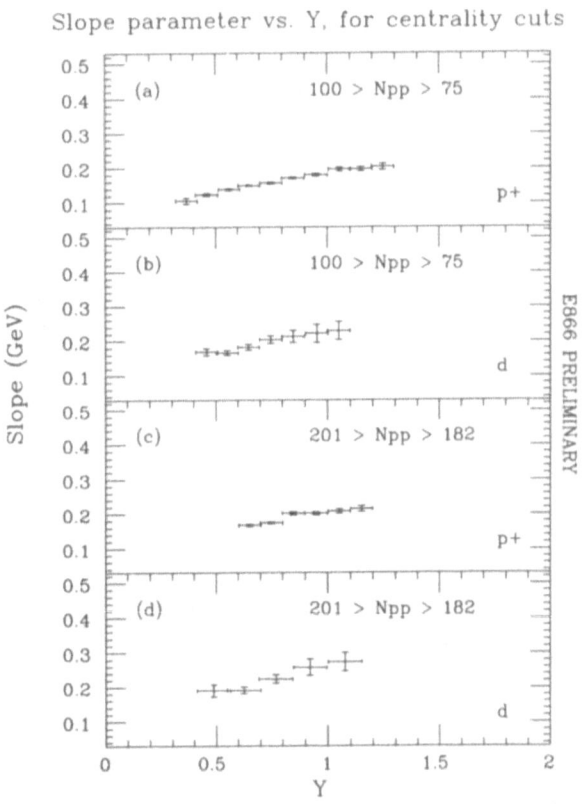

Figure 4. Slope parameter for protons (a), (c) and deuterons (b), (d) as a function of the rapidity for different centrality bins.

The yield distributions dn/dy as a function of rapidity are shown in Fig. 5. Even though the yields for the different N_{pp} bins are not normalized, it is possible to observe a different trend in the dn/dy evolution as a function of the rapidity between the less central bin (a) and (b) and the more central bin (c) and (d). For the $100 > N_{pp} > 75$ bin the proton yield decreases slowly as a function of the rapidity, while the deuteron yield decreases much faster up to $y \simeq 0.8$; after this point the distribution remains constant. For the more central bin both the deuteron and proton distributions behave similarly for all rapidity values. The deuterons with rapidities closer to the target in the less central bin are most likely produced in the spectator region; this is the component that produces the decrease of dn/dy observed in Fig. 5 (b).

The d/p ratio is shown as a function of rapidity in Fig. 6 for different centrality bins. For the most central bin, the d/p ratio remains constant as a function of the rapidity. As the bins become less central, the d/p ratio becomes larger for rapidities closer to the target rapidity. The effect shown in Fig. 6 (a) for the inclusive centrality cut, where the deuteron-to-proton ratio decreases as a function of centrality, is presumably due to the contribution in the yield by deuteron spectators. This effect is also shown in Fig. 7, where the d/p ratio is plotted as a function of centrality for different rapidity cuts. For $N_{pp} > 140$, the d/p ratio does not change as the centrality increases. If the d/p ratio is a measure of the freeze-out baryon density, this stays constant after a certain centrality.

In order to further improve this study it would be very illustrative to look at the d/p ratio in the most central bin as a function of the global particle multiplicity.

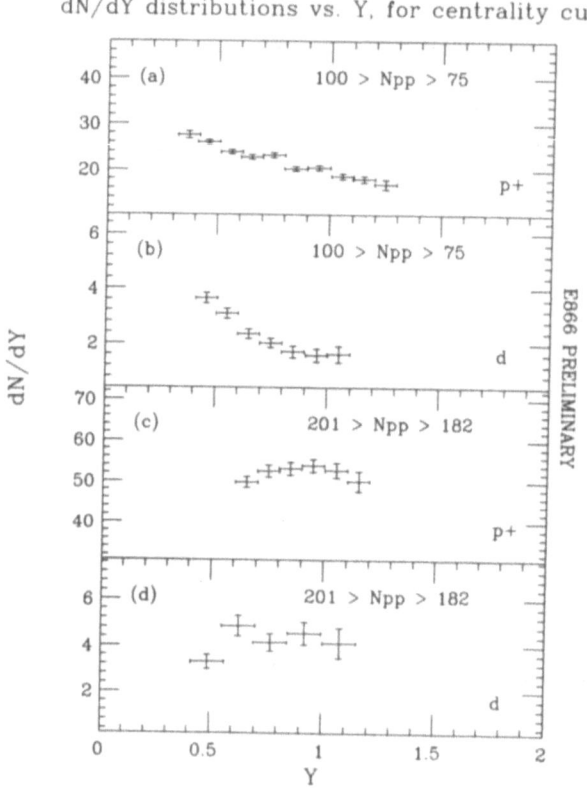

Figure 5. Yield distributions for protons (a), (c) and deuterons (b), (d) as a function of the rapidity for different centrality bins.

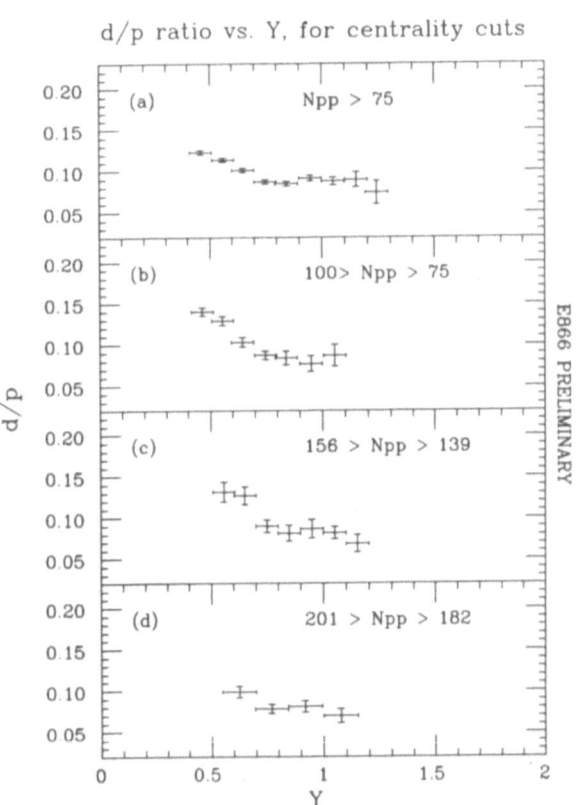

Figure 6. Ratio d/p as a function of the rapidity for different centrality bins.

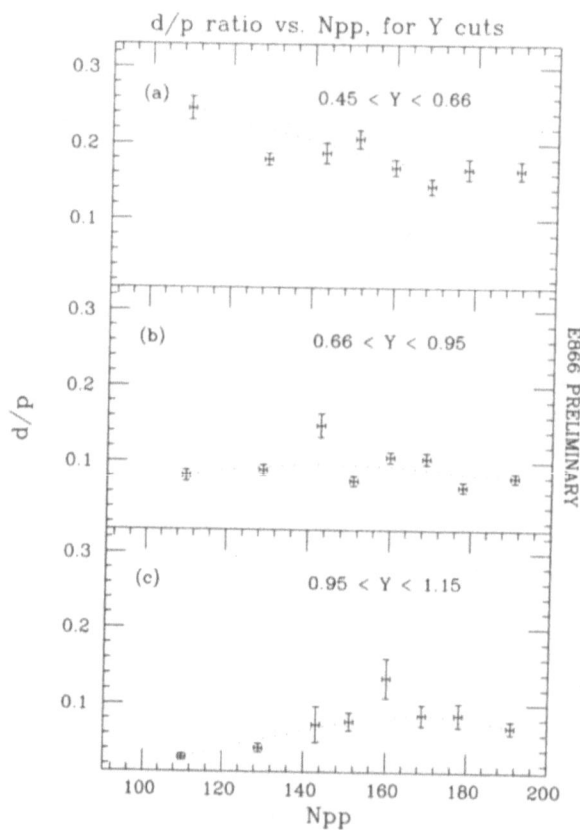

Figure 7. Ratio d/p as a function of the centrality for different rapidity bins. The dashed line are the smoothed points in the plot.

Unfortunately, the 1993 E866 experiment did not have a multiplicity detector and such analysis must wait for the availability of the 1994 data set which includes a multiplicity array the NMA. Another interesting point would be to calculate the cross sections in terms of a variable other than m_t, a variable like $(\gamma\beta)_t$, which scales with the velocity of the particles not with the momentum. Work is in progress to this end.

Acknowledgments

This work was supported by the U.S. Department of Energy under contracts with BNL (DE - AC02 - 76CH00016), Columbia University (DE - FG02 - 86 - ER40281), LLNL (W - 7405 - ENG - 48), University of Maryland (DEFG02 - 93ER40802), MIT (DE - AC02 - 76ER03069), UC Riverside (DE - FG03 - 86ER40271), and by NASA (NGR - 05 - 003 - 513), under contract with the University of California, and by Ministry of Education and KOSEF (951 - 0202 - 032 - 2) in Korea, and by the Ministry of Education, Science, and Culture of Japan.

REFERENCES

1.   U. Heinz *et al.*, Phys. Rev. Lett. **58** (1987) 2292.
2.   W. Weise, Nucl. Phys. **A553** (1993) 59.
3.   Z. Dhen, Proceedings of HIPAGS, edited by C.A. Pruneau, G. Welke, R. Bellweid, S.J. Bennett, J.R. Hall, and W.K. Wilson, Dept. of Physics and Astronomy, Wayne State Univ., Aug. 22-24 1996.
4.   H. Sorge *et al.*, Phys. Lett. **B355** (1995) 27.
5.   R. Mattielo *et al.*, Phys Rev. Lett. **74** (1994) 2180.
6.   A. Z. Mekjian, Phys. Rev. **C 17** (1978) 1051.
7.   J. L. Nage and B. S. Kumar *et. al.*, Phys. Rev. **C 53** (1996) 367.
8.   D.E. Kahana, *et al.*, Phys. Rev. **C 54** (1996) 338.
9.   T. Abbott *et al.*, Nucl. Instrum. Methods **A290** (1990) 41.
10. E-802 Collaboration, T.Abbott *et al.*, Phys. Rev. **C 50**, (1994) 1024.

RECENT RESULTS FROM CERN-WA98

Paul Stankus for the WA98 Collaboration:

M.Aggarwal[13], A.Agnihotri[14], Z.Ahammed[20], A.L.S.Angelis[1], V.Antonenko[15], V.Arefiev[6], V.Astakhov[6], V.Avdeitchikov[6], T.C.Awes[12], S.K.Badyal[5], A.Baldine[6], L.Barabach[6], C.Barlag[10], S.Bathe[10], B.Batiounia[6], T.Bernier[16], K.B.Bhalla[14], V.S.Bhatia[13], C.Blume[10], D.Bock[10], R.Bock[2], E.-M.Bohne[10], D.Bucher[10], A.Buijs[19], E.-J.Buis[19], H.Büsching[10], L.Carlen[8], V.Chalyshev[6], S.Chattopadhyay[20], K.E.Chenawi[8], R.Cherbatchev[15], T.Chujo[18], A.Claussen[10], A.C.Das[20], M.P.Decowski[19], Devanand[5], V.Djordjadze[6], P.Donni[1], I.Doubovik[15], S.Eliseev[11], K.Enosawa[18], H.Feldmann[10], P.Foka[1], S.Fokin[15], V.Frolov[6], M.S.Ganti[20], S.Garpman[8], O.Gavrishchuk[6], F.J.M.Geurts[19], T.K.Ghosh[20], R.Glasow[10], B.Guskov[6], H.A.Gustafsson[8], H.H.Gutbrod[16], R.Higuchi[18], I.Hrivnacova[11], M.Ippolitov[15], H.Kalechofsky[1], R.Kamermans[19], K.-H.Kampert[10], K.Karadjev[15], K.Karpio[3], S.Kato[18], S.Kees[10], H.Kim[12], B.W.Kolb[2], I.Kosarev[6], I.Koutcheryaev[15], A.Kugler[11], P.Kulinich[9], V.Kumar[14], M.Kurata[18], K.Kurita[18], N.Kuzmin[6], I.Langbein[2], A.Lebedev[15], Y.Y.Lee[2], H.Löhner [7], D.P.Mahapatra[4], J.Maharana[4], M.D.D.Majumdar[20], V.Manko[15], M.Martin[1], A.Maximov[6], R.Mehdiyev[6], G.Mgebrichvili[15], Y.Miake[18], D.Mikhalev[6], G.C.Mishra[4], I.S.Mittra[13], Y.Miyamoto[18], S.Mookerjee[14], D.Morrison[17], D.S.Mukhopadhyay[20], V.Myalkovski[6], H.Naef[1], G.K.Nandi[4], S.K.Nayak[4], T.K.Nayak[2], S.Neumaier[2], A.Nianine[15], V.Nikitine[6], S.Nikolaev[15], S.Nishimura[18], P.Nomokov[6], J.Nystrand[8], F.E.Obenshain[17], A.Oskarsson[8], I.Otterlund[8], M.Pachr[11], A.Parfenov[6], S.Pavliouk[6], T.Peitzmann[10], V.Petracek[11], S.C.Phatak[4], F.Plasil[12], M.L.Purschke[2] J.Rak[11], S.Raniwala[14], V.S.Ramamurthy[4], N.K.Rao[5], B.Raven[19], F.Retiere[16], K.Reygers[10], G.Roland[9], L.Rosselet[1], I.Roufanov[6], J.Rubio[1], S.S.Sambyal [5], R.Santo[10], S.Sato[18], P.Saxena[13], H.Schlagheck[10], H.-R.Schmidt[2], G.Shabratova[6], I.Sibariak[15], T.Siemiarczuk[3], K.Singh[13], B.C.Sinha[20], N.Slavine[6], K.Söderström[8], N.Solomey[1], S.P.Sørensen[17], P.Stankus[12], G.Stefanek[3], P.Steinberg[9], E.Stenlund[8], D.Stüken[10], M.Sumbera[11], T.Svensson[8], M.D.Trivedi[20], A.Tsvetkov[15], C.Twenhöfel[19], L.Tykarski[3], J.Urbahn[2], N.v.Eijndhoven[19], W.H.v.Heeringen[19], G.J.v.Nieuwenhuizen[9], A.Vinogradov[15], Y.P.Viyogi[20], A.Vodopianov[6], S.Vörös[1], M.A.Vos[19], B.Wyslouch[9], K.Yagi[18], Y.Yokota[18], G.R.Young[12]

[1]University of Geneva, Switzerland, [2]GSI Darmstadt, Germany, [3]INS Warsaw, Poland, [4]IOP Bhubaneswar, India, [5]University of Jammu, India, [6]JINR Dubna, Russia, [7]KVI Groningen, Netherlands, [8]University of Lund, Sweden, [9]MIT Cambridge, USA, [10]University of Münster, Germany, [11]NPI Rez, Czech Rep., [12]ORNL Oak Ridge, USA, [13]University of Panjab, India, [14]University of Rajasthan, India, [15]RRC (Kurchatov)

Advances in Nuclear Dynamics 3, edited by
Bauer and Mignerey, Plenum Press, New York, 1997

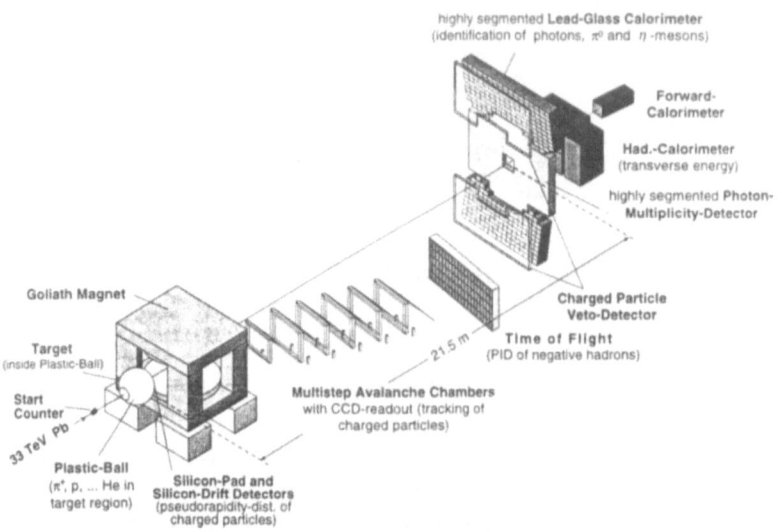

Figure 1. The WA98 experimental setup in 1995.

Moscow, Russia, [16]SUBATECH Nantes, France, [17]University of Tennessee Knoxville, USA, [18]University of Tsukuba, Japan, [19]University of Utrecht, Netherlands, [20]VECC Calcutta, India

She sang beyond the genius of the sea.

...

She measured to the hour its solitude.
She was the single artificer of the world
In which she sang.

Wallace Stevens, *The Idea of Order at Key West*, 1936

Introduction: A Tour of the Experiment

The CERN experiment WA98 is a general-survey, open-spectrometer experiment designed to examine 160 A GeV/c Pb+A collisions at the CERN-SPS. The experiment has a broad physics agenda, as suggested by its many different subsystems. A diagram of the experiment as it stood in 1995 is shown in Figure 1. Detectors whose results are presented here are described briefly:

Surrounding the target is the venerable Plastic Ball detector (PBall), an array of $\sim$400 $E - dE/dx$ telescopes packed in a geodesic-dome-like arrangement. Just downstream from the target, still inside the PBall, the collision is observed by the Silicon Pad Multiplicity Detector (SPMD), a circular annulus with $\sim$4000 charged-particle-sensitive pads covering the range $2.4 < \eta < 3.8$. Charged particles are bent in a wide-aperture dipole magnet and are tracked in six Multi-Step Avalanche Chamber (MSAC) tracking stations. Individual photons are detected in the Photon Multiplicity Detector (PMD),

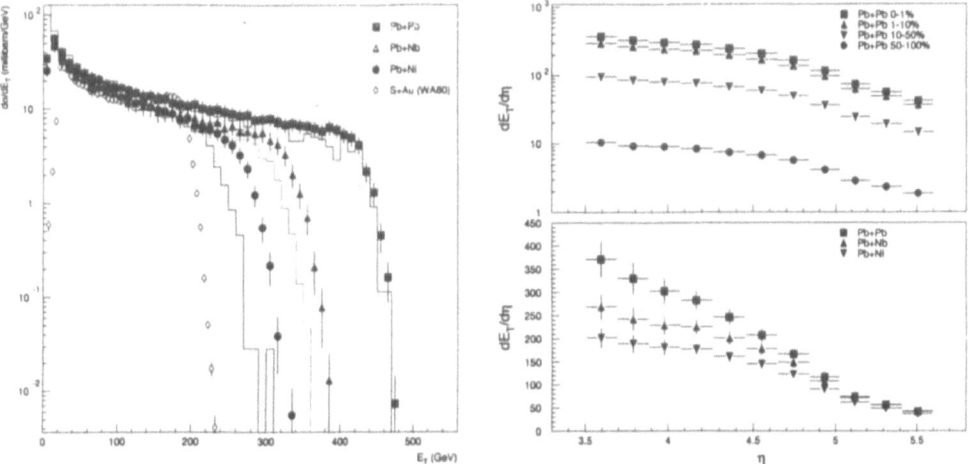

Figure 2. Left: Distribution $d\sigma/dE_t$ of transverse energies measured in the pseudorapidity interval $3.5 < \eta < 5.5$ for different systems in WA98 and WA80; the lines are predictions made using the VENUS 4.12 model. Right: Distributions $dE_t/d\eta$ of transverse energy at different pseudorapidities.

a highly-segmented pixel array of preshower counters. At higher lab angles, energies and positions of individual photons are reconstructed in the highly-laterally-segmented ($\sim$10,000-element) Lead Glass Calorimeter Array (LEDA). At the most forward angles ($\eta > 3.5$) the total hadronic and electromagnetic energy in different η bins is measured by the segmented Mid-Rapidity Calorimeter (MIRAC).

The physics topics addressed by WA98 include: global measurements of E_t and multiplicity in the MIRAC, SPMD and PMD; correlated $N_{charged}/N_{neutral}$ fluctuations as would be produced in a disordered chiral condensate (DCC), using the SPMD and PMD; charged particle singles spectra and pair correlations for HBT using the MSACs; reconstruction of π^0 and η^0 mesons as well as inclusive and direct photons in the LEDA; global measurements of flow patterns using the PBall and the PMD; as well as many others. Preliminary analyses of E_t production, HBT measurements, the DCC search, and reaction plane measurement are discussed here below.

Transverse Energy Production

The spectra $d\sigma/dE_t$ for transverse energy production at forward pseudorapidities, and distributions $dE_t/d\eta$ of transverse energy at different pseudorapidities, are shown in Figure 2 for different Pb+A colliding systems. The monotonically decreasing shapes of the $d\sigma/dE_t$ distributions agree qualitatively with what one would expect for a symmetric collision, or a one with a larger projectile, as opposed to the peak structure observed in S+Au collisions [1]. More revealing are the distributions of E_t over different pseudorapidities. These allow a determination, through Gaussian* fits, of the peak $dE_t/d\eta$, indicating the energy density reached in the collision, and the width of the

---

*Though it is true that the WA98 spectra are measured on only one side of mid-rapidity, confidence in Gaussian fitting is lent after observing that combining the WA98 spectrum for the symmetric Pb+Pb collision, plus its reflection, with the distribution straddling mid-rapidity measured by NA49[2] leads to a smooth and reasonably Gaussian shape overall.

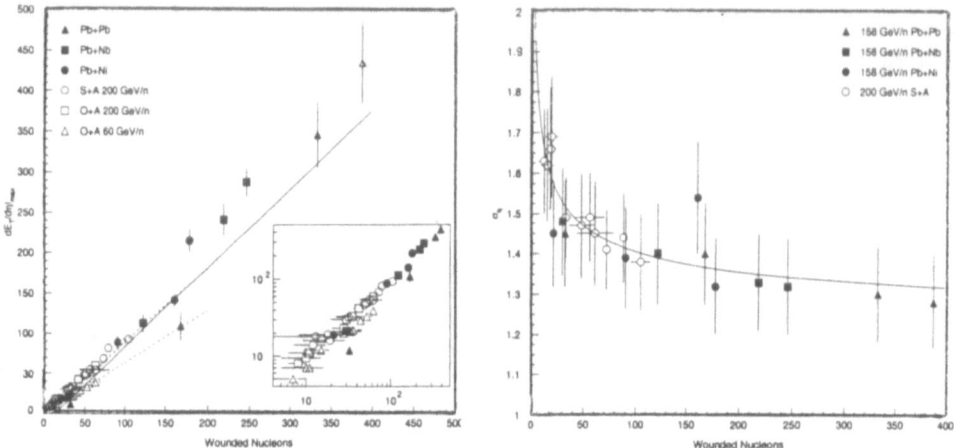

Figure 3. Left: Values of the maxima of $dE_t/d\eta$ distributions for different centrality classes in different collisions are shown, plotted against the average number of wounded nucleons in each collision class as determined through a Glauber-style calculation; the inset graph is the same data on a log-log scale. Right: Values of the widths of the $dE_t/d\eta$ distributions.

$dE_t/d\eta$ distributions which indicates the degree of stopping in the collision.

The behavior of the peaks and widths of the $dE_t/d\eta$ distributions for events at different centralities in a variety of colliding systems are plotted in Figure 3. Here the different classes in the different systems are each described by the average number of wounded nucleons per event, as determined through a Glauber-style calculation. Plotted this way, it is clear that the trend seen[1] in the lighter systems continues in the present data. The peak $dE_t/d\eta$'s are proportional to the number of wounded nucleons over a wide range and a variety of systems. The widths continue to decrease with increasing system size, indicating a greater degree of stopping in larger systems. The corresponding Bjorken-style estimates for initial energy densities also rise from ~ 2 GeV/fm^3 in central S+Au to ~ 3 GeV/fm^3 in central Pb+Pb.

Particle Tracking and HBT

Charged particles are detected and tracked in WA98 through the series of six MSAC tracking stations. Each of these consists of a multi-step avalanche chamber to amplify the particles' primary ionization, followed by a converter to generate visible light at location of the ionization, whose pattern is then recorded in CCD cameras. In effect, each station is a pixel plane with on the order of a million elements. The tracking arm is backed up by a scintillator-slat time-of-flight array to allow particle identification, though the spectrometer is primarily used to track negative particles which are in the great majority pions.

Figure 4 shows the measured single-particle spectra for negative particles in central collisions compared with the spectra for π^0's as reconstructed in the LEDA array[3]; the shapes of the two spectra continue match very smoothly. The figure also shows the C_2 correlation function for pairs of negative particles, along with parameters from a one-dimensional Gaussian functional fit[4]. The fit radius parameter of 6.7 fm is similar

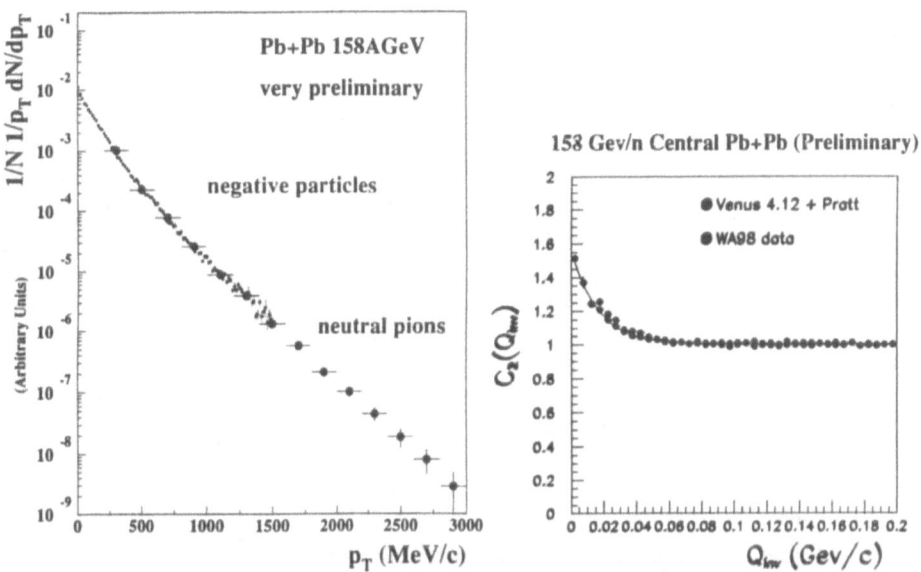

Figure 4. Left: Comparison of single-particle Pt spectrum shapes; small symbols are negative charged particles as tracked in the MSAC stations, large symbols are π^0's reconstructed in the LEDA (relative normalization is arbitrary). Right: The C_2 correlation function for negative pairs produced in central Pb+Pb collisions as a function of Q_{inv}; the parameters resulting from a functional fit are shown, and compared to similar analyses from other experiments.

to those seen in other SPS Pb+Pb experiments. While Gaussian functional fits are traditional in HBT analyses, the WA98 data are noticeably better[4] described by a simple exponential in Q_{inv} than by a Gaussian.

Search for Disordered Chiral Condensates

One of the more intriguing physics topics recently discussed has been the possibility that within a high-energy heavy-ion collision there might be formed a *disordered chiral condensate* (DCC). A DCC is an extended domain of excited nuclear matter which exhibits isospin coherence; when the DCC decays/hadronizes, the resulting pions are either preferentially charged or preferentially neutral. The experimental signature of DCC formation, then, would be larger-than-expected fluctuations in the $\pi^0/\pi^{\pm}$ ratio measured event-by-event. WA98 can observe such fluctuations, event by event, by comparing $N^{charged}$ as measured in the SPMD to N^{photon} as measured in the PMD (within their common pseudorapidity coverage).

Of course, in order to make a sensitive search one needs to do two things: (1) to credibly model what size the "expected" fluctuations would take, and (2) to predict the size (and frequency) of fluctuations that would result from a DCC being formed. In the WA98 preliminary DCC analysis these were carried out as follows. The VENUS simulation was used to model the basic Pb+Pb central collisions; parameters were tuned until the predicted spectra for individual $N^{charged}$ and N^{photon} per event agreed very well with the measured spectra, after which the joint distribution was presumed to model collisions in which no DCC is formed. The simulated formation of DCC's was modeled according two parameters: frequency, i.e. on what fraction of events is a DCC formed; and intensity, i.e. what fraction of the total pions produced in that event come from the DCC domain. Figure 5 shows the simulated distributions with and without a DCC component, and the distribution in the WA98 data that were used for the preliminary analysis. The distribution in the data clearly resembles that of the simulation with no DCC formation, and there were no events in the designated search region for DCC candidate events.

Within the two-parameter model of DCC production, these data can be used to set limits on the intensity and frequency with which DCC's are produced in central Pb+Pb collisions. The resulting limits are illustrated in Figure 6. In the future, of course, the sensitivity of even this simple DCC search can be improved. Substantially increased statistics from a period of dedicated DCC-search running in 1996 will improve sensitivity to the frequency; a more sophisticated model for the background fluctuations will increase sensitivity to the intensity. A new approach to modeling the background is currently being investigated, in which the fluctuations within limited regions of azimuthal angle ϕ are modeled by rotating the pattern of charged hits relative to neutrals within single events. In searching for DCC's which are localized in ϕ this method can obviate the need for a sophisticated simulation and in principle lower the uncertainties of modeling background fluctuations to a different order altogether.

Reaction Plane Measurement

It has been a long-standing question in relativistic heavy-ion physics as to whether any artifact of the *orientation* of the original impact parameter – the reaction plane

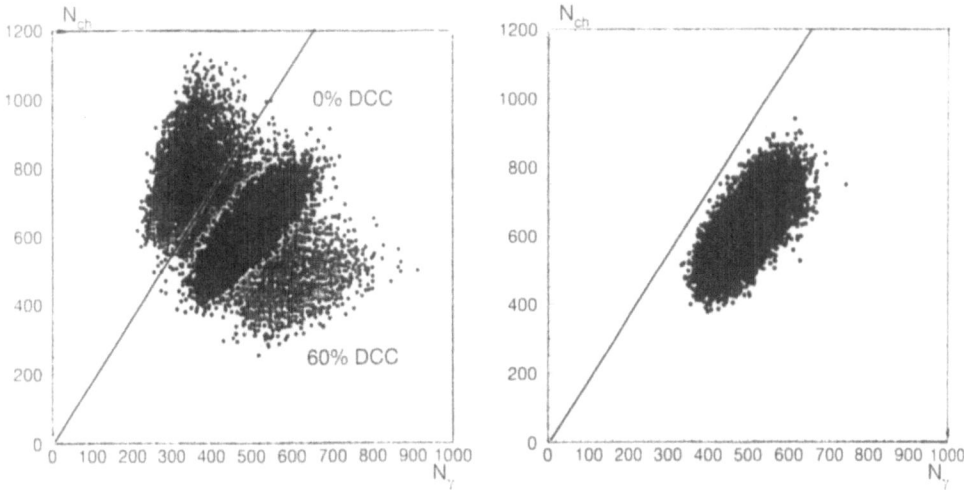

Figure 5. Left: Event-by-event $N^{charged}$ vs N^{photon} joint distributions for simulated Pb+Pb collisions in which no DCC's are formed (dark points) compared to the case in which 60% of the pions in each event come from a DCC (light points). Right: The joint distribution in the WA98 data used in the preliminary analysis. The region to the upper left of the line is the search region for DCC candidate events.

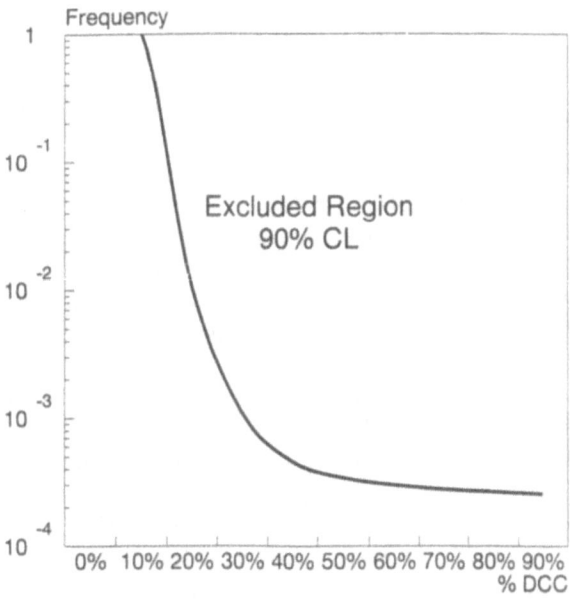

Figure 6. Limits, in a simplified model, on the intensity and frequency of DCC production in central Pb+Pb collisions from the preliminary WA98 analysis.

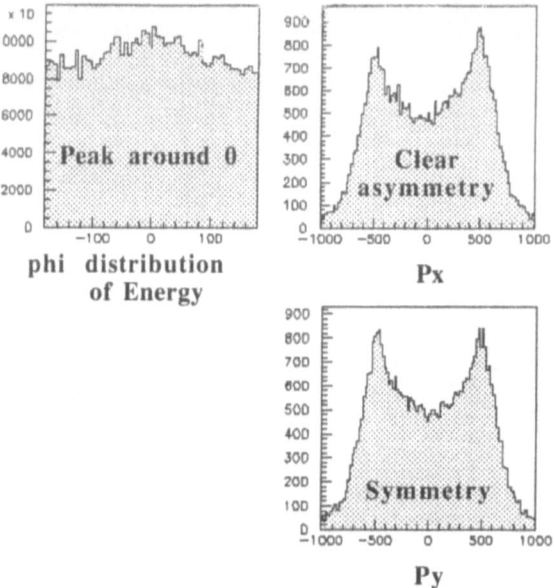

Figure 7. Distribution of the direction of secondary light fragments relative to the net proton direction, weighted by fragment kinetic energy. Left: distribution of fragment azimuthal angles; Right: projection of fragment directions along the net proton direction (the "X" axis) and cross to it (the "Y" axis). These are compiled from events of intermediate centrality.

direction – remains in the pattern of the particles' final states. Accordingly, much effort has gone into looking for azimuthal asymmetries using detectors with wide coverage, and then looking for correlations between the axes of asymmetries measured with different kinds of particles or in different rapidity ranges.

As a first step in such a program, a technique has been investigated in WA98 to identify a reaction plane direction using secondaries near target rapidity observed with the PBall. Each of the PBall's elements can identify the relatively low-energy (in the lab) secondaries by measuring their total kinetic energy in a thick scintillator and their velocity through ionization in a thin scintillator – the well-known E *vs* dE/dx match. Of particular concern here are the identification of protons *vs* light fragments, for which the technique works well with few complications. The candidate for a reaction plane direction on a given event is taken as the average azimuthal direction of all the protons in the target region, weighted by their kinetic energy.

In order to check that this candidate had the right properties for a reaction plane direction, two checks were made. First, if the direction of the summed proton momentum is not simply a random fluctuation, then it should be correlated with the directions of other secondary particles. To study this, the direction of secondary light fragments in the target region, *relative* to the net proton direction in that event, was compiled; the results are shown in Figure 7. There correlation is clear: light fragments are emitted preferentially in the same direction as the protons. Further, the extent of the fragment–proton correlation provides a measure of how extant the reaction plane is for a given class of events, which leads to the second check. If this correlation is really a remnant of the original collision geometry then it should weaken in the most

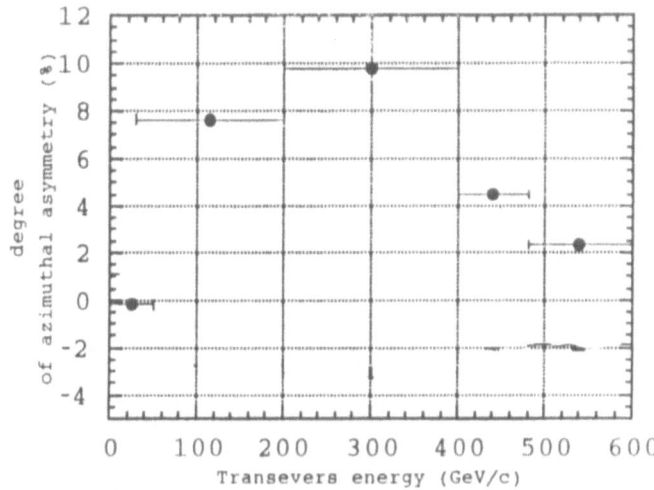

Figure 8. The strength of the proton–fragment correlation *vs* centrality for different event classes; the strongest effect is seen for intermediate centralities as would be expected for a remnant of the original collision geometry. (The ordinate is the slope resulting from fitting the $|\phi_{fragment} - \phi_{protons}|$ distribution with a simple linear function.)

peripheral collisions, which most strongly resemble p+p collisions in which there are fewer re-interactions with spectator matter; and it should greatly weaken or disappear completely in the most central collisions where there is little or no original asymmetry. Figure 8 bears this expectation out: the proton–fragment correlation is strongest for events of intermediate centrality.

With these observations, the net proton direction as measured with the PBall is a reasonable candidate for an event-by-event reaction plane direction. It will be interesting to see in later analyses how the patterns of particle and energy flow at higher rapidities correlate with this direction; and how individual particle/pair measurements such as HBT and P_t spectra differ with and across this direction.

Conclusions and Outlook

The WA98 experiment is a multi-purpose survey experiment with a broad physics agenda to investigate Pb+A collisions at the CERN-SPS. A sample of preliminary findings have been described here. Distributions of transverse energy produced at forward rapidities continues the trend seen with lighter projectiles: the larger interacting systems (as measured by number of wounded nucleons) show a higher peak energy and a greater degree of stopping. HBT-style correlations of negative particles indicate a source size in central collisions similar to that of the lead nuclei, and similar to those observed in other SPS experiments. A search has been made for excess event-by-event charged/neutral fluctuations as might result from formation of the disordered chiral condensate (DCC). Within a simple model this search has ruled out DCC formation at the level of $\sim 1/10^3$ collisions; greatly increased statistics from dedicated running, as well as more sophisticated background simulations and production models are expected to extend the sensitivity of the search in the future by several orders of magnitude. Target-rapidity protons measured in the Plastic Ball reveal a reasonable candidate for an event-by-event reaction plane direction in collisions of intermediate centrality. Future analyses of wide-acceptance detectors at mid rapidities will measure the correlation of various particles' production with this reaction plane direction, potentially shedding light on the relationship between geometry and dynamics in these collisions.

Acknowledgments

The speaker would like to thank Peter Steinberg, Hal Kalechofsky, David Morrison and Mizuki Kurata for their help in preparing the presentation (and for doing the original work!).

This work was supported jointly by the German BMBF and DFG, the U.S. DOE, the Swedish NFR, the Dutch Stichting FOM, the Swiss National Fund, the Humboldt Foundation, the Stiftung für deutsch-polnische Zusammenarbeit, the Department of Atomic Energy, the Department of Science and Technology and the University Grants Commission of the Government of India, the Indo-FRG Exchange Programme, the PPE division of CERN, the International Science Foundation under Contract N8Y000, the INTAS under Contract INTAS-93-2773, and ORISE. ORNL is managed by Lockheed Martin Energy Research Corporation under contract DE-AC05-96OR22464 with the U.S. Department of Energy.

REFERENCES

1. Albrecht, et al., Phys Rev **C44** (1991) 2736
2. Alber, et al., Phys Rev Lett **75** (1995) 3814
3. Peitzmann, et al., Nucl Phys **A610** (1996) 200c–212c
4. Rosselet, et al., Nucl Phys **A610** (1996) 256c–263c

EVENT-BY-EVENT PHYSICS AT THE CERN SPS

Thomas A. Trainor

Nuclear Physics Laboratory 354290
University of Washington
Seattle, WA 98195

INTRODUCTION

Event-by-event physics in nuclear collisions has been an active field of research for more than a decade. Jet production in p-p and e-e collisions and flow in heavy-ion collisions have been extensively studied at LEP, CSPS, TEVATRON, BEVALAC and AGS among other facilities. Jets and flow are large-amplitude and/or large-scale effects which are detectable despite the rather small particle multiplicities per event from these collision systems. Multiplicities have been limited by small system size (p-p, e-e) or low energy densities (heavy ions). Both jet and flow phenomena are manifestations of symmetry reduction or increased correlation with respect to a nominally thermalized system of produced particles.

With the recent availability of a 158 GeV/nucleon lead beam at the CERN SPS one obtains (due to larger system size and increased energy density) substantially increased event multiplicities which can provide sensitivity to smaller-amplitude symmetry reductions over a finite scale interval. This increased statistical power permits event-by-event study of global thermodynamic variables and the possibility to extend this analysis program beyond global or large-scale variables to a scaling analysis approach. This additional analysis capability is needed to explore fully so-called soft or nonperturbative QCD phenomena.

With anticipated RHIC and LHC particle multiplicities 5 and 50 times greater respectively than those at SPS, future event-by-event analysis should become sensitive to small-amplitude symmetry reductions over an extensive scale range, permitting much more elaborate experimental QCD studies.

EVENT-BY-EVENT PHYSICS

We seek evidence for color deconfinement and chiral symmetry restoration in heavy ion collisions by a correlation analysis of the distribution of produced particles. These

Advances in Nuclear Dynamics 3, edited by
Bauer and Mignerey, Plenum Press, New York, 1997

QCD symmetry enhancements are expected to occur with some observable frequency at CERN SPS energies according to lattice gauge theory predictions [1]. We search for evidence in the hadronic sector for such symmetry increases by looking for unusual variations (dynamical fluctuations) in and correlations among global thermodynamic variables [2] and departures from thermal symmetry over a range of scale in momentum space.

The search for large-scale symmetry increases associated with QGP formation by detecting departures from thermodynamic equilibrium introduces the question of local symmetry reduction. One frequently finds in the dynamics of complex systems situations in which the large-scale symmetry of a system is increased by a strategy of lowered local symmetry. Rayleigh-Bénard convection is an example. By analogy we search for dynamical fluctuations on an event-by-event basis which are manifestations of these local symmetry reductions during the collision process. An important question is whether evidence for these symmetry reductions survives until the decoupling of the multiparticle hadronic system.

Among the proposed phenomena associated with QCD symmetry restoration the disoriented chiral condensate (DCC) serves as a paradigm. If chiral symmetry restoration occurs within the highest energy-density regions of the collision volume the distinction among pion species is predicted to be lost there [3]. In the subsequent cooling process this symmetric state may decay preferentially into a particular isospin state over some finite volume, resulting in nonstatistical deviations of the neutral pion fraction from its usual value (1/3) in some regions of phase space.

The observability of the DCC phenomenon hinges on both the thermodynamic trajectory of the collision process and the observability of the soft-pion component of the particle spectrum. Observable DCC effects seem to depend on significant departure from an equilibrium thermodynamic trajectory (during the cooling phase) according to current model calculations [4]. And manifestation of a DCC in particle spectra is expected to occur only for transverse momenta near or below the pion mass. In general, neutral and charged pions are not observed with the same detector components, there are significant low-momentum detection thresholds, and phase-space acceptance overlap for these two species may be incomplete, giving rise to the possibility of significant systematic error. Nevertheless, theoretical and experimental studies of this phenomenon are being vigorously pursued.

Other soft-QCD effects may result from the nonabelian nature of the QCD color field and may produce various symmetry reductions in the collision system (e.g., color ropes, 'quark-matter' droplets and other transverse string-coupling schemes) [5]. This may produce fluctuations in the momentum-space distribution on an event-by-event basis: for example, fluctuations in the degree of participant stopping reflected in event-by-event variations in the rapidity distribution of net baryons. The baryon-to-meson ratios may also be affected by the higher local energy densities resulting from these cross couplings [6].

At the perturbative level, fluctuations in the propagation of high-p_t partons through hadronic and/or QGP matter may be manifest in modified parton dynamics and subsequent hadronization into jets (jet quenching) [7]. Fluctuations in the early stages of the parton cascade may also result in significant event-by-event fluctuations in produced hadron spectra [8, 9, 10].

In spite of major progress in the theoretical understanding of the QGP in recent years there are only a few specific predictions of measurable phenomena, especially in the hadronic sector. The experimental emphasis therefore has to be to search every aspect of the available data for the possibility of anomalous behavior which could be related to QGP formation. What follows is a summary of progress to date on three major event-by-event analysis programs associated with two large-acceptance SPS experiments, WA98 and NA49.

WA98 DCC ANALYSIS

WA98 is a large-acceptance photon and hadron spectrometer. This dual capability makes a DCC search possible by comparing the yields of photons (85% from π_0 decay) and charged particles (80% $\pi_{\pm}$) detected with corresponding multiplicity arrays (PMD and SPMD). The charged particle acceptance is $2.35 < \eta < 3.75$, and the photon acceptance is $3.0 < \eta < 4.2$, with a 0.4 GeV cutoff and 20% hadron contamination. A calorimetric centrality trigger accepts 10% of the minimum-bias cross section.

The initial WA98 event-by-event DCC analysis [11] utilizes the total yields of neutral and charged particles for each event, distributed on a 2D scatter plot. This distribution shows a significant linear correlation corresponding to correlated variations of the two species with total event multiplicity. The event-by-event signal of interest is the degree of deviation from the correlation axis. This deviation is equivalent to a variation in the N_γ / N_{ch} ratio, or the neutral particle fraction, as a global event variable. Excessive fluctuation (beyond counting statistics) of this deviation from the correlation axis could be an indication of an anomalous neutral-to-charged pion ratio resulting from DCC formation.

With 16k events analyzed there are no events falling outside an established DCC cut at present. A further 16k events have recently been processed. It is planned to extend this analysis to smaller-scale regions of the acceptance and incorporate momentum information to make possible a search on the low-p_t component of the pion spectrum.

NA49 EVENT-BY-EVENT ANALYSIS

NA49 is a large-acceptance (nearly 4π) hadronic spectrometer [12]. It consists of two vertex magnets with an internal or vertex TPC in each magnet and two more external or main TPCs (left-right) covering the forward hemisphere. Triggering is done with a small-angle veto calorimeter at 4% of the inelastic cross section (b<3.5 fm). The tracking analysis software has been developed in two stages. For the work described here a pilot event-by-event analysis was carried out on 160k events (6% of total data volume) with main TPC tracking only.

NA49 event-by-event analysis has proceeded on two complementary fronts: large-scale or global event-variable analysis – based primarily on a thermodynamic approach, and scaled correlation analysis of momentum space, which attempts to extend the event-by-event concept over a range of scales limited only by event multiplicity. There is also a flow analysis program underway.

Global Variable Analysis

Available information in the hadronic sector comes from momentum distributions of various charged and neutral particle species or their decay products. The initial

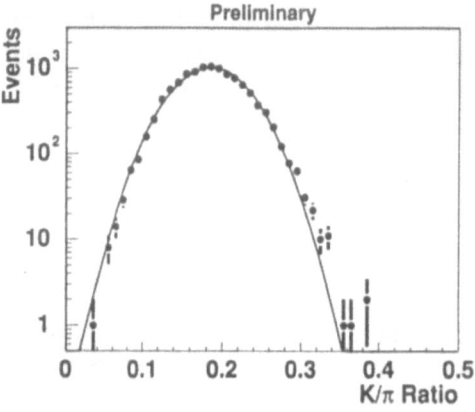

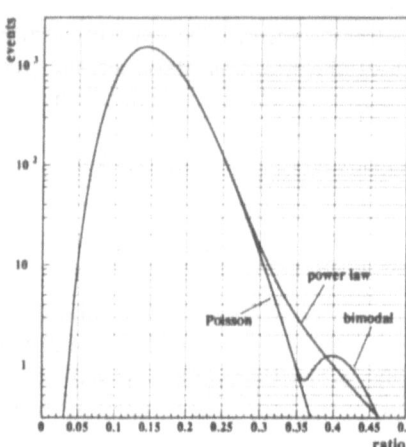

Figure 1. Frequency distribution of uncorrected K/π ratio for 11k events illustrating NA49 global-variables analysis [13].

Figure 2. Illustration of possible deviations from standard Poisson distribution due to power-law or bimodal component.

approach to event-by-event analysis of these spectra relies on so-called 'global' event variables: integral quantities which reflect the largest-scale properties of the multiparticle phase space.

Examples of global variables are $<p_t>$, K/π and other abundance ratios, m_t-spectrum slope parameter and some measure of the width of the rapidity distribution about the CM (for each species). These quantities all measure in some way the 'boundedness' of the phase-space distribution in momentum and flavor composition.

These are the only statistically meaningful quantities at lower multiplicities, and fluctuations in these variables may be dominated there by counting statistics. However, with increasing multiplicity and collision energy density these quantities should carry significant dynamical information in their variances.

In global variable analysis several variables are evaluated on an event-by-event basis. Distributions of these variables are compared to reference systems provided by event-mixing techniques or by Monte Carlo event generators. Mixed events are constructed with the same multiplicity distribution as the original data. As an example the distribution of K/π ratio in the NA49 main TPC acceptance is shown in fig. 1. These data are not acceptance corrected. Thus, the shape of this distribution is relevant here, but not its mean value.

One looks for deviations of the distributions of these variables from an event-mixed or other reference population. An illustration of possible deviation forms is shown in fig. 2. One can also look for correlations among global variables. One seeks a (possibly small) set of events that can be isolated from the general event population as being somehow anomalous, deviating from smooth central-limit behavior. This would signal unusual trajectories through a thermodynamical state space for this event class, perhaps corresponding to color deconfinement phenomena.

Preliminary results of this study [13] relate to rescattering effects and chemical equilibrium. Distributions of global variables are consistent with central limit behavior over three orders of magnitude. Any deviations are consistent with projectile pileup

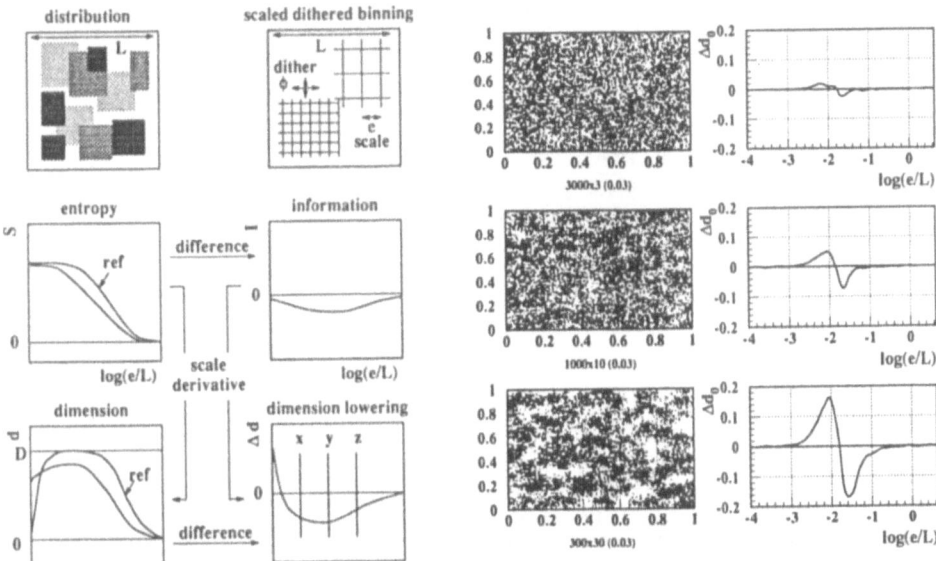

Figure 3. Schematic representation of scaled correlation analysis procedure.

Figure 4. Dimension lowering distributions for three different degrees of correlation.

effects. Comparison of the variance of event-by-event $<p_t>$ with the variance of p_t over the entire event ensemble for Pb-Pb data and simulated p-p collisions indicates that the correlations found in hadronic p_t spectra as produced in individual p-p collisions, and attributed in part to string fragmentation statistics, are significantly attenuated by rescattering in Pb-Pb collisions. A variance analysis of the K/π ratio indicates that fluctuations in the flavor composition of produced particles in Pb-Pb are small compared to the average strangeness enhancement in Pb-Pb vs p-p.

Scaled Correlation Analysis

With larger collision systems and energies, event multiplicity (statistical power) becomes sufficiently great to extend global-variable analysis by using more detailed measures of the correlation content of the phase-space distribution. At the SPS the total pion multiplicity for 158 GeV/nucleon Pb-Pb collisions is more than 1000, making it worthwhile to pursue a more detailed analysis for this particle species. In practice we take all charged hadrons that fall into the detector acceptance for the initial event-by-event analysis.

In constructing a model-independent analysis system sensitive to anomalous events one wants to construct a differential measure of the correlation content of the momentum distribution. The general procedure for this analysis [14] is represented in fig. 3. The object is to extract all available information from a multiparticle distribution in a model-independent way. Scale is all-important. Especially for a system out of equilibrium, the state of the system must be described over scale for the description to be complete.

At each scale value the system is binned, and the primary correlation measure,

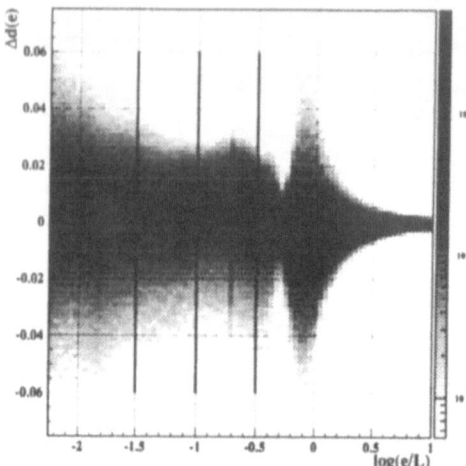

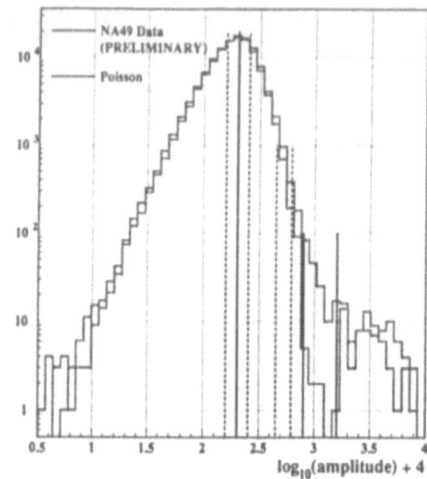

Figure 5. Dimension lowering distribution on scale for real data events showing three scale points used to form an event-spectrum space.

Figure 6. Frequency distribution on 'amplitude' or distance from the event spectrum origin for real data and Poisson events.

the scaled entropy S(e), is determined. A dithering procedure is carried out to reduce aliasing effects. The entropy is then compared with that for some reference system to obtain the scaled information I(e). The reference may be an analytic expression, a Monte Carlo simulation, an ensemble average, mixed events, etc.

The scale derivative of the entropy is the scale-local dimension, a new concept in correlation measurement. The scale derivative of the information is the 'dimension lowering' $\Delta d(e)$, a measure of the transport of dimension on scale, and the central differential measure of scaled correlation for this event-by-event analysis. It measures changes in correlation or symmetry content of the distribution independently at different scales. A simulation illustrating dimension transport on scale for three different degrees of correlation produced by a hierarchical generator are shown in fig. 4.

Because of the limited statistical power of the NA49 multiplicity we have first carried out 1D analyses (here on m_t-m_π). There is a tradeoff between the number of dimensions (degrees of freedom) of the analyzed distribution and the accessible range of scale. With 700 particles in the main TPC acceptance we can analyze a 1D distribution over almost two decades of scale. For a 2D distribution one would be restricted to less than one decade. The concept of 2D scaled analysis may be moot for NA49 data except for large-amplitude effects.

From the values of dimension lowering at three points in scale (-0.5, -1, -1.5) we form a cartesian 'event spectrum.' Each point in this distribution represents an event. Distance from the origin in this space represents deviation from the reference. Direction represents the 'shape' of this deviation. To continue the analysis we compare the event spectrum for real data with that for a set of 'Poisson partners' analyzed in identical fashion. Each element of this Poisson partners set corresponds to a particular event

50

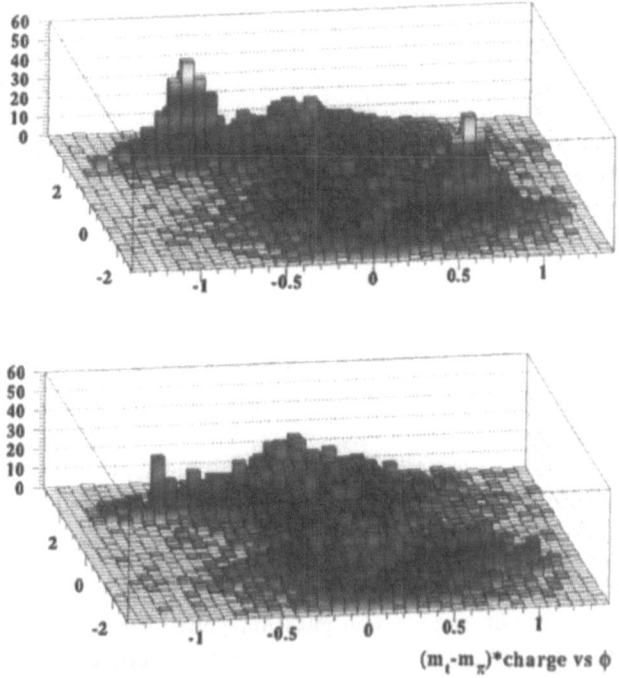

Figure 7. Distributions of positive and negative particles on transverse mass relative to pion mass (GeV/c^2) vs ϕ for anomalous events (top) and reference events (bottom).

in the real data sample, with the same multiplicity and a simple thermal distribution having slope parameter equal to that of the ensemble averaged real data.

Fig. 5 shows the dimension lowering distributions for a number of real-data events and the scale values chosen to construct the event spectrum. Fig. 6 shows the distribution of distances from the event spectrum origin (amplitude) on a log-log plot for real data and Poisson partners. There is an obvious difference between the two distributions near $\log(\text{amp}) = 3$. We select real-data events from this neighborhood and designate them as anomalous events.

Finally, in fig. 7 we plot the m_t-m_π vs ϕ distribution for this ensemble of anomalous events and for an equal number of control events selected from the region of the $\log(\text{amp})$ distribution near the peak. Here one sees in the anomalous event set a prominent peak at m_t-$m_\pi = 0.7$ GeV/c^2, for both signs of particle charge. The fraction of anomalous events is a few permil. The number of additional particles/event that are contained in the peak structure is about 15 out of 350, or 4%. This should be compared with the rms deviation of the normal-event multiplicity, which is about 40 particles. This result demonstrates the high sensitivity of this model-independent method for selection of anomalous events.

Physical interpretation of the anomalous events found in this pilot study is questionable because the main TPCs alone, being outside the magnetic fields of the vertex magnets, cannot uniquely determine the point of origin of a particular particle, whether it is from the main vertex or not, or whether the particle came from a pile-up interaction in the TPC gas. The full data analysis presently underway incorporates unified or global tracking across all NA49 TPCs and should eliminate such ambiguities.

CONCLUSION

The event-by-event physics program at the CERN SPS is just now entering the large-scale data analysis stage after more than a year of pilot studies and software development. Much progress has been made in identifying the observables and procedures that will provide an optimized analysis. We can anticipate a period of intense study of systematic effects as large processed data volumes now become available.

Coupled to this analysis program are the problems of significance and interpretation. By what criteria can we conclude that a class of events is anomalous, and by what means can we interpret the physics content of these anomalous events? The problem of significance is a long-standing one in data reduction. However, standard criteria such as the χ^2 probability distribution may not suffice in the present context because of the complex and nonlinear nature of the problem. This is an area where new statistical concepts are needed.

The standard use of Monte Carlo event generators coupled to GEANT modeling of the experimental apparatus in more traditional event-ensemble physics analysis is not practical for event-by-event analysis, where the number of simulated events needed is typically an order of magnitude or more larger, and the computation time/event is also ten or more times larger due to additional model complexity and larger collision system size. There is simply not enough computation power presently available to pursue this standard approach for event-by-event analysis. Alternatives are presently being explored, including elimination of GEANT modeling for all but a small subpopulation of events and direct analysis of event-generator output for correlation content.

Finally, in order to better anticipate what observable effects may arise in event-by-event analysis there are now several Monte Carlo collision model codes which permit one to explore a model parameter space which includes some anomalous dynamics in order to map out the corresponding fluctuation content of produced events. These 'calibrations' should prove valuable for the interpretation problem.

REFERENCES

1. J.C. Collins and M.J. Perry, Phys. Rev. Lett. **34** (1975) 151; E.V. Shuryak, Phys. Rep. **C61** (1980) 71 and **C115** (1984) 151.
2. R. Stock, Proceedings of the NATO Advanced Research Workshop on *Hot Hadronic Matter: Theory and Experiment*, 1994, Divonne, France.
3. K. Rajagopal and F. Wilczek, Nucl. Phys. **B379** (1993) 395.
4. J. Randrup, LBL-39328, Dec 1996, submitted to Nucl. Phys. A.
5. K. Werner, Phys.Rev.Lett. **73** (1994) 1594.
6. N. Armesto *et al.*, Phys. Lett. **B344** (1995) 301.
7. X.N. Wang and M. Gyulassy, Nucl. Phys. **A544** (1992) 559.
8. J. Harris and B. Müller, Ann. Rev. Nucl. Part. Sci. **46** (1996) 71.
9. K. Mrówczyński, Phys. Lett. **B315** (1993) 118.
10. M. Gyulassy, K. Kajantic and L. McLerran, Nucl. Phys. **B237** (1984) 477.
11. B. Wysłouch, Proceedings of the International Conference on *High Energy Physics*, 1996, Warsaw, Poland.
12. Nucl. Phys. **A610** (1996) 188c.
13. G. Roland, International Workshop XXV on *Gross Properties of Nuclei and Nuclear Excitations*, 1997, Hirschegg, Austria.
14. T.A. Trainor and J.G. Reid, *Introduction to STAR Level-2 Trigger Algorithms*, University of Washington Nuclear Physics Laboratory preprint, Aug. 1995.

Nuclear temperature measurement and secondary decay

Hongfei Xi[1], M.J.Huang,[1] W.G.Lynch,[1] M.B.Tsang[1] J.D.Dinius[1]
S.J.Gaff,[1] C.K.Gelbke[1] T.Glasmacher[1] G.J.Kunde,[1] L.Martin[1]
C.P.Montoya[1] M.Azzano[2] G.V.Margagliotti[2] P.M.Milazzo[2] R.Rui[2]
G.Vannini[2] L.Celano,[3] N.Colonna[3] G.Tagliente[3] M.D'Agostino[4]
M.Bruno [4] M.L.Fiandri[4] F.Gramegna[5] A.Ferrero[6] I.Iori ,[6] A.Moroni[6]
F.Petruzzelli[6] P.F.Mastinu[7]

[1]National Superconducting Cyclotron Laboratory and Department of Physics
and Astronomy, Michigan State University, East Lansing, MI 48824, USA
[2]Dipartmento di Fisica and INFN, Teieste, Italy
[3]INFN, Bari, Italy
[4]Dipartmento di Fisica and INFN, Bblogna, Italy
[5]INFN, Laboratori Nazionali di Legnaro, Italy
[6]Dipartmento di Fisica and INFN, Milano, Italy
[7]Dipartmento di Fisica, Padova and INFN, Bologna, Itally

INTRODUCTION

In heavy ion collisions, the products produced in the reaction are dominated by their phase space. It is quite common to interpret data by statistical physics. Temperature, which is a basic quantity of any statistical physics, needs to be addressed experimentally. Assuming thermal and chemical equilibrium are reached at freeze out time, two method can be used to measure nuclear temperature. One method is to measure the yields of particle unstable states in correlation experiments, from the ratios of the yields of excited states in the same isotope, temperatures are deduced by applying Bolztman factor[1]. Another method to measure temperature is based on the double ratios of two isotope pairs [2], (y_1/y_2) and (y_3/y_4) differing by the number of neutrons and/ or protons. These two methods are only valid when measured yields are the same as the primary yields. The problem of nuclei temperature has been studied extensively in the last decad. Most work focus on the temperature measurement by excited states method[3,4,5,6]. These experimental results show that the temperatures deduced from excited states are around 5 MeV. More recently, nuclear temperature was studied by looking double isotope ratios for Au + Au collisions at E/A=600 MeV. The temperatures deduced from double ratio of $(Y(^6Li)/Y(^7Li))/((Y(^3He)/Y(^4He))$ remain

Advances in Nuclear Dynamics 3, edited by
Bauer and Mignerey, Plenum Press, New York, 1997

relatively constant as a function of deduced excitation energy for $2.5 \leq E/A \leq 10$ MeV but increase rapidly at $E/A \geq 10$ MeV[7], indicates a first order liquid gas phase transition. Since the fragments produced in the reaction normally are highly excited. The final isotope distribution could be quite different from primary distribution due to secondary decays, therefore it is very important to calibrate the thermometers used in the experiment due to secondary decay effect. In this report, we will show experimental results for temperature measurements both form excited states population and double isotope ratios. We will also compare experimental data to a model which take care of secondary decays carefully.

Experimental Results

The experiment was done at National Superconducting Cycltron Laboratory at Michigan State University. A 35 AMeV Au beam from the K1200 Cyclotron was impinged on a 5 mg/cm^2 Au target. Measurement was performed with the Miniball 4π array[8] which provide impact parameter selection via the charge particle multiplicity Nc at $23° \leq \Theta_{lab} \leq 160°$ and the Multics high resolution hodoscope which measured isotope yields at $2° \leq \Theta_{lab} \leq 23°$[9]. Within this latter angular domain, isotope yields were determined for 1) fragments emitted from the central source at $v_{lab} \geq 0.13c$ in central collision($N_c \geq 22$). and 2) fragments emitted at $v_{lab} \geq 0.27c$ from the projectile-like residue in peripheral collisions($4 \leq N_c \leq 9$). Temperatures from double isotope ratio was determined via albego formula[2]

$$T_{iso} = \frac{B}{ln(aR_{iso})} \tag{1}$$

where $R_{iso} = \frac{Y(A_i,Z_i)/Y(A_i+1,Z_i)}{Y(A_j,Z_j)/Y(A_j+1,Z_j)}$ is the double yield ratio of the specific pairs of isotopes, $B = B(A_i, Z_i) - B(A_j, Z_j) + B(A_j + 1, Z_j) - B(A_i + 1, Z_i)$ is the difference between corresponding ground state binding energies, and **a** is a statistical factor involving the ground state spins[2]. Similarly for equilibrated systems, temperatures can be obtained from the yields of two states in the same isotope according to the expression

$$T_{ex} = \frac{E_2^* - E_1^*}{ln(a_{ex} \cdot R_{ex})} \tag{2}$$

where $R_{ex} = Y_1/Y_2$ is the ratio of the yield of the upper state divided by the yield of the lower state and $a_{ex} = (2J_2 + 1)/(2J_1 + 1)$ is a statistical factor involving the spins J_1 and J_2 of the two states with excitation energies E_1^* and E_2^*.

Relative population of specific states of ^5Li, ^4He and ^{10}B fragments were measured by detecting the coincident decay products and an apparent temperature T_{app} was obtained for each ratio by eq.2. Solid points in the left panel of Fig1 indicate the measured temperature from excited states population from $^5Li(E^* = 16.66MeV, J = 3/2^+)$ and $^5Li(E^* = 0MeV, J = 3/2^-)$, $^4He(E^* = 20.1MeV, J = 0^+)$ and ^4He ground states, and also $^{10}B(7.34MeV, 2^-; 7.467MeV, 1^+, 7.487MeV, 2^+, 7.5599MeV, 0^+)$ with $(4.77MeV, 3^+)$. The error of the apparent temperatures reflect both from the statistical uncertainty and the uncertainty due to background subtraction[17]. The solid point in the right panel of fig.1 show the isotope temperature deduced from $(^{13,14}C, ^{6,7}Li, ^{9,10}Be, ^{2,3}H, ^{12,13}C, ^{12,13}B, ^{8,9}Li, ^{11,12}B, ^{7,8}Li)/^{3,4}He$ as a function of binding energy difference. The apparent temperatures from different isotope ratios show a fluctuation behavior. Since the measured yields are influenced by secondary decays, so secondary decay effect has to be corrected before we can compare different thermometers

secondary decay model

Assume a parent nucleus was formed in heavy ion collision at high excitation energy, this excited nucleus will deexcite by emission particles and gamma rays. We consider the emission of nuclei with $1 \leq Z \leq 20$ in their ground states or in any of their excited states. The spectrum of allowed excited states includes both the known and tabulated[10] excited states as well as an empirically based extrapolation of the level density into the continuum as described in ref[11]. We approximate the emission by two stages: 1) a first stage where these states are initially populated when the fragments are emitted from the system, and 2) a second stage during which the excited fragments decay according to standard statistical theory.

We assume that the first stage of emission can be described by statistical decay mechanisms; possible candidates range from the evaporation from a heavy residue to the complete vaporization of the system. For simplicity, we approximate the initial population of an excited state of an emitted nucleus with excitation energy E_i^*, spin J_i, neutron number N_i and charge number Z_i with the expression

$$P_i(N_i, Z_i, E_i^*, \mu_p, \mu_n, T_{em}) \propto$$

$$(2J_i+1) \cdot (N_i+Z_i)^{1.5} \exp(-\frac{V_i}{T_{em}}+\frac{Q_i}{T_{em}}) \exp(-\frac{E_i^*}{T_{em}}) \exp(-\frac{Z_i\mu_p + N_i\mu_n}{T_{em}}) \exp(-t_b/t_i) \quad (3)$$

where V_i is the Coulomb barrier, $-Q_i$ is the separation energy for emission of this nucleus from a residue of mass number A_o and charge number Z_o, and T_{em} is the emission temperature.

In our calculations, μ_p and μ_n in eq 3. were not priori given the values assigned to them as "chemical potentials " or "free excitation energies" within specific statistical models[12,13,14]. Instead, they were adjusted to reproduce the measured charge distributions. This requirement put a constraint on the calculation. When we calculate for a specific reaction, we parameterized the charge distribution obtained from experiment by the power law. By changing the μ_p and μ_n, we adjusted the final calculation charge distribution from the calculation has the same power law feature[15].

As the excitation energy is increased into the continuum, the calculations must consider decays of short-lived states with no barrier to particle emission; however, it is likely that many such short-lived states will decay before break up. To take this pre-breakup cooling effect into account, we include in the initial population a factor, $exp(-t_b'/t_i)$. Here, $t_i = t(E^*/A)$ is the mean lifetime of the emitted fragment calculated according to the Weisskopf model[16] for statistical decay. t_b is the breakup timescale, in our calculation we take t_b as 100 fm/c[15].

For the second (decay) stage of the calculation, we focus on the decay of nuclei from both the tabulated low lying discrete states and continuum states. Each decay was calculated using tabulated branching ratios when available, or by using the Hauser-Feshbach formalism[17]. Unknown spins and parities of tabulated discrete states were

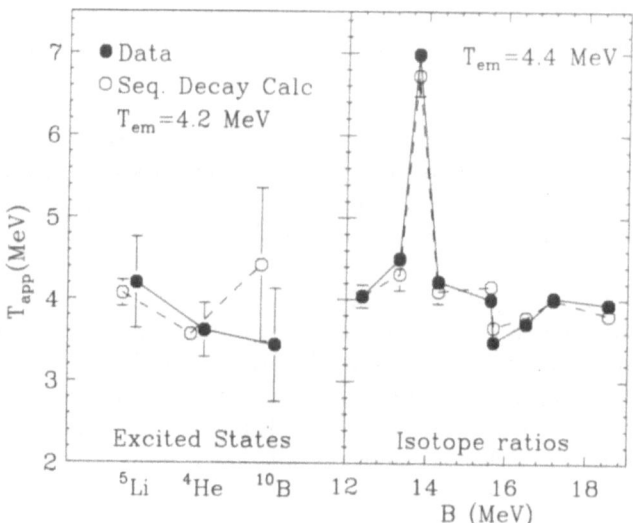

Figure 1. Apparent temperatures obtained from relative populations of excited states for $^5Li,^4He,^{10}B$ (left panel) and from Isotope ratios(right panel). The closed point are the data and the open point are the results from secondary decay calculations

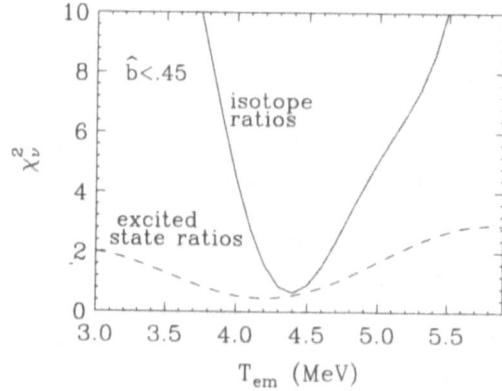

Figure 2. Results of the least squares analysis for the relative population of excited states of $^5Li,^4He,^{10}B$, (dashed line) and for the isotope double ratio temperatures (solid line).

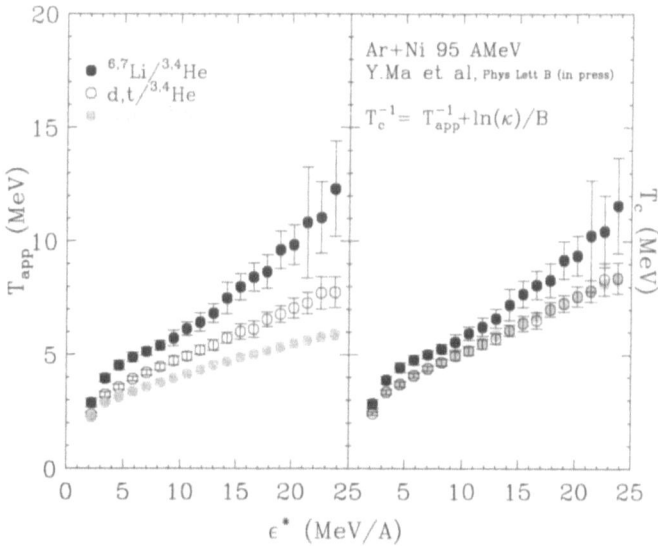

Figure 3. Apparent temperatures for the isotope thermometers for central Au + Au at 35AMeV(top) and peripheral collision(bottom). the left panel show the corrected tempertures

randomly assigned in these calculations and then changed in subsequent calculations to assess the corresponding uncertainties which are of the order of 5% [15].

In order to compare secondary decay model calculation to data the calculation we performed for intial temperatures ranging from 2 Mev to 6 MeV and the agreement between theory and experiment was assessed by calculating the corresponding values for the reduced χ^2 using the expressing,

$$\chi^2_\nu(T_{em}) = \frac{1}{\nu} \sum_{i=1}^{\nu} \frac{[R_{expt,i} - R_{calc,i}(T_{em})]^2}{\sigma^2_{expt,i} + \sigma^2_{calc,i}} \qquad (4)$$

independently for the isotope ratios and for the excited states population. here the σ_{expt} and σ_{cal} are the experimental and theoretical uncertainties and the summation runs over the relevant excited state population or isotope ratios[17]. Fig.2 show the results from this χ^2 fit. best fit values of 4.4 ± 0.2 MeV and 4.2 ± 0.6 MeV are determined for isotope ratio temperature and excited states population temperatures. In Fig.1 the open point are from the secondary decay model calculation. it can be seem that the calculation reproduce the data very well.

Universal fluctuation behavior for isotope temperatures

The good agreement between secondary decay calculation and the isotope ratio temperatures indicates that the fluctuation behavior among isotope temperatures might be universal. To explore the question,we assume the measured ratio R can be written as $R = \kappa \cdot R_0$, R_0 is the ratio before secondary decay. We can rewrite the expression for isotope temperature as

$$\frac{1}{T_{app}} = \frac{1}{T_o} + \frac{\ln \kappa}{B} \qquad (5)$$

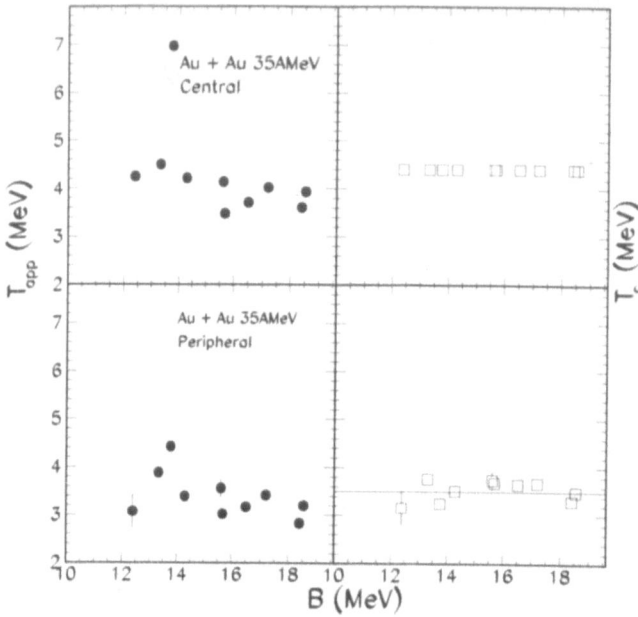

Figure 4. Raw (left panel) and corrected (right panel) temperatures for the three isotope yield ratios as a function of excitation energies for the system Ar + Ni at 95 AMeV[19]

here T_0 is the real temperature and T_{app} is the apparent temperature. For Au+Au at 35 AMeV. we take the value 4.4MeV as T_0 which come from secondary decay calculation, then we can obtained a set of $ln\kappa/B$ for different isotope thermometers. then using this set $ln\kappa/B$, we can get the corrected temperature for other system.

The lower part of Fig.3 show the results for Au+Au peripheral collision. The fluctuation after corrected $ln\kappa/B$ are reduced compare to the apparent temperature. Left panel in Fig.4 shows the results from ref[19], the right panel show the corrected temperatures. It can be seen that after corrected $ln\kappa/B$, three different thermometers are merge together. This indicates the fluctuation among different isotope thermometers are mainly come from secondary decay effect.

The correction factors in eq.5, however, are predicted to be temperature dependent[15] An examination of the temperature dependence of the individual yields that comprise these ratios may suggest experimental test whereby the predicted temperature dependence may be confirmed or rejected. To illustrate the various factors that govern this temperature dependence, we have calculated the yield ratio used in fig.3 for temperature form 3 to 11MeV. and we obtained a $(ln\kappa/B)_{cal}$ by using the apparent temperature from isotope ratios and emission temperature at 4.4 MeV. we found by applying this set of $(ln\kappa/B)_{cal}$ obtained from one temperature point. the fluctuation among different isotope temperatures are reduced and the remaining average temperature was plotted as function of emission temperature in fig.5 as the solid line. it can be seen that below 5 MeV, the corrected temperature are nearly indentical to emission temperature and start flatten out beyond 6 Mev. This is because at high temperature contribution from feeding to a alpha particles cancels the increase of ^3He yield. so any thermometers

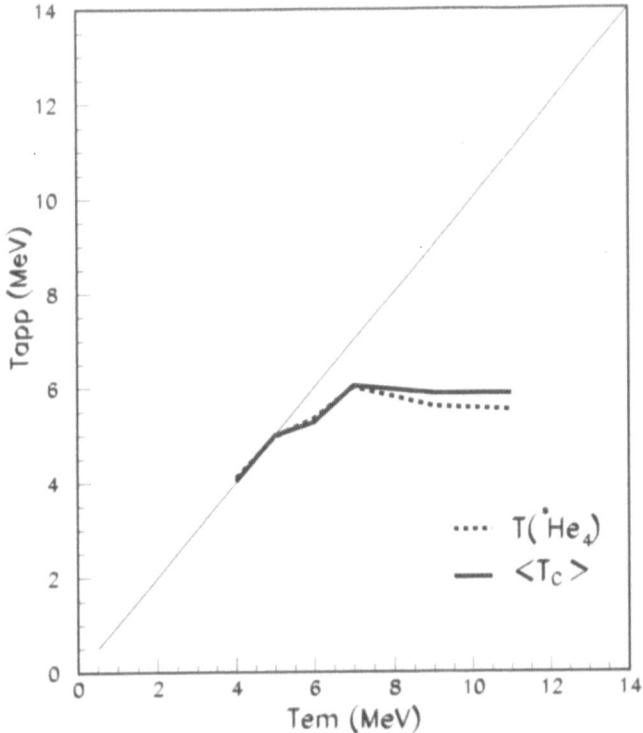

Figure 5. Results of secondary decay calculation. T_{iso} (solid line) and $T(^4He^*)$ (dashed line).

involving $^3He/^4He$ will nolonger good thermometer for temperature beyond 6 Mev. In the same calculation, we have also studied the temperature from the ratio of yield of ^4He ground state to excited states at 20.1 Mev. We again normalize $T(^4He^*)$ to 4.4 Mev. then the normalize $T(^4He^*)$ was plotted as dash line in fig.5. We can see the temperature dependent trend of T_{iso} and $T_{^4He^*}$ are quite indentical. Since the binding energy of ^3He is close to $^4He^*$, so all calculation which take the secondary effect will predict identical trends for T_{iso} and $T(^4He^*)$. If measurements reveal strong differences between these two quantities as suggested in for central Au+Au collisions at incident energies above 50 AMeV[20], the resolution for this difference is not likely to be found in equilibrium physics. rather there must be strong non-equilibrium contributions to ^3He yield, ^4He yields or both.

SUMMARY

In summary, we have measured breakup temperatures for Au+Au reactions at 35A MeV. Temperatures extracted from isotope ratios and excited state populations are virtually the same after correct secondary decay. fluctuation among different isotope temperatures are mainly come from secondary decay. We have obtained ast of empirical correction factors for 35A MeV Au+Au which is able to remove the fluctuation among different isotope temperatures for most of the other reactions. Sequential decay calculation reproduce these correction factors and show that an identical saturation in the apparent temperatures is expected for the isotope temperatures based on $^{3,4}He$ pairs as upon the ratio of the ground to excited state of ^4He. If subsequent measurements do not show these two thermometers to be consistent, the explanation does not lie in equilibrium physics. instead there might be strong non-equilibrium contributions to ^3He or ^4He or both.

This work was supported by the National Science Foundation under Grant Nos. PHY-95-28844 and PHY-93-14131.

REFERENCES

1.   D.J. Morrissey, W. Benenson, W.A. Friedman, Ann. Rev. Nucl. Part. Sci. **44**, 27 (1994) and references therein.
2.   S. Albergo et al., Nuovo Cimento **89**, 1 (1985).
3.   H.M. Xu et al., Phys. Rev. C 40, 186 (1989)
4.   T.K. Nayak et al., Phys. Rev. C 45, 132 (1992).
5.   F. Zhu et al., Phys. Rev. C 52, 784 (1995).
6.   C. Schwarz et al., Phys. Rev. **C48**, 676 (1993) and references therein.
7.   J. Pochodzalla et al., Phys. Rev. Lett. **75**, 1040 (1995).
8.   R.T. de Souza et al., Nucl. Instr. Meth. **A 295,** 109 (1990).
9.   P.F. Mastinu et al., Nucl. Instr. and Meth. Phys. Res. **A338**, 419 (1994) and references therein.
10. F. Ajzenberg-Selove, Nucl. Phys. **A392**, 1(1983); **A413**, (1984); **A433**, 1 (1985); **A449**, 1 (1985); **A460**, 1 (1986).
11. Z. Chen, C.K. Gelbke, Phys. Rev. **C38**, 2630 (1988).
12. J. Randrup, S.E. Koonin, Nucl. Phys. **A356**, 223 (1981).
13. W.A. Friedman and W.G. Lynch, Phys. Rev. **C28** (1983) 16; (1983) 950.
14. D.Hahn and H.Stoker, Nucl. Phys. **A476** 718(1988)
15. H. Xi, W.G. Lynch, M.B. Tsang, W.A. Friedman, Phys. Rev. C R2163, (1996)
16. V.F.Weisskopf, Phys. Rev. **52**, 295 (1937).
17. W. Hauser and H. Feshbach, Phys. Rev. **87**, 366 (1952).
18. M.J.Hunag, H. Xi, W.G. Lynch, M.B. Tsang, et al, Phys. Rev. Letter. in press
19. Y. Ma et. al, Phys. Lett. B, (1996) in press
20. V. Serfling, Ph. D. Thesis, University of Frankfurt, (1996); J. Pochodzalla, Proceedings of First Catania Relativistic Ion Studies, Acicastello, Italy, May 27-31, (1996); G. Raciti et al, Proceedings of First Catania Relativistic Ion Studies, Acicastello, Italy, May 27-31, (1996).

NET PROTON AND NEGATIVELY-CHARGED HADRON SPECTRA FROM THE NA49 EXPERIMENT

Milton Toy [1,2] for the NA49 Collaboration

[1]University of California, Los Angeles
[2]Lawrence Berkeley National Laboratory

INTRODUCTION

The collision of heavy nuclei at ultrarelativistic energies provides an opportunity to study a highly excited system with an energy density of several GeV/fm^3 in a volume about the size of the ion. Quantum Chromodynamics (QCD) predicts that hadronic matter at a sufficiently high energy density will be transformed into a state of deconfined quarks and gluons called the quark-gluon plasma (QGP). In a nucleus-nucleus collision, a portion of the initial longitudinal energy that is carried by the beam and target baryons is transformed to other degrees of freedom, namely produced particles and transverse momentum. The rapidity shift of the baryons, or 'stopping', is estimated by the final state rapidity distribution of the net protons ($p - \bar{p}$). The final state negatively-charged hadrons (h^-) consisting of π^-, K^-, and $\bar{p}$ are all created by the collision and are a measure of the total particle production under the assumption that the positive, negative, and neutral particle production are roughly equal. While these measurements alone cannot show that a phase transition to a QGP has occured, they characterize the basic features of the collisions.

THE NA49 EXPERIMENT

At the CERN SPS, lead ions are accelerated to an energy of 158 GeV per nucleon and are then used to bombard a stationary nuclear target. The NA49 experiment aims to measure over 60% of the approximately 1300 charged particles produced in a central impact Pb+Pb collision.

To achieve this goal, large volume time projection chambers (TPC) are used as the tracking detectors in the experiment. The TPCs also provide a measurement of the energy loss (dE/dx) of charged particles with momenta above 3 GeV/c for particle identification purposes. Two TPCs (VTPC) are inside magnets of 4.5 T-m bending power each and detect charged particles produced directly from the collision in both

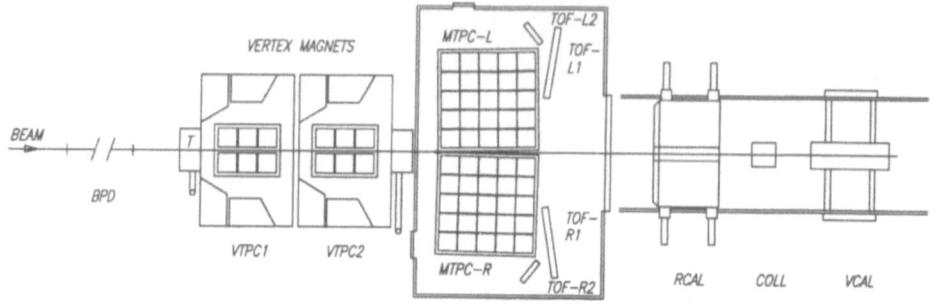

Figure 1. Layout of the NA49 experiment. The target position is indicated by T. The detector components are the time projection chambers VTPC-(1,2) and MTPC-(L,R), time-of-flight system TOF, and calorimeters RCAL and VCAL.

target and projectile rapidity regions and the charged decay products from neutral strange particles[1]. The magnets also provide momentum analysis and spatial separation of the charged particles to facilitate tracking. Two TPCs (MTPC) of size 13 m^2 each are located downstream of the magnets and have nearly full phase-space coverage from mid-rapidity (2.9) to near beam rapidity. The TPC data are read out by over 180,000 channels of electronics performing 512 samples in time. After data compression, the typical event size of a central Pb+Pb collision is 8 MB. In the MTPCs, over 400 charged particles are tracked in each chamber. The momentum resolution is $\Delta p/p = 0.3\%$ and the dE/dx resolution is $\sigma_{dE/dx} = 5\%$. Time-of-flight detectors (TOF) along with momentum and dE/dx information from the TPCs provide particle identification at mid-rapidity. The time resolution is $\sigma_{TOF} = 60$ ps. A forward zero degree calorimeter measures the energy of the projectile spectator nucleons and is used to characterize the centrality of the collision for triggering purposes. A segmented ring calorimeter was used to measure transverse energy and event anisotropy during dedicated calorimetry-only runs[2] [3].

Construction of NA49 was completed in 1995. The data-taking rate is approximately 1 Hz and as of 1997 over 2 million events have been collected for a number of different centrality triggers and target nuclei.

DATA

MTPC, or 'Main' TPC, data from the most central collisions that make up 5% of the total inelastic cross section were used for this analysis. The coverage of the MTPCs is in the projectile rapidity region ($y > 2.9$). For particle identification, up to 90 samples of a particle's energy loss in the TPC gas can be taken over 3.5 m of track length. The dE/dx resolution is at the level of about one standard deviation between p and K as well as K and π. With this resolution, single-particle identification cannot be achieved but particle yields can be obtained statistically by fitting an ensemble of particles in a momentum interval with Gaussian distributions for p, K, π, e. An analysis specific to extracting a proton yield was employed similar to the 'plus minus minus' method where the distribution of net protons in y-p_T space are approximated by the charge difference

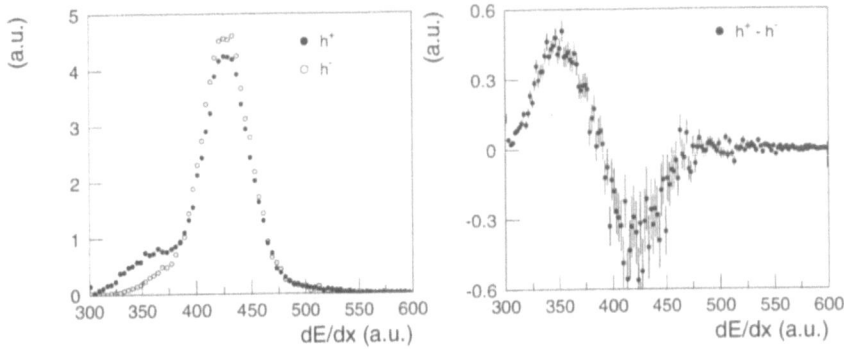

Figure 2. dE/dx distributions for positively- and negatively-charged hadrons (left) for a small interval in y-p_T near mid-rapidity. The main peaks are from pions and the difference between π^+ and π^- is due to the net isospin of the lead nucleus. The $h^+ - h^-$ distribution (right) has positive $(p - \bar{p}) + (K^+ - K^-)$ and negative $\pi^+ - \pi^-$ components.

of all hadrons ($h^+ - h^-$). In this case, the dE/dx information is also used. The energy loss distributions for h^+ and h^- are shown in Fig. 2 for a small interval of 'rapidity' (calculated with an assumption of the proton mass) and transverse momentum. While there is no clear p-K-π separation, there is a prominent pion peak for both charges and the protons can be seen in the left shoulder of the h^+ distribution. Also shown is the difference of the charge distributions, which has a peak corresponding to $p - \bar{p}$ (with $K^+ - K^-$) and another peak from $\pi^+ - \pi^-$. A two Gaussian fit to this distribution and subsequent integration of the positive component gives the net protons with a small kaon contamination that is corrected for.

The MTPCs track particles far away from the target-vertex and, without a 'seen' vertex within the chamber, are subject to misidentifying the trajectories of secondary particles as primary-vertex tracks. All data were corrected for e^+ and e^-, secondary hadronic interactions with detector materials, and weak decays of K_S^0 and Λ by results from simulations.

The sensitivity of the method by which the net protons are measured is affected by h^+-to-h^- mismatches in dE/dx and differences in dE/dx resolution, as well as the uncertainties in the other corrections. These factors combine to result in a systematic error of 15%. The h^- data are from a simple particle counting measurement and has a smaller estimated systematic error less than 10%.

The preliminary semi-inclusive rapidity distributions of $p - \bar{p}$ and h^- are shown in Fig. 3. The integrated number of net protons from the data is 65. There are approximately 10 spectator protons on average measured in the forward calorimeter and an extrapolation of the measured data to beam rapidity can yield another 5 protons for an approximate total of 80 protons in the projectile region of rapidity. Also shown are data from S+S reactions at 200 GeV/N from the NA35 experiment[4]. In order to compare the data on a similar scale, the S+S data are scaled by the ratio of participating nucleons (6.4) which is based on trigger models from each experiment. The Pb+Pb system has a greater degree of stopping than S+S as evidenced by the plateau in the rapidity distribution around mid-rapidity and thus results in a larger energy deposition at mid-rapidity. The broad shape of the net proton distribution suggests that the

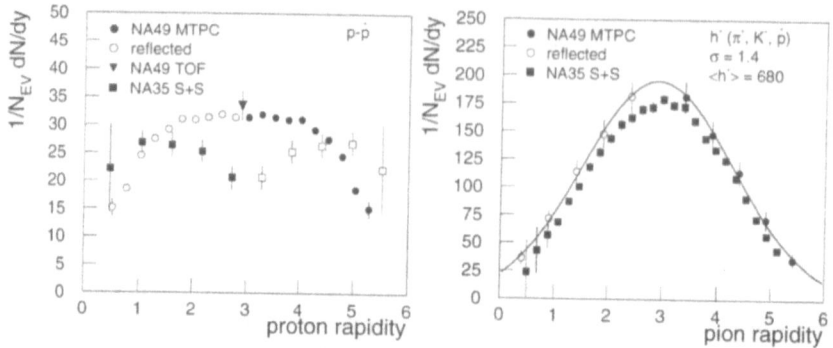

Figure 3. Preliminary rapidity distributions of protons–anti-protons (left) and negatively-charged hadrons (right) from central Pb+Pb collisions. Scaled (see text) S+S data for comparison.

protons experience a strong longitudinal expansion[6]. Data from Au+Au collisions at BEVALAC[7] and AGS[8] energies show that as the rapidity gap between the beam and target increases, the proton spectra becomes flatter and wider. The protons 'fill in' the available phase-space and the proton density at mid-rapidity decreases from over 160 protons per unit rapidity at the BEVALAC down to nearly 30 at CERN. This change in proton density shows that the baryochemical potential of the system depends on the beam energy and will be reflected in the final state particle yields.

The h^- data in Fig. 3 are plotted as a function of rapidity calculated with the assumption of the pion mass. A Gaussian fit to the distribution returns $\sigma = 1.4$ units of rapidity and an integrated yield of 680 particles. A purely thermal source of temperature 200 MeV would produce a distribution of only $\sigma = 0.6$ and this difference can be attributed to longitudinal expansion of the reaction volume. The scaled S+S data show good agreement with the Pb+Pb data and indicate that there is little contribution to particle production from any sort of collective nuclear effect.

Displayed in Fig. 4 are the preliminary transverse momentum inverse slope parameters for the net protons and the mean transverse momenta of the h^-. The inverse slope

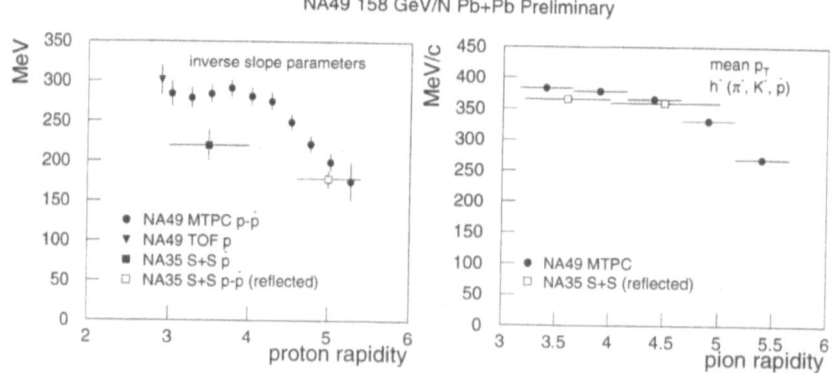

Figure 4. Preliminary transverse momentum distributions. Inverse slope parameters of protons–antiprotons (left) and mean transverse momentum of negatively-charged hadrons (right).

parameters come from fitting the net proton p_T spectra in intervals of rapidity to an exponential function in transverse mass ($m_T = \sqrt{p_T^2 + m^2}$): $dN/dp_T = Cp_T exp(-m_T/T)$, where T is the inverse slope parameter. The Pb data slopes are nearly constant at 280 MeV over a range from mid-rapidity to 1.5 units of rapidity from mid-rapidity. S+S data of $p - \bar{p}$ near target rapidity[4] and $\bar{p}$ near mid-rapidity[5] have lower slope parameters of around 200 MeV. The h^- data are composed of three particle species and consequently the p_T spectra cannot be fit by a simple single exponential function. Instead, the mean p_T are calculated and in this case, the Pb and S results are similar. These results may be due to a collective transverse flow of particles, which is larger in Pb+Pb reactions than in S+S. Because the h^- are predominantly pions, any increase in a common velocity will affect the heavier protons more than the pions.

SUMMARY

The transverse momentum and rapidity distributions of negative hadrons and net protons have been measured in the projectile rapidity region from central Pb+Pb collisions at 158 GeV/nucleon. The net proton rapidity distribution shows greater stopping than what has been seen in the lighter S+S system while the produced negative hadron rapidity distribution agrees with the S+S result after scaling for the difference in the number of participating baryons. The transverse momentum spectra indicate an increase of transverse flow with system size, which takes up some of the additional energy deposited as evidenced by the stopping result.

ACKNOWLEGEMENTS

This work was supported by the U.S. Department of Energy, the Bundesministerium für Bildung und Forschung, Germany, the Research Secretariat of the University of Athens, the Polish State Committee for Scientific Research, the Polish-German Foundation, the Hungarian Scientific Research Foundation, and EPSRC, U.K.

REFERENCES

1. G. Cooper et al. (NA49 Collaboration), these proceedings.
2. T. Alber et al. (NA49 Collaboration), Phys. Rev. Lett. **75**, 3814 (1995).
3. T. Wienold et al. (NA49 Collaboration), Proceedings of the 12th Winter Workshop on Nuclear Dynamics, Snowbird UT, Feb. 3-10, 1996.
4. J. Bächler et al. (NA35 Collaboration), Phys. Rev. Lett. **72**, 1419 (1994).
5. T. Alber et al. (NA35 Collaboration), Phys. Lett. B **366**, 56 (1996).
6. E. Schnedermann, J. Sollfrank, and U. Heinz, Phys. Rev. C **48**, 2462 (1993).
7. T. Wienold et al. (EOS Collaboration), preliminary data, private communication.
8. Y. Akiba et al. (E866 Collaboration), Nuc. Phys. A **610**, 193c (1996).

SEQUENTIAL AND PRE-EQUILIBRIUM NUCLEON EMISSION IN Sn + Ca REACTIONS AT 35A MEV

D.K. Agnihotri, B. Djerroud, J. Tõke, W. Skulski, and W.U. Schröder,
B. Davin,[1] E. Cornell,[1] and R.T. DeSouza[1]

Dept. of Chemistry and NSRL, Univ. of Rochester, Rochester, NY, 14627
[1]Dept. of Chemistry, Indiana University, Bloomington, IN, 47401

Introduction

Over the past decade, an enormous amount of work has been carried out to gain an understanding of heavy-ion reaction mechanisms in the Fermi-energy regime (E/A > 20MeV). At low energies, the phenomenology of heavy-ion reactions is reasonably well understood in terms of binary dissipative reaction mechanisms[1]. For Fermi bombarding energies, an essentially dissipative mechanism is still observed[2, 3, 4, 5], although the dissipation and equilibration of energy is less effective than at lower bombarding energies. New reaction phenomena discovered include incomplete energy dissipation[2, 4, 6], dynamical emission of multiple intermediate-mass fragments[7], and the nearly complete disassembly of projectile- and target-like fragments[8].

In the present work, the emission probabilities and patterns of light particles have been used as tools to probe the reaction mechanism. Statistically evaporated nucleons carry important information about the reaction scenario, e.g., about the number of massive primary fragments, the total excitation energy and its division among the fragments, and the rate or degree of N/Z equilibration of the system. On the other hand, the emission patterns of instantaneous, non-statistical (pre-equilibrium) particles are expected to exhibit characteristics of the dynamics of the early stages of the interaction. For example, the emission probabilities and energy spectra of pre-equilibrium particles carry information of importance for the prospects of forming nuclear systems at critically high temperatures. It has also been pointed out recently[9] that pre-equilibrium nucleon emission can provide information about the isospin dependence of the equation of state of nuclear matter. Pre-equilibrium particle emission processes at Fermi energies have been studied[10, 11] experimentally to some extent for fusion and fusion-like heavy-ion reactions, while such information is still very scant for dissipative reactions[12, 13, 14].

In this paper, results of an exclusive measurement of neutron and proton yields are presented for the two reactions ^{112}Sn+^{48,40}Ca at E/A=35MeV. The particle energies and emission angles have been measured in coincidence with projectile-like fragments

Advances in Nuclear Dynamics 3, edited by
Bauer and Mignerey, Plenum Press, New York, 1997

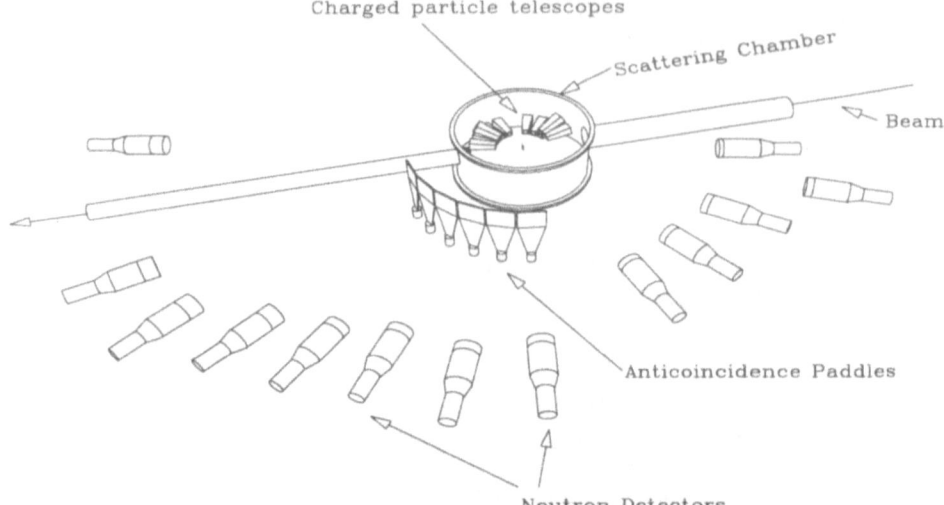

Figure 1. A pictorial view of the experimental setup.

(PLF). The yields of neutrons and protons have been measured as functions of the dissipated energy, for statistical as well as pre-equilibrium nucleons. Of particular interest is the influence of the projectile N/Z ratio that is possibly "remembered" in the yields of the emitted nucleons. For the ^{40}Ca and ^{48}Ca projectiles, these ratios are N/Z=1.0 and 1.4, respectively, while N/Z=1.24 for the target, ^{112}Sn. The N/Z ratios for the composite systems for two cases are N/Z=1.17 and 1.28, respectively. In the following section, experimental details are given, followed by a discussion of experimental results and conclusions.

Experimental Details

The experiments were performed at the K1200 Cyclotron of the National Superconducting Cyclotron Laboratory (NSCL) at Michigan State University. A schematic view of the experimental setup is shown in Fig 1. A 2.5-mg/cm^2 thick ^{112}Sn self-supporting target was bombarded with 35-MeV/A ^{40}Ca and ^{48}Ca beams. A thin-walled scattering chamber was used, in order to minimize neutron scattering. Two position-sensitive, solid-state detector telescopes (SSB) with three elements, placed near the grazing angles ($\Theta_{1/4} \approx 4.2^o$ and 5.2^o for the reactions ^{112}Sn + ^{48,40}Ca, respectively), detected projectile-like fragments (PLF), along with intermediate-mass fragments (IMF). Light charged particles (LCP) and intermediate-mass fragments (IMF) were detected with a set of nine, three-element (gas/Si/CsI) detector telescopes [19] placed at different laboratory angles on one side of the beam.

A 12-detector neutron time-of-flight spectrometer was set up, as shown in Fig.1, to measure the energy and the angular distributions of neutrons. These neutron detectors consisted of 5"-diameter NE213 liquid-scintillator cells of thicknesses between 1.5" and 3.0", viewed by Amperex XP2041 photomultipliers. They were placed on the side of the beam opposite to the charged-particle telescopes, in order to avoid "shadowing" effects. An effective pulse-shape discrimination technique[20] was used to distinguish between neutrons and γ-rays. Fast plastic anticoincidence counters ("paddles") were

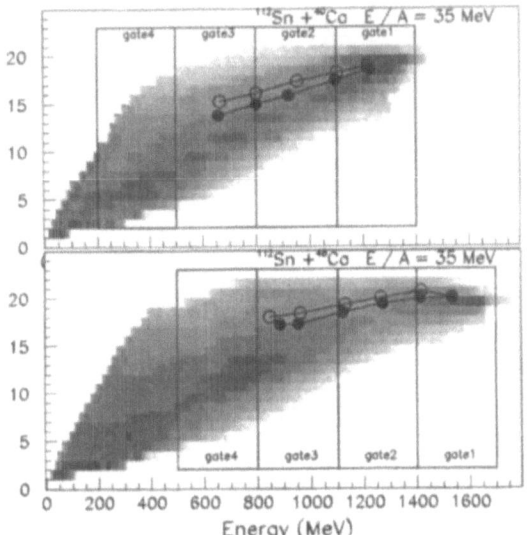

Figure 2. Two-dimensional distribution of the charged yield, measured with the PLF telescope at $< \Theta > = 5.9°$ plotted vs. atomic number(Z) and kinetic energy(E) for ^{112}Sn+^{40}Ca and ^{112}Sn+^{48}Ca. Symbols represent model calculations.

placed in front of the forward-angle neutron detectors, in order to tag the fast, light charged particles (LCP) that passed through the chamber walls and were stopped in the neutron detectors. In the off-line analysis, such events were not considered.

Results and Discussion

In Fig.2, two-dimensional distributions of the charged yields, measured in the PLF telescope at $< \Theta > = 5.9°$, are plotted *vs.* atomic number (Z) and kinetic energy (E), for the ^{112}Sn+^{40}Ca (upper panel), and the ^{112}Sn+^{48}Ca (lower panel) reactions. The circles joined by lines are quantitative model predictions for fragment E-Z correlations. Open circles indicate results of calculations[22] with the nucleon exchange model[23] (NEM) for representative primary reaction fragments, corrected for sequential evaporation using the code EVAP[24]. Solid circles represent calculations taking the projectile as representative of the primary, pre-evaporation PLFs. The overall character of the two distributions is quite similar, with a yield concentrated along a well defined ridge in the E-Z plane. The trends of this E-Z correlation are qualitatively reproduced by the model calculations. The experimental E-Z distributions for the two reactions ^{112}Sn+^{40}Ca and ^{112}Sn+^{48}Ca differ for quasi-elastic and partially damped events. For the ^{40}Ca-induced reaction, the ridge defines a strong correlation between Z and E of the PLF, consistent with that due to an evaporative decay of excited primary ^{40}Ca fragments. On the other hand, for the ^{48}Ca-induced reaction, the correlation between Z and E of the PLFs is weaker, in particular for partially damped events. This latter feature suggests a better preservation of the primary Z-distribution of the neutron-rich PLFs. Neutron and proton data have been analyzed in more detail as functions of PLF energy in the relatively broad regions indicated by the "gates" defined by the frames in Fig.2. In this fashion, the degree of dissipation was varied in the analysis.

Plotted in Fig.3 are the energy spectra of neutrons from the ^{112}Sn + ^{48}Ca reaction for different laboratory angles, fitted with a schematic model assuming sequential emis-

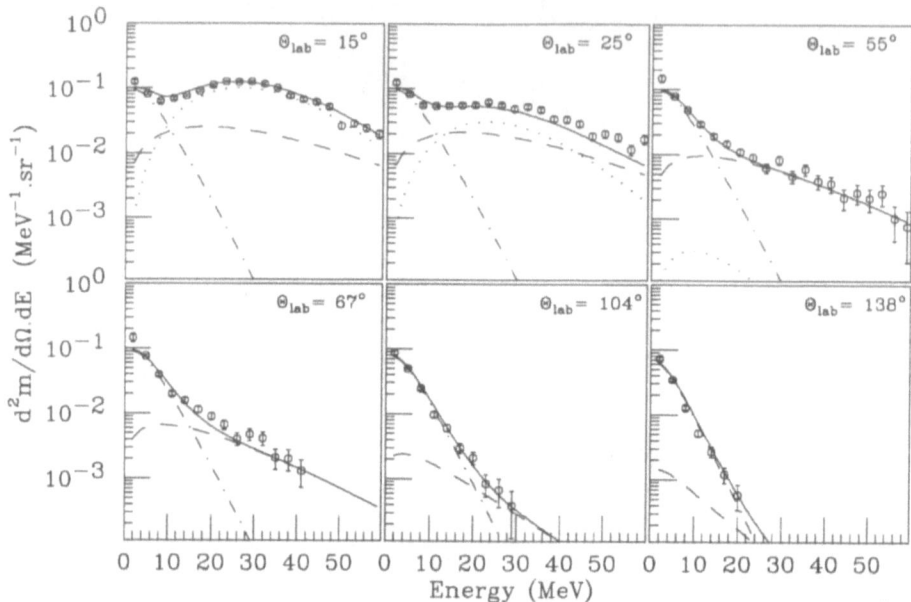

Figure 3. Energy spectra of neutrons from the ^{112}Sn + ^{48}Ca reaction for different laboratory angles. Data are represented by the symbols, while the curves represent a three-source fit.

sion from three moving emitters ("sources"). Data are represented by open symbols with error bars, while solid curves represent a three-source model fit. The minimal model used in the parameterization of data combined the contributions of neutrons evaporated from projectile-like (dotted curves) or target-like (dot-dashed curves) fragments. The remaining yield was found to be well reproduced by isotropic emission from a hypothetical third emitter (dashed curves) moving with approximately 55% of the beam velocity. The data can definitely not be reproduced by fits with only two moving (PLF and TLF) sources. Neutrons from this third source exhibit an energy spectrum with a logarithmic slope of the order of 10MeV and have been attributed to pre-equilibrium processes. Similar energy spectra have been observed for protons, with a pre-equilibrium component which, at intermediate angles, is well separated from thermal distributions. Average multiplicities of neutrons and protons have been obtained as functions of excitation energy from an angular extrapolation of the model fits to the data. The evaporative multiplicities are quantitatively reproduced by statistical-model calculations for the decay of the primary PLF and TLF using the code GEMINI[21].

Fig.4 shows neutron (crosses) and proton (circles) energy spectra, measured at an angle of 20° w.r.t. the PLF detection angle. The top panels are for ^{112}Sn+^{40}Ca, while the bottom panels represent ^{112}Sn+^{48}Ca. The four columns from left to right, as selected by gates in PLF kinetic energy (cf. Fig.2), correspond to an increase in dissipated energy. At forward angles, e.g., at 20°, most of the nucleons with energies above 15MeV are sequentially evaporated from the PLF, as suggested by a well defined bump in the energy spectrum. For peripheral ^{40}Ca-induced reactions, one notices that proton emission from the PLF is more probable than neutron emission. On the other hand, for mid-peripheral collisions, proton emission competes with neutrons on an equal footing, while for more central collisions, the neutron yield exceeds the proton yield. This behavior suggests that, with increasing energy loss, the primary PLF becomes more neutron rich. In ^{48}Ca-induced reactions, the neutron yields dominates

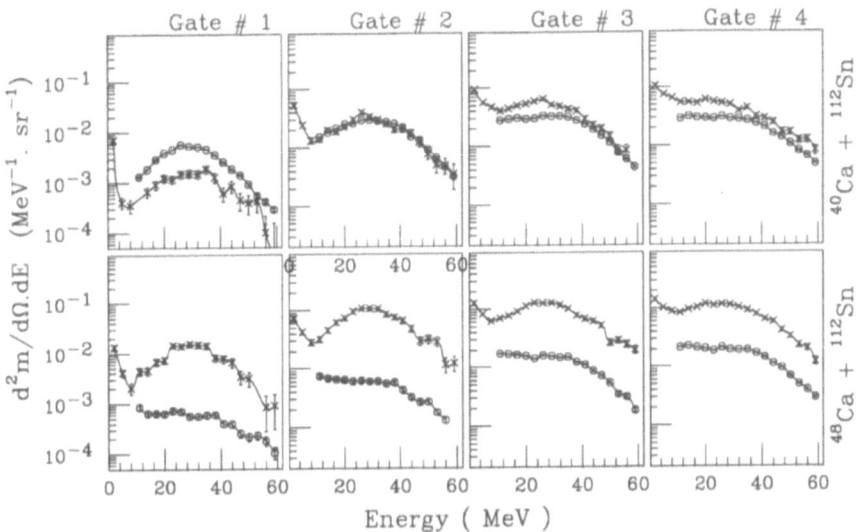

Figure 4. Neutron (crosses) and proton (circles) energy spectra, measured at an angle of $20°$ w.r.t. the PLF detection angle for ^{112}Sn+^{40}Ca (top) and ^{112}Sn+^{48}Ca (bottom). The dissipated energy increases from left to right, as selected by PLF kinetic energy (cf. Fig.2).

over the proton yield for the entire range of energy dissipation. This behavior indicates preservation of the neutron richness of the projectile by the primary PLF fragments. However, the difference in the neutron and proton yields decreases for higher degrees of energy dissipation. This does not imply an equilibration of the N/Z asymmetry degree of freedom, as shown further below. The above observations are consistent with the PLF Z distributions shown in Fig.2. For ^{48}Ca-induced reactions, most of the PLF excitation energy is carried away by neutrons, leading to a better preservation of the PLF Z distributions than for ^{112}Sn+^{40}Ca.

Shown in Fig.5 are the neutron-to-proton multiplicity ratios (M_n/M_p) for the PLF components for the two reactions ^{112}Sn+^{48}Ca and ^{112}Sn+^{40}Ca, plotted vs. excitation energy. The solid and dashed curves represent neutron-to-proton multiplicity ratios, calculated for emission from projectile-like fragments with N/Z ratios equal to those of the beams(1.40 and 1.00) and those of the composite systems (1.28 and 1.17), respectively. The experimental trends are very different for these two reactions. For the ^{48}Ca-induced reaction, the data are consistent with a unique ratio of N/Z=1.40 for the primary PLFs, which is equal to that of the projectile. However, for the ^{40}Ca-induced reaction, the fragment N/Z ratio appears to evolve gradually with excitation energy, from that of the beam to the average N/Z of the composite system. From such trends, one learns that the equilibration of the N/Z degree of freedom is a gradual process, just like at lower bombarding energies[1]. As expected also, the degree of mass-to-charge equilibration attained in a reaction depends strongly on the N/Z asymmetry in the entrance channel. Calculations for the potential energy surface driving mass and charge exchange in the reaction have been performed for the two systems. For the ^{40}Ca-induced reaction, the surface shows a steep gradient towards larger PLF-N/Z values. For the reaction ^{112}Sn+^{48}Ca, the injection point (entrance channel) is located already close to the local minimum of the potential energy surface. Hence, the potential energy surfaces are consistent with the very different trends observed in the data for the two different systems, both as far as PLF Z-distributions (cf. Fig.2) and M_n/M_p ratios (Fig.5) are

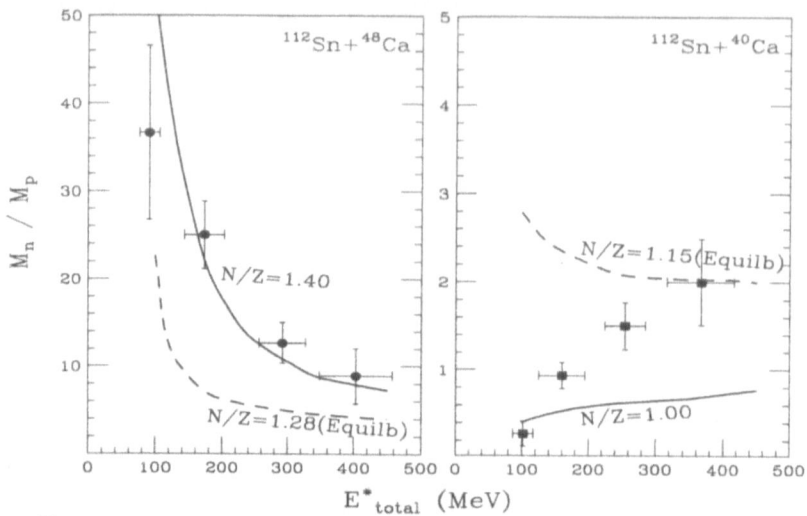

Figure 5. The neutron-to-proton multiplicity ratio M_n/M_p for the PLF components vs. total excitation energy, for ^{112}Sn + ^{48}Ca (circles, left panel) and ^{112}Sn + ^{40}Ca (squares, right panel). Curves represent M_n/M_p ratios calculated for emission from projectile-like fragments with N/Z equal to that of the beams or the composite systems.

concerned. It is interesting to notice that the effect of the potential energy surface still contributes to the dynamics of collision at Fermi-energies.

Shown in Fig.6 (left panel) are the pre-equilibrium neutron multiplicity (squares) and the total evaporative (PLF+TLF) neutron multiplicity (circles), plotted as functions of total excitation energy. Data for ^{40}Ca- and ^{48}Ca-induced reactions are joined by dashed and solid lines, respectively. The multiplicity for evaporated neutrons is higher for the ^{48}Ca-induced reaction than for the ^{40}Ca induced reaction, as expected, because of the additional 8 neutrons in the former case. From the ratio of the pre-equilibrium multiplicity to the total evaporative multiplicity, one concludes that pre-equilibrium nucleons can carry away not more than 20-25% of the total dissipated energy. On the right panel of Fig.6, the neutron-to-proton multiplicity ratios (M_n/M_p) for the pre-equilibrium components are plotted as functions of excitation energy, for the reactions ^{112}Sn + ^{48}Ca (circles) and ^{112}Sn + ^{40}Ca (squares). As in the case of evaporative particles, one notices a strong dependence of the pre-equilibrium multiplicity ratios on the projectile N/Z ratio. These trends also suggest that the pre-equilibrium nucleons originate mostly in the projectile. One may expect that the in-medium nucleon-nucleon scattering cross-section plays an important role in determining the N/Z ratio of the fast particles emitted early in a nuclear collision. This expectation is, however, not borne out in mean-field (BUU) calculations [9]. Instead, these calculations suggest that effects of the kind observed may be caused by the isospin-dependence of the nuclear equation of state.

Experimental results were compared with Fermi-jet model[15−18] calculations. In Fig.7, pre-equilibrium neutron multiplicities are plotted vs. the total excitation energy. The symbols represent data for ^{48}Ca (circles) and ^{40}Ca (squares), while the dashed and dashed-dotted curves represent results of the corresponding Fermi-jet model calculations. As is obvious from Fig.7, the model is unable to reproduce the absolute pre-equilibrium neutron multiplicities. The experimental angular distributions of such neutrons are also in contradiction to the prediction of the Fermi-jet model. Experimentally, fast particle-jets at backward angles suggested by the model are not observed, either.

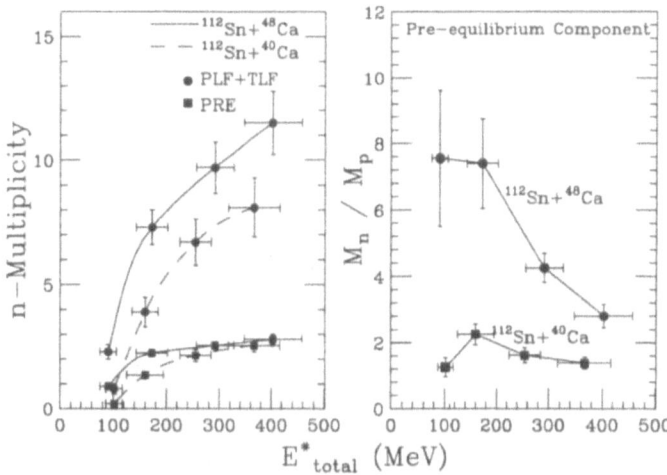

Figure 6. Pre-equilibrium (squares) and evaporative (circles) (PLF+TLF) neutron multiplicities plotted vs. total excitation energy (left panel). In the right panel, the neutron to proton multiplicity ratios M_n/M_p for pre-equilibrium components are plotted vs. total excitation energy for the two reactions.

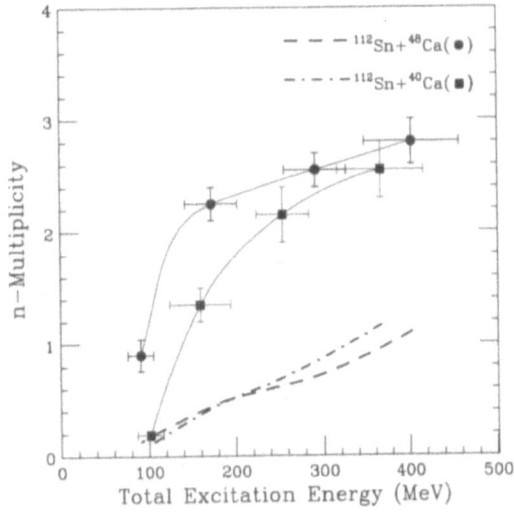

Figure 7. Pre-equilibrium neutron multiplicity is plotted vs. the total excitation energy. The solid symbols represent data for ^{48}Ca- (circle) and ^{40}Ca- (squares) induced reactions while the dashed and dashed-dotted curves represent results of the Fermi-jet model calculations for respective projectiles.

Conclusions

In summary, an exclusive measurement of the neutron and proton yields from the reactions ^{112}Sn + ^{40,48}Ca was performed in coincidence with projectile-like fragments. These nucleons are mostly evaporated sequentially from the hot primary PLFs and TLFs. However, the measured energy and angular distributions of neutrons and protons cannot be explained in terms of statistical emission from PLFs and TLFs alone. An additional component is observed, most clearly at intermediate angles, attributed to pre-equilibrium processes. This component can be modeled with an effective source, moving with approximately half the beam velocity and emitting isotropically nucleons with energy spectra characterized by a logarithmic slope parameter of approximately 10MeV. Average energies and multiplicities of evaporated nucleons suggest that at most 50-60% of the total kinetic energy available in the entrance channel is dissipated in the reaction. Of this dissipated energy, approximately 20-25% is carried away by pre-equilibrium nucleon emission. The trends of the neutron-to-proton ratios M_n/M_p of the evaporated nucleons have been found to be very different for the two reactions. These trends are strongly correlated with the entrance channel N/Z asymmetry and are indicative of a gradual equilibration of the N/Z asymmetry. The M_n/M_p ratios for pre-equilibrium particles exhibit a strong memory of the projectile N/Z ratios, a possible consequence of the isospin dependence of the nuclear equation of state. Fermi-jet model calculations fail to reproduce the total multiplicities or angular distributions of pre-equilibrium nucleons observed in the present measurements.

This work was supported by the U.S. Department of Energy under Grant No. DE–FG02–88ER40414 and DE–FG02–87ER40316.

REFERENCES

1. W.U. Schröder and J.R. Huizenga, in Treatise on Heavy-Ion Science, ed. D.A. Bromley (Plenum, New York and London, 1984) Vol. 2, p. 113, and references therein.
2. B. Lott et al., Phys. Rev. Lett. **68**, 3141 (1992)
3. W.U. Schröder Nuclear Physics **A538**, 439 (1992)
4. B.M. Quednau et al., Physics Lett. **B309**, 10 (1993)
5. S.P. Baldwin et al., Phys. Rev. Lett. **74**, 1299 (1995)
6. B. Djerroud et al., Advances in Nucl. Dyn. 2, (Plenum, New York and London, 1996), p333
7. J. Töke et al., Nucl. Phys. **A583** 519 (1995), and Phys. Rev. Lett. **77**, 3514 (1996)
8. E. Piasecki et al., Phys. Rev. Lett. **66**, 1291 (1991)
9. Bao-An Li et al., Phys. Rev. Lett. **78**, 1644 (1997); Also see contribution to this proceeding
10. D. Hilscher, Proc. Spec. Meeting Pre-equil. Nucl. Reactions, (B. Strohmaier, Editor), OECD, Paris 1988, NEANDC-245 'u', p.245; and references therein
11. E. Holub et al., Phys. Rev. **C33** (1986) 143
12. J.L. Wile et al., Phys. Rev. Lett. **63** (1989) 2551, and Phys. Rev. **C39** (1989) 1845
13. G.A. Petitt et al., Phys. Rev. **C40** (1989) 692
14. B.A. Remington et al., Phys. Rev. **C38** (1988) 1746
15. M.C. Robel, Ph.D. thesis, LBL, Berkely, preprint LBL-8181 (1979), unpublished
16. J. P. Bondorf et al., Phys. Lett. **B84** (1979) 162, and Nucl. Phys. **A333** (1980) 285
17. K. Möhring et al., Nucl. Phys. **A440** (1985) 89
18. J. Randrup and R. Vandenbosch, Nucl. Phy. **A474** (1987)219, and Phy. Rev. **C48** (1993)857
19. D. Fox et al., Nuclear Instrumentation and Methods, **A368** (1996) 709
20. J. Töke et al., Nuclear Instrumentation and Methods, **A334** (1993) 653
21. R.J. Charity, GEMINI code available via anonymous ftp from wunmr.wustl.edu/pub/gemini
22. W.U. Schröder et. al., Nucl. Science Research Conference Series, Vol.11, (1986) 255
23. J. Randrup, Nucl. Phys. **A307** (1978) 319; **A327** (1979)490; **A383** (1982) 468.
24. N.G. Nicolis, D.G. Sarantites, and J.R. Beene, computer code EVAP (unpublished); derived from the computer code PACE by A. Gavron. Phys. Rev. **C21**, 230 (1980).

DISSIPATIVE COLLISIONS AND MULTIFRAGMENTATION IN THE FERMI ENERGY DOMAIN

W. Skulski,[1] J. Dempsey,[2] D.K. Agnihotri,[1] S.P. Baldwin,[1] B. Djerroud,[1] J. Tõke,[1] W.U. Schröder,[1] R.J. Charity,[2] L.G. Sobotka[2]

[1]Dept. of Chemistry and NSRL, University of Rochester, Rochester New York 14627
[2]Dept. of Chemistry, Washington University, St. Louis, Missouri 63130

Introduction

Multiple emission of intermediate–mass fragments (IMFs) is one of the salient reaction modes in heavy–ion collisions in the Fermi–energy domain ($E/A \approx 10$ to 100 MeV).[1] It has been established that for heavy systems, and at the lower boundary of the Fermi energy regime ($E/A \approx 30$ MeV), sequential *thermal* IMF emission from either the projectile–like fragment (PLF) or the target–like fragment (TLF) is rather weak.[3, 5] IMFs are rather emitted from the "neck zone" formed between PLF and TLF, both in peripheral[2, 3, 4] and in central[6] collisions. It has also been shown that in peripheral collisions the "neck zone" is transiently formed from the neutron–rich projectile and target surface matter.[3, 7] These findings are consistent with a *dynamical* IMF production scenario, such as that of microscopic molecular dynamics models.[8, 9, 10] Since protons and α-particles are emitted mostly sequentially from PLF and TLF,[3, 5] while in the same events, IMFs are emitted mostly from a midrapidity "neck–like" source, a single fused system can be ruled out as a possible source of all particles and fragments, for nearly the entire observed cross section.

It will be shown in this paper that, in Bi+Xe collisions at $E/A = 55$ MeV, a dissipative reaction mechanism can be identified with a large cross section. Multiple IMFs can be observed in coincidence with projectile–like fragments, consistent with dynamical production mechanisms. It will also be demonstrated that an "Arrhenius plot" constructed from multi–IMF emission probabilities is consistent with such dynamical reaction scenarios.

Advances in Nuclear Dynamics 3, edited by
Bauer and Mignerey, Plenum Press, New York, 1997

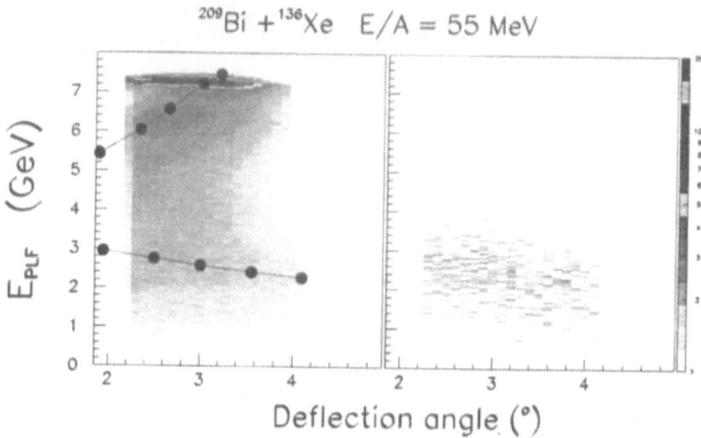

Figure 1. The yield of massive charged fragments with $Z \geq 10$, observed with a silicon–CsI(Tl) telescope at forward angles, for the reaction Bi+Xe at E/A=55 MeV, plotted versus laboratory deflection angle and fragment energy. *Left panel:* inclusive data. The line represents results of nucleon exchange model calculations.[11] *Right panel:* data in coincidence with four intermediate–mass fragments (IMFs).

Experimental procedure

The experiment was performed at the K1200 Cyclotron of the National Superconducting Cyclotron Laboratory at Michigan State University. A 55–MeV/nucleon ^{136}Xe beam bombarded a 1–mg/cm^2 thick ^{209}Bi target. The experimental setup consisted of the Rochester SuperBall, the St. Louis MiniWall, the MSU MiniBall, and a Si-CsI forward array. Massive projectile–like fragments were detected with an annular Si–CsI(Tl) telescope covering the angular range between 2.2° and 4.5°. The telescope consisted of a position–sensitive annular Si detector backed by an array of 16, two-centimeter thick CsI(Tl) detectors with photodiode readout. Light and intermediate–mass charged products were measured with the Miniball and Miniwall detectors, which covered the angular range from 5.3° to 160°. These devices were able to resolve elements of hydrogen, helium, and lithium, as well as three isotopes of hydrogen over part of the angular range. Fragments with Z>3 were counted, but not resolved according to their atomic number, through the whole angular range covered by the detectors.

All charged–particle detectors were placed inside the scattering chamber of the University of Rochester SuperBall, a 16–m^3 gadolinium–loaded liquid–scintillator detector. The SuperBall measured, with almost 4π coverage, the multiplicity of neutrons, and provided information on the summed neutron kinetic energy. Together, these detection systems provide the means to correlate the energy and charge of the projectile remnant with the charged–particle emission characteristics and the multiplicity and total kinetic energy of the associated neutrons.

Results

Dissipative collisions and multifragmentation

The yield of massive charged fragments with $Z \geq 10$, observed with the silicon–CsI(Tl) telescope at forward angles, is plotted in Fig. 1 versus laboratory deflection angle and fragment energy. Inclusive data and data taken in coincidence with four

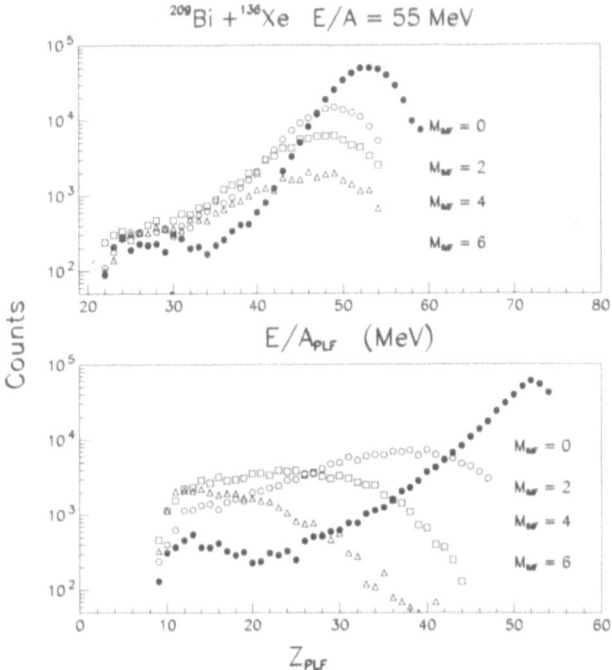

Figure 2. Yield of massive charged fragments from the reaction Bi+Xe at E/A=55 MeV, observed with a silicon–CsI(Tl) telescope at forward angles, plotted versus fragment energy per nucleon (upper panel) and fragment charge (lower panel). Different sets represent data in coincidence with various numbers of IMFs.

IMFs are plotted in the left and right panels, respectively. Also plotted in the figure are the predictions of the nucleon-exchange model[11], corrected for evaporation. Most of the yield in the left panel is distributed along a characteristic Z–shaped ridge. From the small–angle limit of the telescope aperture to the grazing angle of $\theta \approx 4°$, a ridge of mainly elastic yield runs horizontally at $E \approx 7$ GeV. As energy is dissipated in the reaction, the ridge bends back towards smaller angles and turns to smaller E values. For high degrees of dissipation, a third part of this ridge is discernible at $E \approx 3$ GeV. The correlation of fragment energy with angle, shown in Fig. 1, is characteristic of a dissipative orbiting process,[12] where the two reaction partners rotate about their common center of gravity for a fraction of a revolution, while kinetic energy of relative motion is dissipated. Subsequently, the fragments disengage as primary TLF and PLF and deexcite, while proceeding along Coulomb trajectories. As demonstrated by the good agreement between the trajectory calculations[11] and the data, the orbiting feature, known from a similar Bi+Xe experiment at $E/A = 28$,[13] persists at a beam energy which is almost twice as large. The right panel of the same figure shows, that multi-fragmentation events are concentrated in the third, returning branch of the Wilczyński diagram, consistent with significant energy dissipation necessary to cause multiple IMF emission from the system.

In the upper and lower panels of Fig. 2, the yield of massive PLFs is plotted versus the fragment energy–to–mass ratio E/A_{plf} (top), and fragment charge Z_{plf} (bottom), respectively, as a function of IMF multiplicity m_{imf}. The $m_{imf} = 0$ data sets are peaked at values $E/A \approx 53$ MeV and $Z \approx 52$, which are close to that of the beam. The peaks of $m_{imf} > 0$ data sets are shifted towards smaller E/A and Z values, consistent with increasing energy damping. Fig. 2 shows that multifragmentation occurs in coincidence with the production of PLFs and is thus accompanied by dissipative orbiting. Mul-

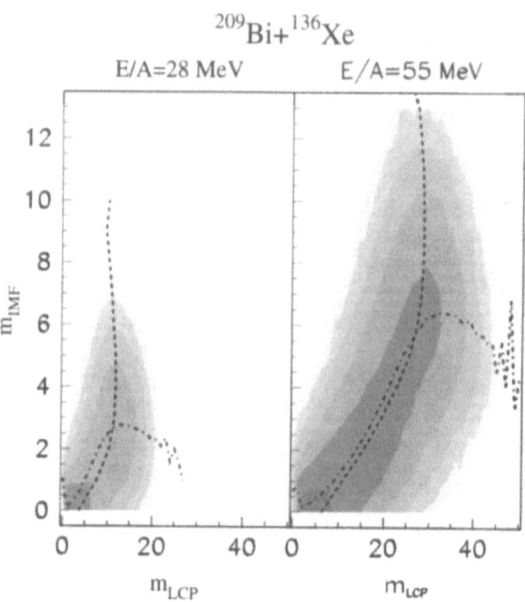

Figure 3. Logarithmic contour plot of the yield for the reaction Bi+Xe at E/A=28 and 55 MeV, plotted versus m_{imf} and m_{lcp}. Dotted lines depict the two average correlations: $\overline{m}_{lcp} = f(m_{imf})$ and $\overline{m}_{imf} = g(m_{lcp})$.

tifragmentation associated with the disintegration of a fused projectile–target system has not been detected in the present experiment.

In Fig. 3, correlations between m_{lcp} and m_{imf} are depicted as logarithmic contour plots. Data from the $E/A = 28$ MeV experiment[13] are shown in this figure for comparison. Both the 28 and 55 MeV data sets look qualitatively similar to each other, with m_{imf} correlated with m_{lcp} for relatively small values of m_{imf}, i.e., up to $m_{imf} \approx 2$ at 28 MeV/u, and up to $m_{imf} \approx 6$ at 55 MeV/u. For larger values of m_{imf}, the two variables appear uncorrelated, which leads to vertical "saturation" ridges of m_{imf} as a function of m_{lcp}. In Fig. 3, both average correlations, $\overline{m}_{lcp} = f(m_{imf})$ and $\overline{m}_{imf} = g(m_{lcp})$, are indicated as dashed and dash–dotted lines, respectively. These average correlations do not coincide with each other over the whole range of m_{lcp} and m_{imf} values. On both panels, the average $\overline{m}_{lcp} = f(m_{imf})$ follows the cross section ridges on the two-dimensional plots. The other average deviates from the ridges and enters a region of fluctuations and background (to the right of the ridges), for both the 28 MeV/u and 55 MeV/u data sets. Beyond their respective maxima, the one-dimensional correlations $\overline{m}_{imf} = g(m_{lcp})$ are strongly influenced by fluctuations, pileup, and other background events. Consequently, the m_{imf} observable has been adopted as a "sorting variable" throughout this work.

Resolution of an apparent conflict between observations of dynamical and "thermal" properties of multifragmentation

At a first glance, there seems to be a conflict between the experimental evidence for a dynamical nature of multifragmentation, as established in several recent papers,[3, 6] and thermal features claimed[14] for the same process. Therefore, it is important to examine the applicability of the concepts of "thermal reducibility" to the Bi+Xe reaction ($E/A = 28$ MeV), where the dynamical nature of multifragmentation has been well established.[3, 6] As shown in Fig. 4, this data set can be presented in a form similar

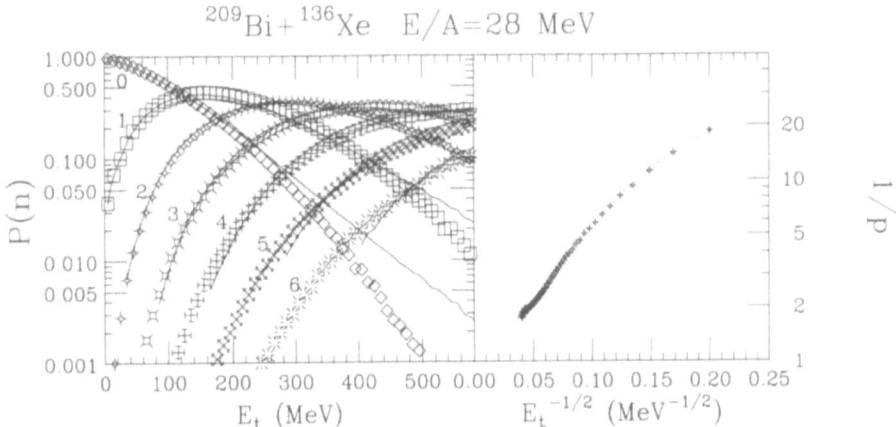

Figure 4. *Left panel:* Experimental probabilities (symbols) for the emission of n IMFs vs. total transverse energy E_t. The lines represent binomial fits to the data. *Right panel:* The reciprocal of the reduced IMF emission probability $1/p$ plotted vs. $1/\sqrt{E_t}$, for the reaction Bi+Xe at E/A=28 MeV. The E_t observable contains of the LCP and IMF contributions.

to the representation chosen in a series of papers[14] in an attempt to demonstrate an essentially "thermal and reducible" nature of multifragmentation. In the left panel of this figure, relative yields for multi–IMF events are plotted as "excitation functions" $P(m_{imf}) = f(E_t)$, where the total transverse energy E_t is defined in terms of particle energy E_i and its polar angle θ_i, as $E_t = \sum_{i=1}^{N} E_i \cdot \sin^2 \theta_i$. Solid curves in this figure represent the binomial formula $P_N^M(p) = \binom{M}{N} \cdot p^N \cdot (1-p)^{M-N}$ fitted to the data. Here, M and $N = m_{imf}$ are formal parameters of the binomial distribution, and $p = f(E_t)$ is an "elementary IMF emission probability", calculated as a function of E_t from the mean values and variances of m_{imf} distributions.[14] The data are well described by the binomial formula, except for higher energies and low values of m_{imf}, i.e., $m_{imf} = 0$ at $E_t > 300$ MeV and $m_{imf} = 1$ at $E_t > 500$ MeV. The right panel of Fig. 4 shows the dependence of $1/p$ on $E_t^{-1/2}$, i.e., an Arrhenius–type plot. The rationale behind such a representation is to emphasize a possibly thermal dependence of p on temperature T, $p \propto \exp\left(-B_{imf}/T\right)$, where B_{imf} denotes an average conditional barrier of the emitter–IMF configuration. In this approach, the transverse energy E_t is assumed to be proportional to the excitation energy E^* of the IMF emitter, and an approximate linearity of the resulting "Arrhenius plot" is taken to represent an experimental proof of a thermal scaling of the IMF emission probability p.

If the assumed proportionality $E_t \propto E^*$ were true, then it is clear from the left panel of Fig. 4 that, on the average, progressively larger numbers of IMFs are emitted from systems of increasing excitation energy and temperature. The proportionality between E_t and E^* can thus be verified by examining other experimental data for different IMF coincidence folds, in search of independent evidence supporting the excitation energy dependence seemingly present in the data depicted in the left panel of Fig. 4.

Figure 5 presents average values of the neutron multiplicity $\overline{m}_n$, the light–charged–particle multiplicity $\overline{m}_{lcp}$, the average transverse LCP energy $\overline{E}_t^{lcp}$, and the average IMF transverse energy $\overline{E}_t^{imf}$, all plotted as functions of m_{imf}. It is clear that none of the plotted variables, *except* $\overline{E}_t^{imf}$, increases monotonically with increasing m_{imf}. These observables level off at about $m_{imf} = 2$, contrary to expectations for a thermal scenario outlined above. Clearly, should the average excitation energy of the system increase

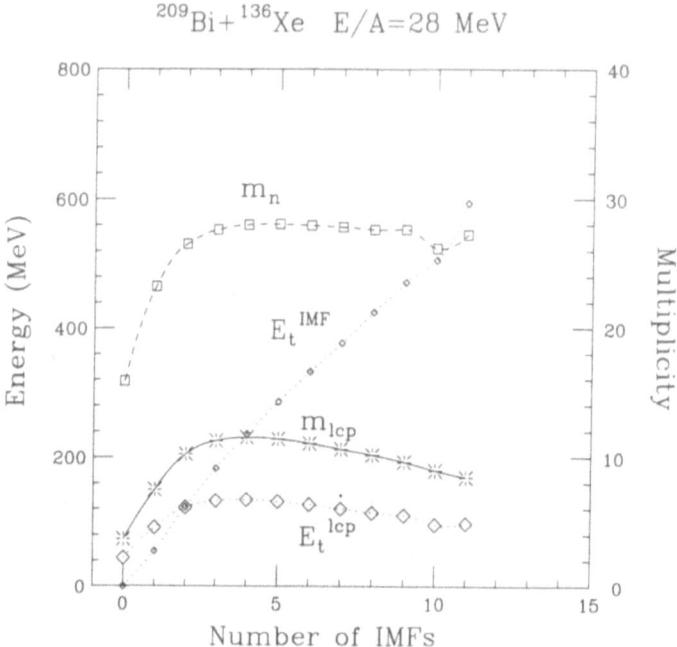

Figure 5. Average values of the neutron multiplicity $\overline{m}_n$ and LCP multiplicity $\overline{m}_{lcp}$, the transverse LCP energy $\overline{E}_t^{lcp}$, and the transverse IMF energy $\overline{E}_t^{imf}$, plotted vs. m_{imf}.

with m_{imf}, one should also observe a corresponding increase of the light–particle multiplicities and/or of the light–particle transverse energies. However, this is not the case. The average thermal excitation energy of the system appears to be approximately constant for $m_{imf} > 2$, as evidenced by constant values of both average LP multiplicities and LCP transverse energies. One should note that, for large m_{imf} values, significant amounts of energy are carried away from the system by IMFs themselves rather than by light particles. Transverse energy of neutrons was measured at $E/A = 35$ in Au+Kr reaction,[5, 15] and conclusions similar to the ones presented in this paper were reached.[5]

Even though the thermal energy deposit does not change significantly for $m_{imf} > 2$, the *total* transverse energy of *all* charged particles continues to increase with every additional IMF, since the IMF energy E_t^{imf} is of course included in the definition of $E_t = E_t^{lcp} + E_t^{imf}$. As shown in Fig. 5, the transverse energy E_t^{imf} indeed increases linearly with m_{imf}, due to a trivial correlation between m_{imf} and E_t^{imf}. The probability distributions for given numbers of IMFs, $P(m_{imf}, E_t)$, are thus displaced with respect to one another by shifts in the E_t variable, that are roughly proportional to m_{imf} (see Fig. 4). These shifts produce a visual pattern shown in the left panel of Fig. 4, which has been interpreted[14] to reflect the equilibrium–statistical (thermal) nature of multifragmentation. However, in the present case, this pattern is almost entirely due to the trivial correlation between m_{imf} and E_t, inherent in the definition of E_t.

Recently it has been demonstrated,[16] that the finite width Γ of the excitation energy distribution, which contributes to a given bin in E_t, causes significant distortions of the experimental "Arrhenius plot". This "binning effect" is illustrated in Fig. 6. Even if IMF fold probabilities were distributed according to a binomial formula, with parameters p, M, and N depending on exitation energy E^*, due to finite width of the E^* vs. E_t correlation, the experimentally observed distributions would not be of a binomial form, when plotted as functions of E_t. Based on numerical simulations of this effect

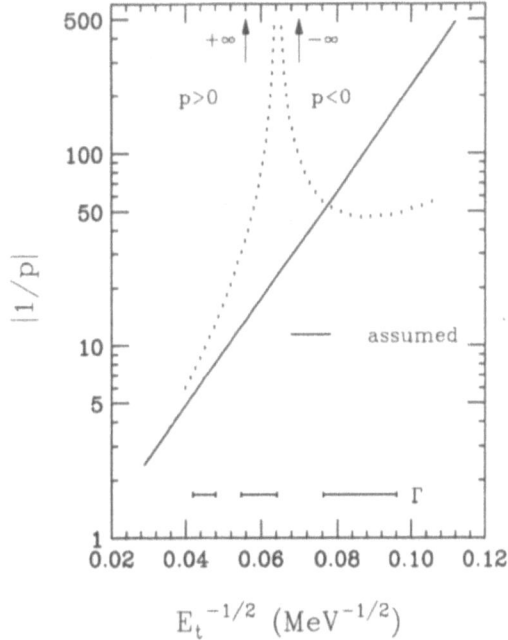

Figure 6. The reciprocal of the assumed "true" IMF emission probability $1/p$ (solid straight line "Arrhenius plot") and of the corresponding experimental observable $1/p_{apparent}$ (dotted line), plotted vs. $1/\sqrt{E_t}$. The latter curve was calculated assuming finite widths Γ of the $E^* \rightarrow E_t$ correlation, predicted by the GEMINI evaporation code.[17] These widths are illustrated at the bottom of the figure as horizontal bars labeled "Γ".

illustrated in Fig. 6, conclusions regarding thermal scaling and binomial reducibility[14] have been refuted.[16]

A convincing interpretation of the data, presented in Fig. 4 in terms of equilibrium statistical models of the kind proposed in the literature, can only be based on the following three key prerequisites: (1) that a thermally equilibrated source was indeed formed in the experiment; (2) that excitation energy of the system can be gauged as $E_t \propto E^*$, and (3) that the width of the E^* vs. E_t correlation is small and can be neglected. None of these assumptions appears to be justified. (1) In the present energy regime, multifragmentation is *not* due to the decay of a single thermalized source, but is rather driven by the *dynamics* of the collision process.[2, 3, 4, 5, 6, 7] (2) Due to its autocorrelation with the m_{imf} observable, the total transverse energy E_t is *not* the best variable to study the IMF emission process. (3) Finally, the finite width of the E^* vs. E_t correlation produces a highly distorted image of the true correlation $P(m_{imf}, E^*)$. Therefore, the linearity of experimentally observed "Arrhenius plots" is largely a result of the representation of data and does not provide decisive support to equilibrium–statistical multifragmentation mechanism. Thus, the Arrhenius–type of representation does not contradict other evidence for a dominant dynamical mechanism of multifragmentation.

Conclusions

The present work has demonstrated that the dissipative reaction mechanism persists and dominates in the Bi+Xe reaction at beam energies as high as 55 MeV/nucleon. Multifragmentation was found concurrent with the production of projectile–like fragments, and is thus tightly related to the dissipative reaction mechanism. Apparent "thermal and reducible" properties of multifragmentation were investigated in cases, where IMFs are known to be emitted dynamically rather than statistically. The IMF multiplicity was used as a "sorting variable" to investigate, whether the intrinsic excitation energy of the Bi+Xe system at 28 MeV/u depends on m_{imf} in a way consistent with statistical equilibrium. It was found, that this excitation energy remains approximately constant for $m_{imf} > 2$. On the other hand. the *total* transverse energy $\overline{E}_t$ of *all* charged particles increases with every detected IMF, due to the contribution by IMFs themselves. This trivial correlation between the observables E_t and m_{imf} leads to a false statistical signature of the IMF production mechanism, in the form of apparent excitation functions for multi–IMF events, and an Arrhenius–type plot of deduced model parameters. In case of the present Bi+Xe reaction, the IMFs are in fact emitted dynamically from the projectile–target interaction zone. The Arrhenius–type representation of the observed IMF probability distributions does not represent a contradiction with the dynamical nature of multifragmentation.

Help of the MSU Miniball group during the experiment is appreciated. This work was supported by the U. S. Dept. of Energy under Grant No. DE–FG02–88ER40414 and DE–FG02–87ER40316.

REFERENCES

1. L.G.Moretto and G.J.Wozniak, Ann. Rev. Part. and Nucl. Sci. **43**, 379 (1993), and references therein.
2. B. Lott et al., in Proc. 9th Winter Workshop on Nuclear Dynamics, Key West, World Scientific. Singapore (1993), p. 159.
3. J. Tõke et al., Nucl. Phys. A **583**, 519 (1995).
   J. Tõke et al., Phys. Rev. Lett. **75**, 2920 (1995).
4. C. Montoya et al., Phys. Rev. Lett. **73**, 3070 (1994).
5. B. Djerroud et al., (to be submitted to Phys. Rev. **C**.) See also B. Djerroud et. al., preprint UR–NSRL–421.
6. J. Tõke et al., Phys. Rev. Lett. **77**, 3514 (1996).
7. J. Dempsey et al. Phys. Rev. **C54**, 1710 (1996).
8. A. Ono, H. Horiuchi, Phys. Rev. **C53**, 2958 (1996).
9. H.Feldmeier, contribution to this Workshop.
10. J. Aichelin Phys. Rep. 202, 233 (1991).
11. W. U. Schröder et al., Nucl. Sci. Res. Conf. Ser. **11**, 255 (1987).
12. W.U. Schröder and J.R. Huizenga, in *Treatise in Heavy–Ion Science*, ed. D.A. Bromley, Plenum Press, New York and London, 1984, Vol. 2, p. 113, and references therein.
13. S. P. Baldwin et al., in Proc. 9th Winter Workshop on Nuclear Dynamics, Key West, World Scientific, Singapore (1993), p. 36.
    S. P. Baldwin, et al., Phys. Rev. Lett. **74**, 1299 (1995).
14. L. G. Moretto et al., Phys. Rev. Lett. **74**, 1530 (1995).
    L. Phair et al., Phys. Rev. Lett. **75**, 213 (1995).
    K. Tso et al., Phys. Lett. **B361**, 25 (1995).
    L. G. Moretto et al., submitted to Phys. Rep., LBL preprint LBLNL–39388 (1996).
15. W. Skulski, B. Djerroud, et al., Phys. Rev. **C53**, R2594 (1996).
16. J. Tõke et al., submitted to Phys. Rev. C.
17. R. J. Charity, computer code GEMINI, available from ftp://wunmr.wustl.edu/pub/gemini.

FERMIONIC MOLECULAR DYNAMICS: MULTIFRAGMENTATION IN HEAVY–ION COLLISIONS AND IN EXCITED NUCLEI

Hans Feldmeier [1] and Jürgen Schnack[2]

Gesellschaft für Schwerionenforschung, D–64220 Darmstadt

INTRODUCTION AND SUMMARY

How multifragmentation happens in heavy ion collisions is still a matter of debate. Explanations reach from nucleation over self organization, spinodal decomposition to cold break–up and survival of initial correlations. For an overview see ref. [1]. A key question is the time scale of the reaction. Slow processes like nucleation or self organization are hindered if the expansion of the whole system is too fast. Another issue is the relaxation time for thermal and chemical equilibrium which is important when statistical models are used to explain multifragmentation [2, 3, 4, 5].

If one considers the decay of excited spectator matter which has not been compressed, the expansion preceding the fragmentation might be slow enough to allow for global equilibration. However, below excitations energies of about 6 MeV per nucleon an equilibrated nucleus will most likely cool down by evaporating nucleons but not by breaking into many intermediate mass fragments (IMF, $Z \geq 3$). The reason is that the barrier for multifragmentation is too high and the nucleus has cooled already by evaporation before it fragments into pieces. In order to drive an expansion across the barrier into the spinodal region just by means of thermal excitation at normal density, one needs excitation energies between 6 and 10 MeV per particle [5]. For these energies the question arises: Can in a heavy–ion collision, which chops off a large fraction of the nucleus, the excitation of the remaining spectator part be distributed fast enough among the kinetic degrees of freedom such that the expansion sets in only after thermal equilibration? Or is it more likely that a peripheral or semi–peripheral collision creates a non–equilibrium object with strong local fluctuations, which drive the system across the barrier towards multifragmentation right away without first thermalizing and then expanding? In such a situation a common mean field would not establish anymore and the system could not be regarded as an equilibrated Fermi gas in a mean field.

---

[1]email: h.feldmeier@gsi.de, WWW: http://www.gsi.de/~feldm
[2]email: j.schnack@gsi.de, WWW: http://www.gsi.de/~schnack

Advances in Nuclear Dynamics 3, edited by
Bauer and Mignerey, Plenum Press, New York, 1997

We want to argue here that a non–thermal situation is actually very helpful in getting the nucleus across the barrier for multifragmentation which exists in the equilibrium potential–energy surface. On the way to equilibrium the system can then easily cross the equilibrium barriers or is already behind them. A non–equilibrium system can feed into all parts of the coexistence region in the phase diagram, large and small volumes, i.e. many or few fragments. And thus the isotopic ratios in the ensemble can reflect the properties of the coexistence region and the liquid–gas phase transition [6]. Therefore, we see the possibility that the phase diagram and the liquid–gas phase transition can also be investigated by non–equilibrium multifragmention of a wounded spectator and may be even better than by thermally excited nuclei (hot Fermi gas in a mean–field) which go through thermal expansion and subsequent formation of fragments.

For the participant matter created for example in central collisions the compression is much stronger and the excitation energy much higher. A transient pressure exists during the time of instreaming matter and thereafter due to recoil of promptly emitted particles. This together with the short mean free path at high excitations is in favour of multifragmentation originating from a more thermalized source. But the whole system is expanding and cooling fast so that it is questionable if there are enough collisions to ensure local equilibrium until freeze out. Experiments show for example that a large part of the excitation energy is converted into radial flow [7].

There is little hope to decide upon these questions experimentally in a unique way, because it is very difficult to measure temperature and flow profiles [6, 7, 8] and even harder or impossible to infer experimentally on the time scale of the evolution of the system. Therefore, microscopic transport models which do not assume equilibration are needed for a better understanding. These models should go beyond the mean field approach, which is a kind of equilibrium assumption in itself, so that in principle they are capable to describe many–body correlations like the formation of fragments. QMD, AMD and FMD are molecular dynamics models which assert this claim. How equilibrium is achieved can then be studied by comparing distributions, for example of mass, charge, kinetic energy etc, with equilibrium distributions.

In the following two sections we investigate within Fermionic Molecular Dynamics (FMD) the decay of a compound system with 46 or 80 nucleons which was created in a heavy–ion collision at a beam energy of about 35 AMeV and the decay of ^{56}Fe which we put in an excited state by scaling the whole many–body wave function and/or randomly moving the centroids of the wave packets. For the definition of the model see ref. [9, 10, 11, 12, 13].

The succeeding section shows FMD collisions of ^{19}F + ^{27}Al and ^{40}Ca + ^{40}Ca in the Fermi energy domain where multifragmentation is the dominant reaction mechanism. The system, however, does not go through a thermalized situation. In section 'Decay of excited nuclei' FMD evolutions of randomly excited nuclei (artificially thermalized source inside the multifragmentation barrier) are investigated. They do not show multifragmentation within a set of about 20 runs. Either they vapourize into individual nucleons, or after expanding and blowing off outer layers the inner part contracts again and an evaporation residue, which can be rather small, is left over. Only if not all correlations are destroyed the excited nucleus expands and decays into fragments and single nucleons, quite similar to the decay following the collision.

MULTIFRAGMENTATION IN COLLISIONS

Central and semi–peripheral collisions of ^{19}F + ^{27}Al and ^{40}Ca + ^{40}Ca calculated within FMD are shown in fig. 1 and 2. We choose an energy of 32 AMeV and 35 AMeV for which the relative velocity between the two nuclei is about the Fermi velocity. Here we expect the break down of the mean–field picture which prevails in the dissipative regime (up to about $E_{beam} = 15A$MeV) where the system either fuses or undergoes a strongly damped binary collision [12]. When the collective velocity becomes comparable to the internal velocities of the nucleons, a common mean field cannot be established any longer and non–equilibrium effects will be important. The picture of a hot Fermi gas in a mean field will no longer be true.

The following figures show a variety of events as contour plots of the one–body density in coordinate space. This density is integrated over the z–direction. Figure 1 and 2 present runs at two impact parameters and two initial orientations of the intrinsically deformed ground states.

In fig. 1 one sees for both impact parameters the creation of a source which lives for about 100 fm/c before it fragments into pieces of all sizes. The time is given in the upper right corners. For the larger impact parameter the source is more stretched and the final momenta of the fragments are not as isotropic as in central collision.

For the larger system ^{40}Ca + ^{40}Ca (fig. 2) the situation is similar except that the number of IMFs is larger and at impact parameter $b = 2.75$ fm the outgoing fragments have still a more isotropic momentum distribution. Pronounced flow sets in for larger impact parameters.

Although the one–body densities at $t = 100$ fm/c in fig. 1 and at $t = 120$ fm/c in fig. 2 look very thermalized, they are not. There are still strong many–body correlations which just cannot be seen in a one–body distribution. Analyzing the time evolution of a cluster one sees that the correlations between the wave packets which finally compose the fragment can be followed back for rather long time. The heavy–ion collision does not completely randomize the many–body state. The important role of correlations for multifragmentation in FMD will be discussed further in the following section where we destroy these correlations artificially.

DECAY OF EXCITED NUCLEI

In order to study the influence of many–body correlations on the decay pattern of an excited nuclear system we create in this section various initial states by exciting a ^{56}Fe nucleus through a combination of scaling the ground–state density and randomly displacing the mean positions of the wave packets without changing the density. Both create excitation, but while the scaling does not destroy the inter–particle correlations, the random displacement does.

In the FMD ground state the wave packets arrange in phase space such that the total energy is minimized. This ground state is an intrinsic state in which the relative positions and orientations of the wave packets in coordinate and momentum space reflect many–body correlations. If one destroys these correlations by randomly displacing all mean position parameters $\vec{r}_k$ around their original positions within a small circle of radius 0.2 fm perpendicular to $\vec{r}_k$ the ^{56}Fe nucleus achieves already 1.6 MeV excitation energy per particle, although the one–body density remains practically unchanged. The

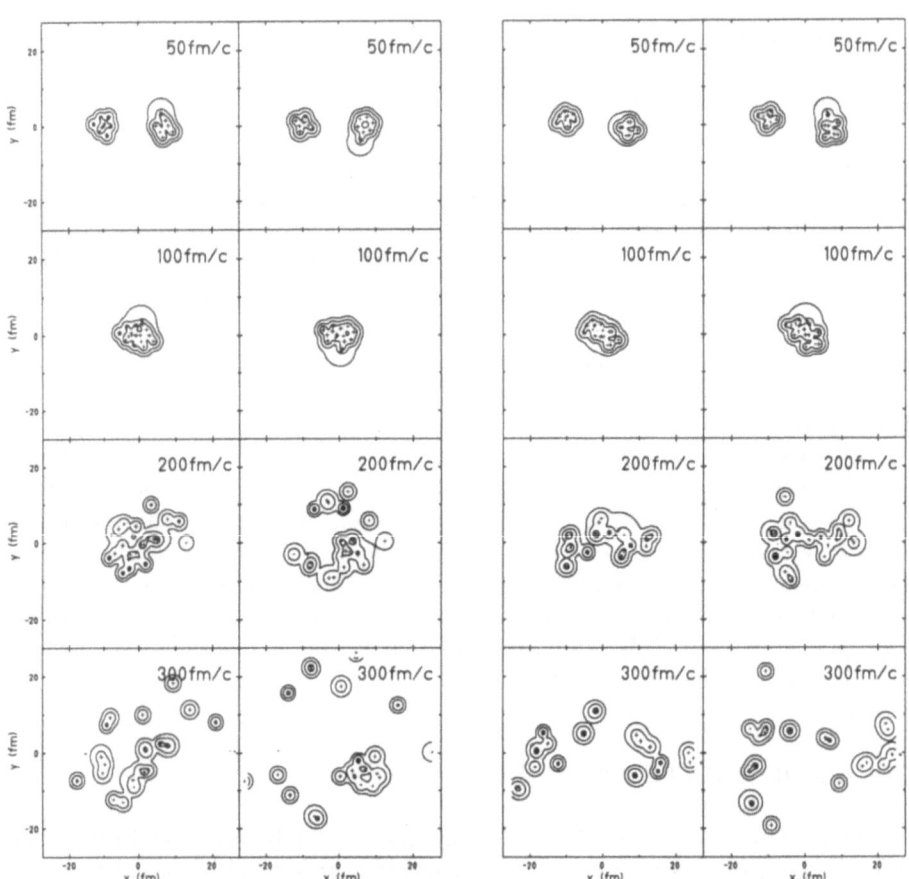

Figure 1. One–body density in coordinate space integrated over z for ^{19}F$+^{27}$Al collisions at 32 AMeV, $b = 0.5$ fm (l.h.s.) and $b = 2.5$ fm (r.h.s.). The contour lines depict the density at 0.01, 0.1, 0.5 fm^{-2}. Crosses indicate the mean positions $\vec{r}_k$ of the wave packets.

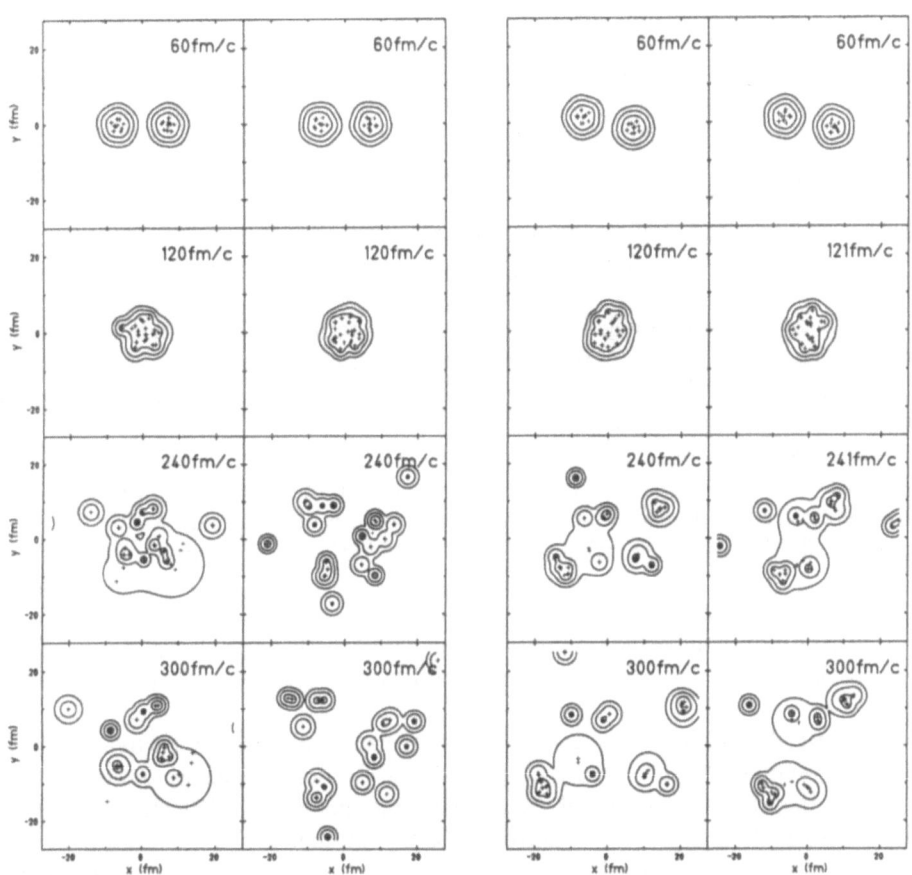

Figure 2. Same as fig. 1 but for ^{40}Ca+^{40}Ca collisions at 35 AMeV, $b = 0.25$ fm (l.h.s) and $b = 2.75$ fm (r.h.s).

first row in fig. 3 shows the time evolution of an excited state with 9.7 $AMeV$ excitation energy which was achieved by random displacements within 2 fm. One sees a fast expansion of the outer density layers which carries away a lot of energy and the survival of a residue in the center which evaporates nucleons on a longer time scale (not shown here).

In the second row the same magnitude of random displacements leads to 11.2 $AMeV$ excitation energy. Here the expansion is faster and the residue is smaller. But there are also other cases around 11.2 $AMeV$ where no residue is left over. Already above $E^* = 12$ $AMeV$ no residues are observed anymore and the nucleus vaporizes completely. There is a sharp transition around $E^* = 11$ $AMeV$ where the thermal expansion can not be brought to a halt by the attractive interaction anymore.

In the fourth row we scaled the density in coordinate space by a factor 2.2, which implies a scaling of the momentum density by 1/2.2, and in addition displaced the mean position vectors within 0.5 fm. The result is an excitation energy of $E^* = 11.3$ $AMeV$ similar to the randomized case in the second row. Up to 200 fm/c also the density develops rather similar to this case. The main difference to random excitation is that the compressed nuclei (by scaling) vaporize mainly by spreading of the wave packets and less by radial motion of the centroids (crosses in fig. 3) whereas in the randomized nuclei the centroids move out faster. At lower excitation energies, i.e. below 10 $AMeV$ (not shown here), a few particles are emitted fast, then the residue goes through damped monopole vibrations while thermalizing, and finally evaporates nucleons after some delay. From this and other runs we conclude that concerning multifragmentation the coherent compression does not change the picture too much compared to random excitation.

The last row shows the evolution of a system in which only the mean position vectors are scaled down by 0.6, but the width parameters and the mean momentum parameters are not changed. On top of that a random displacement within 0.3 fm was applied. The resulting excitation energy $E^* = 10$ $AMeV$ is comparable with the cases discussed above. But unlike in the first row the system expands without forming a residue in the centre and then undergoes multifragmentation. The reason is that this way of exciting the nucleus does not destroy as much the spatial correlations. Those wave packets which are grouped together in the ground state will still be close after the excitation and the correlations survive to a large extend the expansion. The density develops wrinkles rather early (see frame at 50 fm/c) which rapidly become cracks (see 100 fm/c) and at 200 fm/c, where in the first row the randomized system is still rather compact, the fragments are widely separated. One should note that although the one-body density of the initial state seems perfectly symmetric and no extra momentum has been given to the wave packets the expansion amplifies the correlations and the initial symmetry of the one–body density is broken by rapidly growing fluctuations.

This shows that in FMD many–body correlations play an important role in the formation of fragments. It seems that the time, which is set by the expansion, is too short to allow the many–body state to develop the special kind of many–body correlations which are needed to make a cluster with low enough excitation energy, so that it can survive.

The absence of multifragmentation in other calculations of the FMD type [14, 15] is interesting. Especially in [14] the effect of decoupling the center–of–mass degrees of freedom from the internal degrees of freedom of a fragment was investigated because in a many–body state, which is a (antisymmetrized) product of single–particle packets, a fragment has always about 10 MeV kinetic energy in the localization of the c.m.–

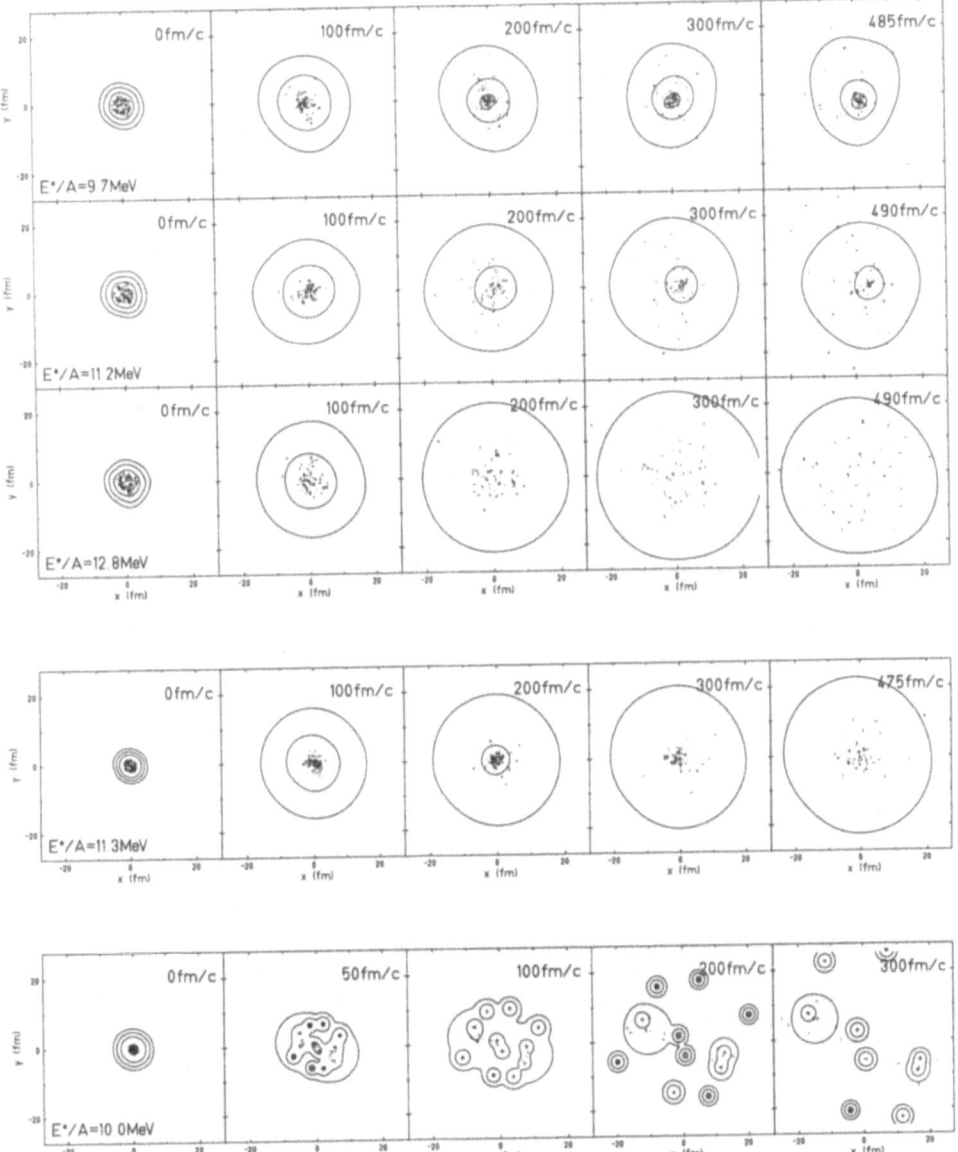

Figure 3. Same as fig. 1 but for ^{56}Fe deexcitations at various excitations. First three rows: random displacement of mean positions $\vec{r}_k$ (crosses). Fourth row: density scaled by 2.2 and small random displacements. Fifth row: $\vec{r}_k \rightarrow 0.6\vec{r}_k$ and small random displacements.

coordinate, which for small fragments might decrease the production probability [16]. Even with the additional degrees of freedom no multifragmentation was found in ref. [14].

One must however be very careful in comparing different FMD type calculations because we found that the result of a heavy–ion reaction is very sensitive to the two-body Hamiltonian employed. In particular if the ground state properties of the different fragments are not described well (with the very same Hamiltonian) the outcome of a collision is unpredictable. But even for interactions which reproduce ground states equally well it can happen that for example two light nuclei do not fuse anymore at low energy (a must for any model). Therefore one may not yet conclude that the FMD trial state is too much restricted in its degrees of freedom since we see multifragmentation, dissipative binary reactions and fusion with the same interaction [12].

REFERENCES

1. *Multifragmentation*, Proceedings of the International Workshop XXII on Gross Properties of Nuclei and Nuclear Excitations, eds. H. Feldmeier & W. Nörenberg, Hirschegg, Austria (1994)
2. J.P. Bondorf, A.S. Botvina, A.S. Ilinov, I.N. Mishustin, K. Sneppen, Phys. Rep. **257** (1995) 133
3. D.H.E. Gross, Rep. Prog. Phys. **53** (1990) 605
4. W.A. Friedman, Phys. Rev. Letts. **60** (1988) 2125;
    W.A. Friedman, Phys. Rev. **C42** (1990) 667
5. G. Papp, W. Nörenberg, APH Heavy Ion Physics 1 (1995) 241
6. J. Pochodzalla, T. Möhlenkamp, T. Rubehn, A. Schüttauf, A. Wörner, E. Zude, M. Begemann-Blaich, Th. Blaich, C. Gross, H. Emling, A. Ferrero, G. Imme, I. Iori, G.J. Kunde, W.D. Kunze, V. Lindenstruth, U. Lynen, A. Morini, W.F.J. Müller, B. Ocker, G. Raciti, H. Sann, C. Schwarz, W. Seidel, V. Serfling, J. Stroth, A. Trzcinski, W. Trautmann, A. Tucholski, G. Verde, B. Zwieglinski, Phys. Rev. Lett. **75** (1995) 1040
7. W. Reisdorf et al. (FOPI collaboration), in press, Nucl. Phys. **A** (1996); nucl-ex/9610009
8. J.B. Natowitz, K. Hagel, R. Wada, Z. Majka, P. Gonthier, J. Li, N. Mdeiwayeh, B. Xiao, Y. Zhao, Phys. Rev. **C52** (1995) R2322
9. H. Feldmeier, Nucl. Phys. **A515** (1990) 147
10. H. Feldmeier, K. Bieler, J. Schnack, Nucl. Phys. **A586** (1995) 493
11. H. Feldmeier, J. Schnack, Nucl. Phys. **A583** (1995) 347c
12. H. Feldmeier, J. Schnack, to be published in Prog. Part. Nucl. Phys. **39** (1997)
13. J. Schnack, H. Feldmeier, GSI–Preprint–97–18
14. D. Kiederlen, P. Danielewicz, preprint, MSUCL–1047 (1996)
15. Ph. Chomaz, M. Colonna, A. Guarnera, Proc. 12th Winter Workshop on Nuclear Dynamics, Snowbird, ed. W. Bauer and G.D. Westfall, Plenum, New York (1996) 65
16. A. Ono, H. Horiuchi, Toshiki Maruyama, A. Ohnishi, Phys. Rev. Lett. **68** (1992) 2898; Prog. Theor. Phys. **87** (1992); Phys. Rev. **C47** (1993) 2652

APPARENT TEMPERATURES IN HOT QUASI-PROJECTILES AND THE CALORIC CURVE

J. Péter[1], M. Assenard[5], F. Gulminelli[1], Ma Y.-G.[1],A. Siwek[1],G. Auger[2], Ch.O. Bacri[6], J. Benlliure[3], E. Bisquer[4], F. Bocage[1], B. Borderie[6], R. Bougault[1], R. Brou[1], P. Buchet[2], J.L. Charvet[2], A. Chbihi[3], J. Colin[1], D. Cussol[1], R. Dayras[2], E. De Filippo[2], A. Demeyer[4], D. Doré[6], D. Durand[1], P. Eudes[5], J. D. Frankland[6], E. Galichet[4], E. Genouin-Duhamel[1], E. Gerlic[4], M. Germain[5], D. Gourio[5], D. Guinet[4], P. Lautesse[4], J.L. Laville[5], J.F. Lecolley[1], A. Le Fèvre[3], T. Lefort[1], R. Legrain[2], O. Lopez[1], M. Louvel[1], N. Marie[3], V. Métivier[5], L. Nalpas[2], A. D. Nguyen[1], M. Parlog[8], E. Plagnol[6], O. Politi[3], A. Rahmani[5], T. Reposeur[5], M.F. Rivet[6], E. Rosato[7], F. Saint-Laurent[3], S. Salou[3], J.C. Steckmeyer[1], M. Stern[4], G. Tabacaru[8], B. Tamain[1], L. Tassan-Got[6], O. Tirel[3], E. Vient[1], C. Volant[2], J.P. Wieleczko[3]

[1]L.P.C. Caen (IN2P3-CNRS/ISMRA et Université), Caen, France
[2]C.E.A. DAPNIA-SPhN, C.E. Saclay, France
[3]GANIL (DSM-CEA/IN2P3-CNRS), Caen, France
[4]I.P.N. Lyon (IN2P3-CNRS/Université), Villeurbanne, France
[5]SUBATECH (IN2P3-CNRS/Université), Nantes, France
[6]I.P.N. Orsay (IN2P3-CNRS), Orsay, France
[7]Dipartimento di Scienze, Univ. di Napoli, Italy
[8]I.F.I.N., I.F.A., Bucharest, Romania

INTRODUCTION

The dependence of nuclear temperature upon excitation energy has been experimentally studied with increasing values of excitation energy over the years. At excitation energies per nucleon, E^*/A, lower than 6 MeV the temperatures deduced from the kinetic properties of the emitted particles and clusters follow the Fermi gas law : $E^* = (A/k).T^2$. The value of the inverse level density parameter k was found to be in the range 8 to 13[1]. When excitation energies up to 10 MeV per nucleon were reached, temperatures obtained from the relative populations of excited levels in the emitted light nuclei did not overcome 5-6 MeV [2], but this limitation could be explained by side-feeding effects. Such hot nuclei were formed in fusion or deep inelastic reactions. At incident energies above 40-50 MeV/u, binary dissipative collisions dominate and the quasi-projectiles reach excitation energies per nucleon and kinetic temperatures above 10 MeV[4, 5]. The study of projectile "spectators" in reactions at several

Advances in Nuclear Dynamics 3, edited by
Bauer and Mignerey, Plenum Press, New York, 1997

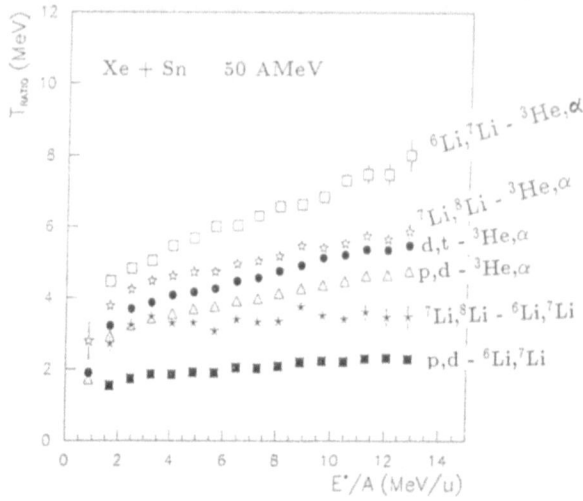

Figure 1. Measured apparent temperature obtained from double isotopic yield ratios versus excitation energy per nucleon of quasi-projectiles with masses 1115±20% formed in ^{129}Xe collisions on Sn at 50 AMeV.

hundreds of MeV/u made it possible to reach similar excitation energies [3]. In this Aladin experiment at GSI, the temperature was obtained via the relative abundances of two isotope pairs [6]. The relation between this temperature Tr^0 and E^*/A was interpreted as indicating a phase transition, with the nuclear gas regime dominating above $E^*/A = 10$ MeV, as predicted [7]. However, a monotonic increase of the temperature with excitation energy was observed in similar conditions[8] and questions were raised about the significance of these caloric curves [9, 10, 11, 12], about the role played by the mass dependence of the decaying nucleus upon E^*/A [13], as well as the strong effects of side-feeding, especially at high temperatures [14]. This point will be discussed by Xi Hong Fei at this meeting.[15].

With Indra at GANIL, the temperature-excitation energy relationship was studied for E^* ranging from 2 to 24 MeV per nucleon in the quasi-projectiles issued from the reaction ^{36}Ar+ ^{58}Ni at 95 MeV/u. Different prescriptions for the determination of temperatures were applied and compared: apparent temperatures extracted from several pairs of isotopes and from the slopes of light charged particle kinetic energy spectra. No indication was found for a fast phase transition [16]. In this paper we show in addition temperatures determined from the population of excited discrete levels, as well as results obtained at 52 and 74 MeV/u and for quasi-projectiles from the reaction ^{129}Xe $+^{124}$Sn at 50 MeV/u. The strong role played by successive de-excitation steps and side-feeding is studied within two very different scenarios.

EXPERIMENTS

The kinetic energies of charged products were measured with the 4π detector array INDRA covering 90% of the 4π solid angle . Isotopic separation was achieved for elements up to Z=4.

As for heavier and lighter systems having the same mass asymmetry studied previously at neighbouring incident energies [4, 5], the dominant process is the formation of a quasi-projectile (QP) and a quasi-target (QT) accompanied by dynamical emission around mid-

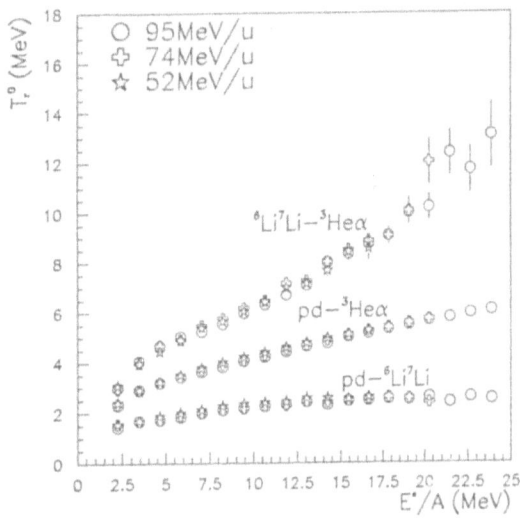

Figure 2. Same as Fig.1 for quasi-projectiles with masses 33±20% formed in ^{36}Ar collisions on Ni at 52, 74 and 95 AMeV.

rapidity. All QP products have velocities well above the detection and isotopic separation thresholds. Therefore, the QP was reconstructed as in [5]. To avoid mass dependence effects, QP's with equal masses (within ±20%) were selected.

Determining a temperature value is justified only if thermal equilibrium was achieved in the source. Experimentally one can check that the angular distributions of various products are isotropic in the source frame (or forward-backward symmetric if the source has a large spin). This is not sufficient to establish that equilibrium had been attained, but this is a necessary condition. The QP products, ranging from protons to ^{8}Li, fulfill this condition at all impact parameters.

The QP excitation energy is equal to the mass balance between the QP mass and its products, added to the sum of kinetic energies of its products in its frame. The kinetic energy of neutrons was taken as the average kinetic energy of protons in the same impact parameter bin, corrected for the absence of Coulomb barrier. Due to large fluctuations in the energy sharing between QP and QT, the excitation energy per nucleon reaches values above the available energy per nucleon in central collisions, as already seen for this system [17]. Details can be found in [16].

APPARENT TEMPERATURES FROM ISOTOPIC RATIOS

Temperatures were calculated from the yields Y of several pairs of light isotopes differing by one neutron, according to [6]:

$$Tr^0_{n-d} = B/\ln(a.\frac{(Y(Z_n, A_n)/Y(Z_n, A_n + 1)}{(Y(Z_d, A_d)/Y(Z_d, A_d + 1)}) \qquad (1)$$

where n (d) stands for the pair of isotopes with the smallest (largest) binding energy difference, and appears at the numerator (denominator).

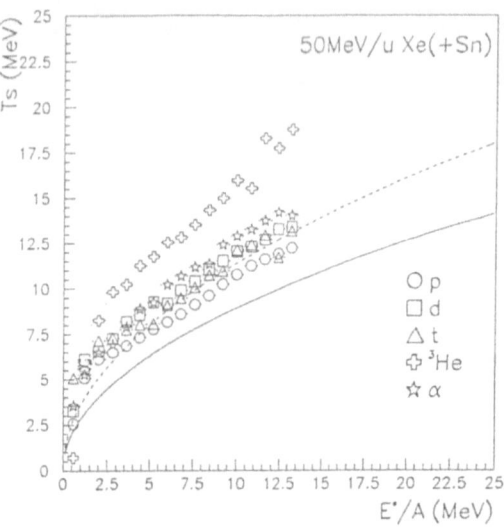

Figure 3. Measured slope parameters from kinetic energy spectra for light particles emitted by quasi-projectiles formed in ^{129}Xe collisions on Sn at 50 AMeV. For orientation, the Fermi gas relationship is shown with A/8 (solid line) and A/12 (dashed line).

By choosing isotopes which do not have low lying levels decaying via γ emission, one can calculate a with the ground states only, as in [3]. This approximation is indicated by the index 0. But at high temperatures, high lying levels contain a significant part of the yield, especially when they have a large spin. When they decay via γ emission, they contribute to the yields in eq. 1. Another problem is due to particle-unbound levels. This reduces the yields of the decaying fragments and increases the yields of daughter fragments (side-feeding) [14]. Therefore Tr^0 is only an apparent temperature.

Let us examine the Tr^0 values obtained with various isotope pairs listed in fig. 1,2. When B is not large (≈ 5 MeV), i.e. p,d-^{6}Li,^{7}Li and ^{7}Li,^{8}Li-^{6}Li,^{7}Li, the curves increase very slowly with E^*/A and saturate at low values. Such ratios are useless. When the isotopes having the largest binding energy difference, $^3He-\alpha$, are involved, larger B values are obtained and Tr^0 increase more with E^*/A. However the values obtained with the isotopes p,d and d,t at the numerator never exceed 6 MeV. Only $T^0(^6Li^7Li-^3He\alpha)$ (used by Aladin group) reaches high values, but instead of a plateau a small and gradual variation of slope is observed.

Since low values of excitation energies are reached in peripheral collisions and high values in central collisions, different collision dynamics at different impact parameters could possibly contribute to the observed behavior. To check this point, Ar + Ni data obtained at 52, 74 and 95 MeV/u were analyzed with the same method, after removing the larger part of the few fusion events present at 52 MeV/u. At the same excitation energy obtained at different impact parameters, the three Tr^0 values are equal: fig. 2.

APPARENT TEMP. FROM KINETIC ENERGY SPECTRA

Kinetic energy spectra in the QP frame were analyzed for light charged particles, since they are less sensitive to possible collective expansion. The (inverse) slope parameters Ts were obtained via the usual fits with Marwell-Boltzmann distributions. Results for quasi-Xe

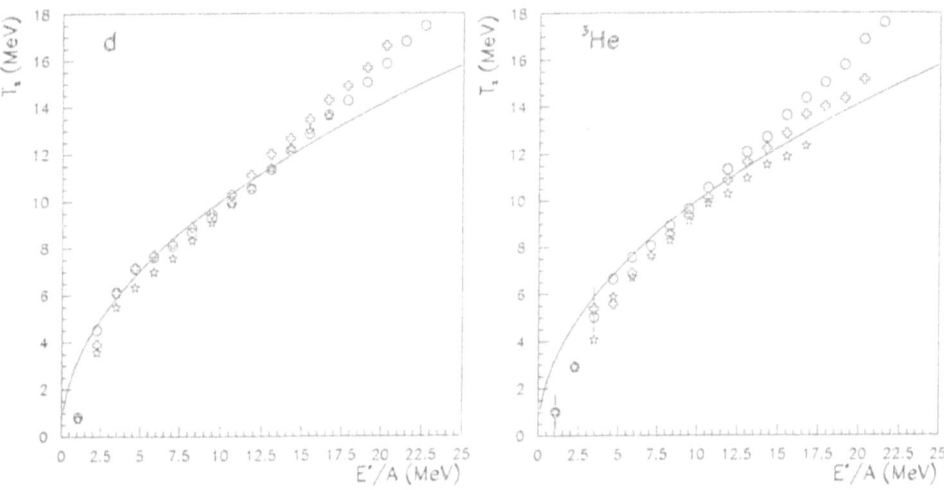

Figure 4. Same as fig. 3 for ^{36}Ar at 52, 74 and 95 AMeV, for deuterons (left) and ^{3}He (right). The symbols are the same as in fig. 2. For orientation, the Fermi gas relationship is shown with A/10 (solid lines).

nuclei are shown in fig. 3. ^{3}He slope parameters are higher than those of other light particles, which may mean they are emitted earlier, but this effect is not seen with quasi-Ar ; a specific study is in progress. The slight increase from p's to α's can be due to a small collective expansion. Each type of particle exhibits a monotonic and fast increase with E^{*}/A, as is also the case for quasi-Ar in Fig. 4. This figure also shows a very good agreement between the values obtained at 3 incident energies for deuterons (left panel). The same independence on incident energy is obtained for p and t. For ^{3}He (right panel) and α, differences in Ts reach 10% at high E^{*}/A values. This is related to the error made on the velocity of the reconstructed QP in central collisions.

APPARENT TEMP. FROM EXCITED STATE POPULATION

The third available method is the relative population of excited states. When a nucleus is emitted from a source with a temperature $Temi$, two levels of this nucleus separated by ΔE are populated in proportion to $e^{-\Delta E/Temi}$ and their spins. The levels which could be identified are those which decay via emission of two charged products detected by two modules. INDRA was not designed for this purpose and only well separated levels can be identified. A detailed discussion of the methods and results will be given [18]. Here is shown only the emission temperature obtained for ^{5}Li between the g. s. and 16.6 MeV : fig.6, bottom panel. The huge statistics needed does not allow to get narrow bins of E^{*}/A. $Temi(^{5}\text{Li})$ does not exceed 6 MeV.

SEQUENTIAL DECAY CALCULATIONS

In order to investigate the difference between the initial temperature, $Tini$, and the various apparent temperatures, we applied the standard evaporation model. Even if its basic

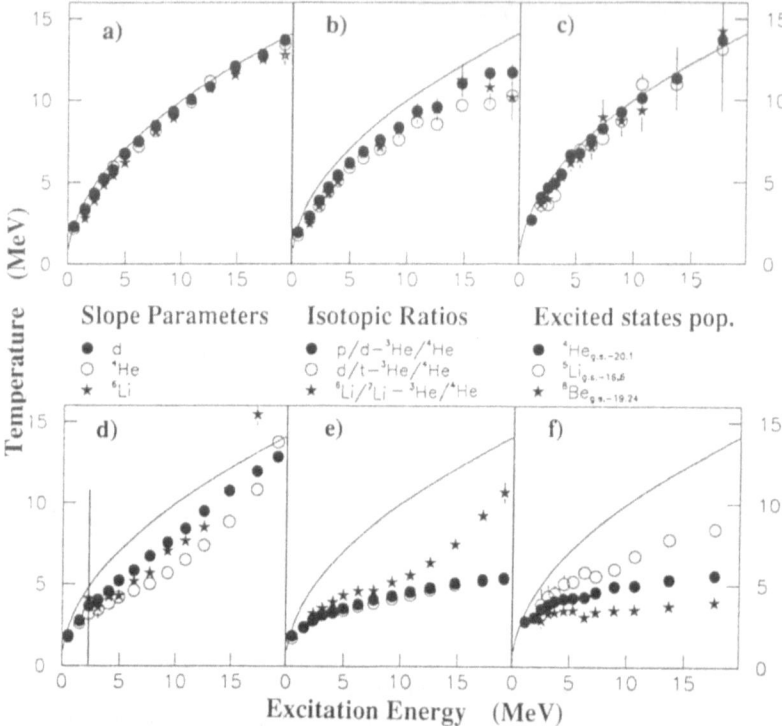

Figure 5. Sequential statistical model calculations. Solid lines: assumed initial temperature versus the excitation energy per nucleon. Points: calculated apparent temperatures obtained via 3 methods. Top panel : particles evaporated first, before decay of discrete levels. Bottom panel : particles emitted at all de-excitation steps, after decay of discrete and continuum states.

assumption (thermal equilibrium re-established between successive independent emissions) is not valid at very high E^*, it allows us to see the effects of several de-excitation steps and side-feeding. The Fermi gas relationship was assumed at all excitation energies.

Firstly, the code was stopped after first chance emission. Ts and $Temi$ have to be equal to the initial temperature : fig. 5a,c. Tr^0 also should be equal, provided we take only fragments emitted in the ground state ; since excited fragments are included (as in the experiment), it is a bit lower at high E^*/A: fig. 5b. When the particle decay of discrete levels is allowed, all temperatures decrease quite moderately.

A very different situation is obtained when the full evaporation chain is studied. Detailed studies were performed by switching off or on particle decay of discrete levels, and taking into account the maximum width of these levels [21]. Fig. 5d,e,f show the final step. Tr^0 values obtained with small values of B (not plotted) are almost constant with E^*/A, like experimental data (fig. 1,2). When B is large, some increase is kept, especially for ^{6}Li ^{7}Li-^{3}Heα which exhibits a change of slope with E^* quite similar to the data (fig. 2) and almost mimics a liquid-gas phase transition ! Even with very large ΔE values, $Temi$ is severely reduced by side-feeding: fig. 5f. The slope parameters are much less influenced by both disturbing processes: fig. 5e.

If the decaying nucleus is thermalized, the various apparent temperatures must be reproduced with the same initial temperature, $Tini$. The calculations in fig. 5 were made

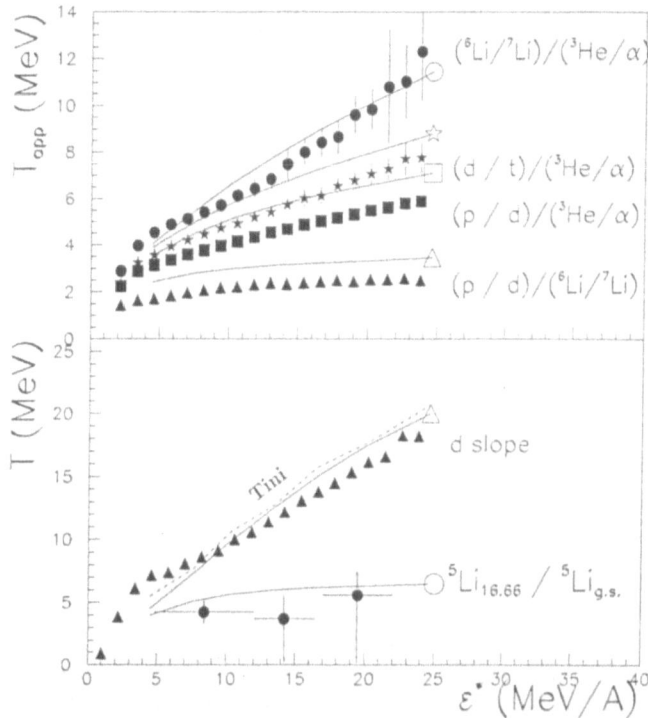

Figure 6. Extended QSM calculations compared to data. Solid symbols: experimental data; upper panel: Tr^0 values; bottom panel: Ts for deuterons, $Temi$ for ^{5}Li (preliminary values). Dashed line: correlation between initial temperature and excitation energy assumed in the calculation. Solid lines and open symbols : calculated apparent temperatures corresponding to the solid symbol with the same shape.

for Ar nuclei with the Fermi gas relationship and can be compared to data (fig. 2, 4 and 6). $Temi(^5$Li$)$ is too large, but a rather good agreement is seen with T^0(p,d-^{3}He,α), T^0(p,d-^{6}Li,^{7}Li) and Ts values, while a correlation assuming a plateau (T_{ini} constant for $E^*/A =$ 3.6-11 MeV,shown in [16]) clearly disagrees with most data.

EXTENDED Q.S.M. AND INITIAL TEMPERATURES

At high E^*/A, sequential decay is not likely, so an opposite assumption was studied[22] : the emitting source is viewed as a nuclear gas in thermal and chemical equilibrium undergoing simultaneous disassembly at a given density ρ and temperature T [19].

Corrections to the ideal gas were included in the form of excluded volume effects [20]. This correction plays an important role. The excitation energy has been calculated as the total energy of the freeze-out configuration.This latter quantity has been fixed to reproduce the experimental excitation energy dependence of the measured ratio between the charge carried by light charged particles (Z=1,2) and the total charge. This leads to a good reproduction of the charge distributions for $E^*/A \geq 10$ MeV. The initial temperatures steadily increase from 2 to 18 MeV.

In fig. 6 the calculated Tr^0 values, Ts for deuterons and $Temi(^5$Li$)$ are compared to the data. The differences between the various apparent temperature are essentially due to side-feeding. These results were obtained by taking into account levels with widths ≤ 2 MeV(source lifetime ≥ 100 fm/c). The sensitivity to this parameter is shown in [16, 22]. Above ≈ 6 MeV per nucleon, all apparent temperatures are rather well reproduced.

CONCLUSION

Apparent temperatures obtained from isotope ratios, kinetic energy distributions and excited state populations were determined for quasi-projectiles formed in collisions of ^{36}Ar on Ni at 95, 74 and 52 MeV/u and ^{129}Xe on Sn at 50 MeV/u. Nearly constant QP masses were selected over the whole range of excitation energies which reaches 24 MeV per nucleon or 13 MeV, depending on the system. Two types of calculations show that apparent temperatures based on isotope yield ratios or excited level populations are very sensitive to the role of high energy resonances and that the initial temperature values are subject to a large uncertainty. At excitation energies above 10 MeV per nucleon, these apparent temperatures are much lower than the initial temperatures. The kinetic temperatures (slope parameters) of light particles keep a good memory of the initial temperature, provided no large collective motion is present. No evidence is found for a plateau followed by a rise (first order liquid-gas phase transition) in the relationship between T and E^*/A, neither in the experimental data (apparent temperatures), nor in the initial temperatures obtained via calculations. The initial temperature grows steadily with excitation energy on the whole range of excitation energies. Up to ≈ 10 MeV per nucleon, this increase is very close to that of a Fermi gas. At higher excitation energies, multifragmentation sets in and the temperature increases faster with excitation energy. Instead of a rather sharp transition, a gradual evolution is found.

REFERENCES

1. R. Wada et al., Phys. Rev. **C 39** (1989) 497 and references therein.
2. J. Pochodzalla et al., Phys. Rev. **C 35** (1987) 1695 and references therein.
3. J. Pochodzalla et al., Phys. Rev. Lett. **75** (1995) 1040.
4. J.C. Steckmeyer et al., Phys.Rev. Lett. **76** (1996) 4895.
5. J. Péter et al., Nucl. Phys. **A593** (1995) 95.
6. S. Albergo et al, Nuovo Cimento **A 89** (1985) 1.
7. J. Bondorf et al, Nucl. Phys. **A 444** (1985)460, **A 448** (1986) 753
8. J. A. Hauger et al., Phys. Rev Lett. 77 (1996) 235.
9. D. H. E. Gross et al., Phys. Lett. B **200** (1988) 397.
10. Sa B.-H., Nucl. Phys. **A 499** (1989) 480.
11. L. G. Moretto et al., Phys. Rev. Lett. **76** (1996) 2822
12. X. Campi et al., Phys. Lett. B **385** (1996) 1
13. J. Natowitz et al., Phys. Rev. C **52** (1995) R2322
14. M. B. Tsang et al., Phys. Rev. C **53** (1996) R1057
15. H. F. Xi et al., preprint MSUCL-1040 (1996)
16. Y.G. Ma et al., Phys. Lett. B **390** (1997) 41
17. M.F. Rivet et al., Phys.Lett. **B388** (1996) 219. B. Borderie et al., Phys.Lett. **B388** (1996) 224
18. M. Assenard, Ph. D. Thesis Nantes, 1997.
19. A.Z. Mekjan, Phys. Rev. C **17** (1978) 1051
20. R. K. Tripathi and L.W. Townsend Phys. Rev. C **50** (1994) R7
21. A. Siwek et al., Proc. XXXI Zakopane School, Poland (1996)
22. F. Gulminelli and D. Durand, preprint LPCC 96-11, Nucl. Phys. A (accepted).

BARYON PRODUCTION IN HIGH ENERGY PB-PB COLLISIONS - RECENT RESULTS FROM NA44

Eva Barbara Holzer for the NA44 Collaboration

I. G. Bearden[1], H. Bøggild[1], K. Bussman[2], J. Boissevain[3], J. Dodd[4],
B. Erazmus[5], S. Esumi[6], C. W. Fabjan[2], D. Ferenc[7], D. E. Fields[3],
A. Franz[2], J. J. Gaardhøje[1], M. Hamelin[8], A. G. Hansen[1],
K. H. Hansen[1], D. Hardtke[9], H. van Hecke[3], E. B. Holzer[8],
T. J. Humanic[9], P. Hummel[2], B. V. Jacak[3], R. Jayanti[9], K. Kaimi[6],
M. Kaneta[6], M. Kopytine[3], Y.Y. Lee[9], M. Leltchouk[4], A. Ljubicic[7],
B. Lörstad[11], N. Maeda[6], A. Medvedev[4], Y. Miake[10], A. Miyabayashi[11],
M. Murray[8], S. Nagamiya[4], S. Nishimura[6], H. Ohnishi[6], G. Paic[7],
S. U. Pandey[9], F. Piuz[2], J. Pluta[5], V. Polychronakos[10], M. Potekhin[4],
G. Poulard[2], A. Sakaguchi[6], M. Sarabura[3], K. Shigaki[6], J. Simon-Gillo[3],
J. Schmidt-Sørensen[11], W. Sondheim[3], M. Spegel[2], T. Sugitate[6],
J. P. Sullivan[3], Y. Sumi[6], W.J. Willis[4], K.Wolf[8], N. Xu[3], and
D. Zachary[9]

[1]Niels Bohr Institute, DK-2100 Copenhagen, Denmark
[2]CERN, CH-1211 Geneva 23, Switzerland
[3]Los Alamos National Laboratory, Los Alamos, NM 87545, USA
[4]Columbia University, New York, NY 10027, USA
[5]Nuclear Physics Laboratory of Nantes, Nantes 44072, France
[6]Hiroshima University, Higashi-Hiroshima 724, Japan
[7]Rudjer Boskovic Institute, Zagreb, Croatia
[8]Texas A&M University, College Station, TX 77843, USA
[9]Ohio State University, Columbus,OH 43210, USA
[10]Brookhaven National Laboratory, Upton, NY 11973, USA
[11]University of Lund, S-22362 Lund, Sweden

Proton and Deuteron transverse mass spectra have been measured by the experiment NA44 at CERN. Preliminary results from 158A· GeV PbPb collisions are presented and compared to 200A· GeV SS and SPb collisions measured in the same experiment. The inverse slope parameter increases with system size and with particle mass providing evidence of flow. Ratios of proton and deuteron rapidity densities are compared for the different systems. In the framework of a coalescence model a radius parameter R can be extracted. R decreases with transverse mass, supporting the hypothesis of radial flow.

Advances in Nuclear Dynamics 3, edited by
Bauer and Mignerey, Plenum Press, New York, 1997

EXPERIMENT NA44

The Focussing Spectrometer, NA44, is designed to measure particle correlations and single particle spectra. It features particle identification at the trigger level and excellent momentum resolution of $\sigma_p/p = 0.2\%$ by dealing with only a few particles at a time. The mean multiplicity within the acceptance of the spectrometer is about 0.5 particles per central PbPb collision. By moving the spectrometer to two different angular settings and by changing the nominal momentum setting (4,6 or 8 GeV/c) a broad region in phase space is scanned. The transverse momentum coverage is $0 \leq p_t \leq 1.6 GeV/c$ near central rapidity.

The SPS accelerator supplies protons, sulphur and lead ions beams at 450, 200 and 158 GeV per nucleon, respectively. Target area detectors include a Cerenkov beam counter with a time resolution of $35 ps$ [1] and an interaction counter made of plastic scintillator used as a centrality trigger. The magnetic arm of the spectrometer consists of 2 dipole magnets and 3 superconducting quadrupoles. The magnets create a magnified image of the target on a focal plane close to the first tracking chamber. The momentum range, Δp is $\pm 20\%$ of the nominal momentum setting. Tracks through the spectrometer are reconstructed with 3 wire chambers with pad and strip cathode readout ($\sigma \approx 200 \mu m$) and 3 hodoscopes with an average time of flight resolution of $100 ps$. Two threshold Cerenkov counters are used to veto electrons and pions.

Data Sample

The data presented here were taken in 1992 for SA collisions and in 1995 for PbPb collisions, using targets of 6.6, 5.9 and 4.6% of a nuclear collision length for SS, SPb and PbPb, respectively. The centrality trigger selects the highest multiplicity events in the pseudorapidity range of approximately $1.3 \leq \eta \leq 3.5$. The fraction of the cross section selected was 8.7, 10.7 and 20% of the total geometrical cross section for SS, SPb and PbPb collisions, respectively. No further selection on higher centrality was performed. The beam energy was $200 A \cdot GeV$ for the sulphur beam and $158 A \cdot GeV$ for the lead beam. This corresponds to a lab rapidity of $y = 6$ for sulphur ions and $y = 5.8$ for lead ions. The data consist of deuterons from the $8 GeV/c$ momentum setting and protons from the $4 Gev/c$ setting as they fall into the same rapidity and transverse velocity window. When taking the ratio of deuterons to protons most of the systematic errors, such as those on life time and acceptance, cancel.

Systematic Errors

The resolution of the transverse momentum is about $10 MeV/c$, the offset at $p_t = 0$ was found to be $\leq 7 MeV/c$, and the p_t scale is known to within 2%. The contamination of the data samples has been estimated to be less than 2.5%. The systematic errors on the absolute normalization are determined by uncertainties of the tracking (2%), particle identification efficiency (1%) and the measurement of the beam flux and centrality (3%).

Correction for Λ Feed-Down

In order to calculate the invariant cross section of protons, their measured number was corrected for Λ feed-down. The correction mainly depends on the Λ/p ratio. Pro-

Table 1. Inverse slopes in MeV/c for protons and deuterons for SS SPb and PbPb. The errors are statistical

y	Fit Range	Centrality	System	p (MeV/c)	d (MeV/c)
		8.7%	SS	155 ± 5	219 ± 45
1.9-2.3	$v_t = 0.41c$ - $0.63c$	10.7%	SPb	199 ± 7	197 ± 37
		20%	PbPb	229 ± 7	318 ± 29

ton and lambda distributions were simulated with the event generator RQMD version 1.08 [2]. The reconstruction probability for a proton arising from the decay of a Λ was evaluated with the help of a GEANT simulation. The lambda correction is highest at low p_t and it increases with system size. For PbPb collisions about 40% of the protons originate from lambdas at $p_t < 0.04 GeV/c$ compared to 30% in SS collisions. Above $p_t = 0.4 GeV/c$ the lambda contribution is constant at about 20% and 15% for PbPb and SS collisions, respectively. The systematic error on the correction factor was estimated by changing the Λ/p ratio from RQMD by a factor of 1.5, which covers the range of published data [3, 4].

TRANSVERSE MASS SPECTRA

Transverse mass, m_t ($m_t = \sqrt{p_t^2 + m^2}$), spectra of particles from pp, pA and AA collisions from a broad range of incident beam energies are generally well described by an exponential function $exp(-m_t/T)$. Here T is the inverse slope parameter. Collective expansion will lead to a boost in transverse velocity and hence to higher inverse slope parameters for the heavier particles. Pion, kaon and proton transverse mass spectra from SS and PbPb from NA44 have been compared to a hydrodynamical model featuring thermalization and flow [5]. Data were found to be consistent with a choice of parameters of freeze-out temperature $T_f = 140 MeV$ and surface velocity β_s of 0.41c and 0.6c for SS and PbPb, respectively [6].

Figure shows the proton and deuteron invariant cross section as a function of transverse kinetic energy, $k_t = m_t - m$, from SS, SPb and PbPb collisions for the 8.7, 10.7 and 20% most central interactions, respectively. The data are from the rapidity interval of $y = 1.9 - 2.3$. Deuteron cross sections are scaled by a factor of 100. The scales on both the axes are the same for protons and deuterons, so that their slopes can be compared. The particle yield increases with system size. A single exponential function was fitted to the data in the k_t range of $0.09 - 0.27 GeV/c$ for protons and $0.18 - 0.54 GeV/c$ for deuterons, which corresponds to the same range in transverse velocity for both particles, ie $v_t = 0.41c$ - $0.63c$. The extracted inverse slope parameters are shown in Figure and listed in Table 1, errors are statistical only. The total systematic errors are less than $10 MeV/c$. The systematic error from the correction of lambda feed down is about 1%.

The inverse slope parameter is higher for deuterons than for protons, furthermore, for a given particle type T increases with system size. A higher degree of nuclear stopping is expected in PbPb than in SS collisions [7], and has been reported [8, 9, 10]. This would lead to higher number of baryons at mid-rapidity and also to an increase in transverse collective flow.

Figure shows ratios of deuteron to proton rapidity densities for the three target-

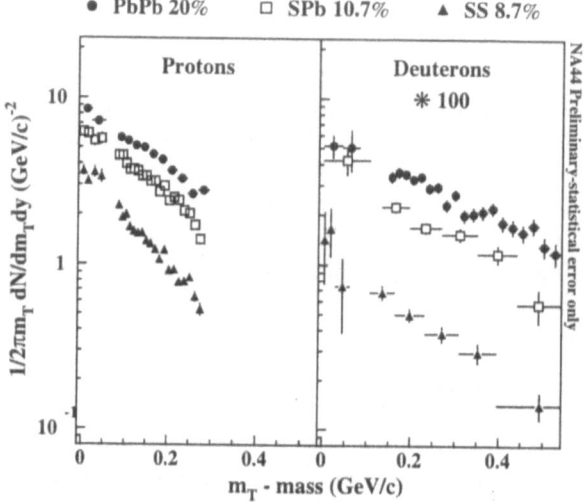

Figure 1. Proton and deuteron invariant cross section as a function of transverse kinetic energy.

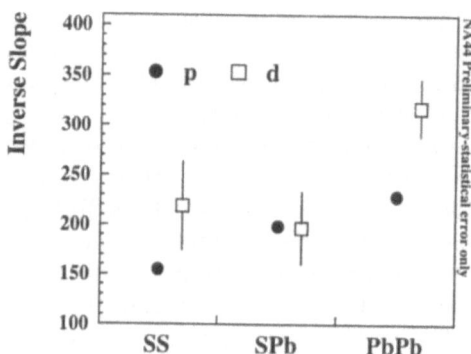

Figure 2. Proton and deuteron inverse slopes in MeV/c for SS, SPb and PbPb. The fit range is $v_t = 0.41c$ - $0.63c$.

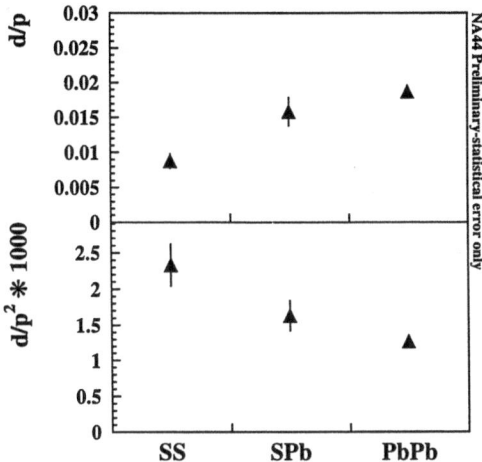

Figure 3. Ratios of deuteron to proton rapidity densities, d/p and d/p^2.

projectile combinations. The number of deuterons and protons per unit rapidity, dN/dy, is denoted by d and p, respectively, on the abscissa. Particle rapidity densities were calculated by summing the yield within the acceptance and using an exponential fit to the m_t spectrum to estimate the yield at high m_t outside of the acceptance. About one third of the cross section is outside of the acceptance.

Proton rapidity densities are corrected for Λ feed down. About 20% of the measured protons originate from lambdas. The ratio of deuterons to protons increases with the size of the colliding system, while the ratio d/p^2 decreases. For SS collisions the neutron to proton ratio is 1, while for PbPb collisions it is 1.53. The factor of 2 increase of the d/p ratio from SS to PbPb collisions cannot be explained by the increase of the n/p ratio.

COALESCENCE RADIUS

Model

In the framework of the coalescence model deuterons are formed from a proton and a neutron close enough in momentum space to coalesce. In a spherically symmetric thermal model by Scott Pratt [11] the single-particle density for protons, $\rho(r)$, is

$$\rho(r) = e^{\frac{-r^2}{2R_G^2}}. \tag{1}$$

Assuming that the invariant cross section of neutrons equals the invariant cross section of protons times $\frac{A-Z}{Z}$, the radius of the source evaluated in the rest frame of the proton-neutron pair that formed the deuteron is given by,

$$R_G^3 = \frac{3}{4}(\pi)^{3/2}(\hbar c)^3 \frac{m_d}{m_p^2} \cdot \frac{(\frac{E_p \cdot d^3 N_p}{dp^3})^2}{\frac{E_d \cdot d^3 N_d}{dp^3}} \cdot \frac{A-Z}{Z} \tag{2}$$

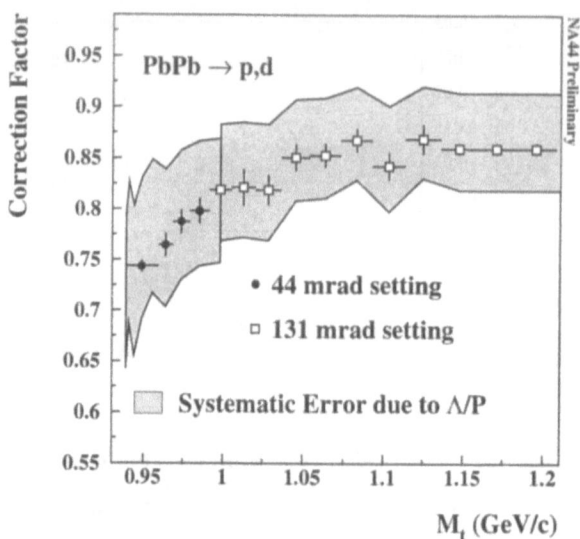

Figure 4. Correction factor to R_G from lambda decay.

The quantity $E \cdot \frac{d^3N}{dp^3}$ is the invariant cross section per central collision. The proton and deuteron invariant cross sections are evaluated at the same velocity.

Radius Parameter as a Function of m_t

Figure gives the correction factor to R_G for lambda feed down in PbPb collisions as a function of m_t. The hatched band around the data gives an estimate of the systematic error due to the Λ/p ratio. The biggest correction occurs at low m_t, leading to a higher systematical uncertainty and model dependence at low m_t. Above $m_t = 1.04 GeV/c$ the Λ correction is below 15% and nearly constant. In this region the m_t dependence of R_G is not sensitive to contamination from Λ decays.

Figure shows the radius parameter R_G, as defined in Equation 2, as a function of the proton transverse mass. The error bars show the statistical errors. The hatched band shows the systematic error due to the Λ/p ratio. R_G decreases with increasing m_t. Assuming that the shapes of the proton and deuteron transverse mass spectra are exponential in m_t,

$$\frac{1}{m_t}\frac{dN_p}{dm_t dy} \propto \exp\left(\frac{-m_t}{T_p}\right),$$
$$\frac{1}{m_t}\frac{dN_d}{dm_t dy} \propto \exp\left(\frac{-m_t}{T_d}\right), \tag{3}$$

with $m_t|_d = 2 \cdot m_t|_p$,
the d/p^2 ratio as a function of the proton m_t has the following form:

$$\frac{d}{p^2}(m_t) \propto \exp\left(2m_t\left(\frac{1}{T_p} - \frac{1}{T_d}\right)\right). \tag{4}$$

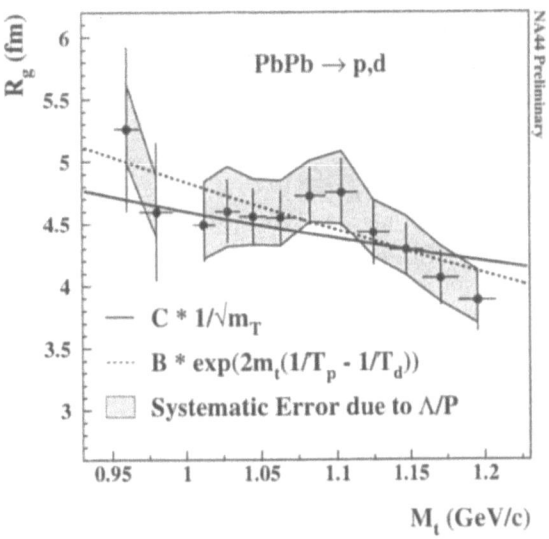

Figure 5. R_G versus the proton transverse mass.

Therefore d/p^2 is a function of m_t, if $T_p \neq T_d$. The dashed line in Figure gives the functional dependence of R_G for strictly exponential proton and deuteron m_t spectra.

Hydrodynamic calculations of a cylindrically symmetric, expanding source show that the source radius as calculated by Bose-Einstein interferometry should have a strong m_t dependence [12, 13]. Collective expansion creates a position-momentum correlation of particles within the source. The higher the transverse momentum of the measured particle pairs, the smaller is the portion of the source that is measured. The extracted HBT radii are predicted to decrease as a function of $1/\sqrt{m_t}$. HBT measurements from NA44 have confirmed this picture [14] for SPb collisions. The solid line, $C \cdot 1/\sqrt{m_t}$, in Figure shows that the data points are consistent with this functional form as well.

SUMMARY

The invariant cross sections of deuterons and protons in the transverse velocity range of $0.41 - 0.63c$ can be represented by an exponential function in m_t. The inverse slope parameters are larger for deuterons than for protons and both of them are bigger in PbPb collisions than in SPb and SS collisions. The deuteron yield increases stronger than the proton yield and less strong than the square of the proton yield going from SS to SPb and PbPb collisions. These effects are not explained by the increase of the n/p ratio from SS to PbPb. The d/p ratio is sensitive to the phase space density of the protons at freeze out, while the d/p^2 ratio is inversely proportional to the volume of the source. Therefore the volume increases from SS to SPb to PbPb. The extracted coalescence radius parameters fall with transverse mass as expected for an expanding, cylindrically symmetric source. The increase of the slope parameters as a function of the particle mass, and the m_t dependence of R_G can be interpreted as signs of collective radial motion of the emitting source.

ACKNOWLEDGMENTS

The author wishes to thank S. Pratt for helpful discussions on the coalescence radius. NA44 is grateful to the staff of the CERN PS–SPS accelerator complex for their excellent work. We thank the technical staff at CERN and the collaborating institutes for their valuable contributions. We are also grateful for the support given by the Austrian Fonds zur Förderung der Wissenschaftlichen Forschung (grant P09586); the Science Research Council of Denmark; the Japanese Society for the Promotion of Science; the Ministry of Education, Science and Culture, Japan; the Science Research Council of Sweden; the US W.M. Keck Foundation; the US National Science Foundation; and the US Department of Energy.

REFERENCES

1. N. Maeda et al., A Cherenkov beam counter for ultra-relativistic heavy-ion experiments, *NIM A 346* 132:136 (1994).
2. H. Sorge et al., Colour rope formation and strange baryon production in ultrarelativistic heavy ion collisions, *Phys. Lett. B289* 6:11 (1992).
3. P. Foca, Strange particle production in S-nucleus collisions at 200 GeV /Nucleon, *in*: "Strangeness in Hadronic Matter, AIP Conference Proceedings 340," J. Rafelski, ed., AIP, Tucson (1995).
4. D. E. Greiner et al., Strangeness Production as a function of centrality, *in*: "Strangeness in Hadronic Matter, AIP Conference Proceedings 340," J. Rafelski, ed., AIP, Tucson (1995).
5. E. Schnedermann et al., Thermal phenomenology of hadrons from $200A$ GeV S+S collisions, *Phys. Rev. C 48* 2462:2475 (1993).
6. I.G. Bearden et al., Collective expansion in high energy heavy ion collisions, *Phys. Rev. Lett.* accepted for publication.
7. A. Capella, Baryon stopping and strange baryon enhancement in heavy ion collisions, *Nucl. Phys. A 610* 132c:138c (1996).
8. H. Boggild et al., Coulomb effect in single particle distributions, *Phys. Lett. B372* 339:342 (1996).
9. I.G. Bearden et al., Mid-rapidity protons in 158A GeV Pb+Pb collisions, *Phys. Lett. B388* 431-436 (1996).
10. S. V. Afanasiev et al., Hadron yields and hadron spectra from the NA49 Experiment, *Nucl. Phys. A 610* 188c:199c (1996).
11. S. Pratt, private communications.
12. T. Csörgő, B. Lörstad, Bose-Einstein correlations for expanding finite systems, *Nucl. Phys. A 590* 465c (1995).
13. Y. M. Sinyukov, Spectra and correlations in locally equilibrium hadron and quark-gluon systems, *Nucl. Phys. A 566* 589c (1994).
14. H. Becker et al., m_t dependence of boson interferometry in heavy ion collisions at the CERN SPS, *Phys. Rev. Lett. 74(17)* 3340:3343 (1995).

STRANGENESS PRODUCTION AND FLOW IN HEAVY-ION COLLISIONS

G. Q. Li,[1] G. E. Brown,[1] C.-H. Lee, [1] and C. M. Ko[2]

[1]Department of Physics, SUNY at Stony Brook
Stony Brook, NY 11794
[2]Cyclotron Institute and Physics Department,
Texas A&M University, College Station, TX 77843

INTRODUCTION

With the development of various heavy-ion facilities, nuclear physics is expanding into many new directions. One of these is the study of the properties of strange particles, namely hyperons and kaons, in dense matter. Strangeness plays a special role in the development of hadronic and nuclear models. The mass of strange quark is about 150 MeV, which is, on the one hand, considerably larger than the mass of light (up and down) quarks ($\sim$ 5 MeV), but on the other hand, much smaller than that of charm quark ($\sim$ 1.5 GeV). In the limit of vanishing quark mass the chiral symmetry is good, and systematic studies can be carried out using chiral perturbation theory for hadrons made of light quarks. On the other hand, if the quark mass is large, one can use the non-relativistic quark model to study the properties of charmed and heavier hadrons. The hadrons with strangeness lie in between these two limits and therefore present a challenge to the theorists.

Nevertheless, chiral perturbation calculations have been extensively and quite successfully carried out in the recent past for the study of kaon-nucleon (KN) and antikaon-nucleon ($\bar{K}N$) scattering [1, 2]. The main reason for the success of the chiral perturbation theory is the existence of a large body of experimental data that can be used to constrain the various parameters in the chiral Lagrangian.

Of great interest to nuclear physicists is the study of the properties of strange hadrons in dense matter. Chiral perturbation calculations in matter are still under development, since a new energy scale, the Fermi momentum, is now involved. For this kind of theoretical development to be successful, a large body of experimental data that can be obtained only from heavy-ion collisions are needed. The study of kaon properties in dense matter is also relevant for astrophysics problems, such as the properties of neutron stars [3].

There have been some studies on strangeness production and flow in heavy-ion collisions at SIS energies [4, 5, 6, 7, 8, 9, 10]. In this contribution we will concentrate on recent development concerning strangeness production in Ni+Ni collisions at 1-2 AGeV. We will show that these observables are sensitive to kaon properties in dense matter and can thus provide useful information for the development of chiral perturbation theory in matter and for the study of neutron star properties. In Section II, we review briefly the current understanding of kaon properties in nuclear matter, the relativistic transport model with strangeness, and the various elementary processes for strangeness production. Our results for strangeness spectra and flow will be reported in Section III, where we will also compare our results for proton and pion spectra to experimental data. Finally a brief summary is given in Section IV.

THE RELATIVISTIC TRANSPORT MODEL WITH STRANGENESS

Our study is based on the relativistic transport model extended to include the strange degrees of freedom [11]. The Lagrangian we use is given by

$$
\begin{aligned}
\mathcal{L} = {} & \bar{N}(i\gamma^{\mu}\partial_{\mu} - m_N + g_{\sigma}\sigma)N - g_{\omega}\bar{N}\gamma^{\mu}N\omega_{\mu} + \mathcal{L}_0(\sigma, \omega_{\mu}) \\
& + \bar{Y}(i\gamma^{\mu}\partial_{\mu} - m_Y + (2/3)g_{\sigma}\sigma)Y - (2/3)g_{\omega}\bar{Y}\gamma^{\mu}Y\omega_{\mu} \\
& + \partial^{\mu}\bar{K}\partial_{\mu}K - \left(m_K^2 - \frac{\Sigma_{KN}}{f^2}\bar{N}N\right)\bar{K}K - \frac{3i}{8f^2}\bar{N}\gamma^0 N\bar{K}\overset{\leftrightarrow}{\partial_t}K.
\end{aligned}
\tag{1}
$$

In the above the first line gives the usual non-linear σ-ω model with the self-interaction of the scalar field. The second line is for the hyperons which couple to the scalar and vector fields with 2/3 of the nucleon strength, as in the constituent quark model. The last line is for the kaon which is derived from the SU(3) chiral Lagrangian. Note that since the hyperon and kaon densities in heavy-ion collisions at SIS energies are very small, their contributions to the scalar and vector fields are neglected.

From this Lagrangian we get the kaon and antikaon in-medium energies

$$
\omega_K = \left[m_K^2 + \mathbf{k}^2 - a_K\rho_S + (b_K\rho_N)^2\right]^{1/2} + b_K\rho_N
\tag{2}
$$

$$
\omega_{\bar{K}} = \left[m_K^2 + \mathbf{k}^2 - a_{\bar{K}}\rho_S + (b_K\rho_N)^2\right]^{1/2} - b_K\rho_N
\tag{3}
$$

where $b_K = 3/(8f_{\pi}^2) \approx 0.333$ GeVfm^3, a_K and $a_{\bar{K}}$ are two parameters that determine the strength of attractive scalar potential for kaon and antikaon, respectively. If one considers only the Kaplan-Nelson term, then $a_K = a_{\bar{K}} = \Sigma_{KN}/f_{\pi}^2$. In the same order, there is also the range term which acts differently on kaon and antikaon, and leads to different scalar attractions. We take the point of view that they can be treated as free parameters and try to constrain them from the experimental observables in heavy-ion collisions.

In chiral perturbation theory, in the same order as the Kaplan-Nelson term there is the range term, which can be taken into account by renormalizing [1]

$$
\Sigma_{KN} \longrightarrow \left(1 - 0.37\frac{\omega_{K,\bar{K}}^2}{m_K^2}\right)\Sigma_{KN}.
\tag{4}
$$

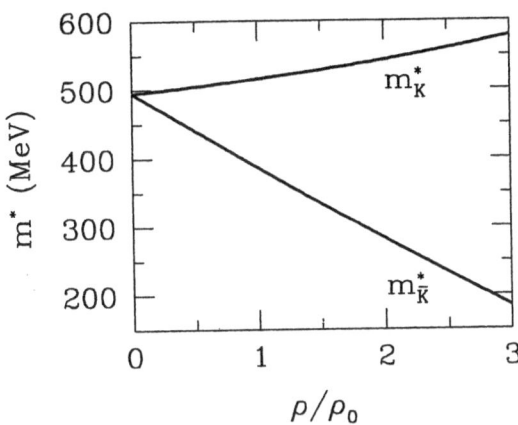

Figure 1. Kaon and antikaon effective mass as a function of nuclear density .

Since $\omega_K \sim m_K$ for the densities considered here, there is a reduction of ~ 0.63 from the range term. However, f_π is also density dependent [12],

$$\frac{f_\pi^{*2}(\rho_0)}{f_\pi^2} \approx 0.68. \tag{5}$$

We do not know quantitatively the behaviour of f_π^*/f_π at higher densities, although we would expect it to continue to decrease with density. We see that, at least in the range of densities $\rho \sim \rho_0$, there is considerable cancellation between effects of the range term and the decrease in f_π. We therefore neglect both for K^+.

In the case of K^- meson, the range term drops rapidly with density as $(\omega_{\bar{K}}/m_K)^2$ decreases. We thus neglect the range term and relate $a_{\bar{K}}$ to a_K, namely, $a_{\bar{K}} \approx (0.63)^{-1}a_K$. Note that f_π^2 in b_K is not scaled with density. The b_K represents the vector interaction, which is $1/3$ of the vector mean field acting on a nucleon. In our recent paper [13], it is shown that for $\rho \sim (2-3)\rho_0$, the region of densities important for kaon and antikaon production and flow, the nucleon vector mean field is estimated to be only about 15% larger than $3b_K = 9/(8f_\pi^2)$. The increase from scaling f_π^{-2} is evidently largely canceled by the decrease from short-range correlations. Our parameterization thus incorporates roughly what we know about the scaling of f_π and of effects from short-range interactions.

Using $a_K = 0.22$ GeV^{2}fm^3 and $a_{\bar{K}} = 0.35$ GeV^{2}fm^3, the kaon and antikaon effective mass, defined as their energies at zero momentum, are shown in Fig. 1. It is seen that the kaon mass increases slightly with density, resulting from near cancellation of the attractive scalar and repulsive vector potential. The mass of antikaon drops substantially. At normal nuclear matter density $\rho_0 \approx 0.16$ fm^{-3}, the kaon mass increases about 4%, in rough agreement with the prediction of impulse approximation based on KN scattering length. The antikaon mass drops by about 22%, which is somewhat smaller than what has been inferred from the kaonic atom data [14], namely, an attractive K^- potential of 200 ± 20 MeV at ρ_0.

From Fig. 1 we find that $m_{\bar{K}}(3\rho_0) \approx 170$ MeV. The correction for neutron rich matetr in using $m_{\bar{K}}$ in neutron stars is about 50 MeV upwards, giving $m_{\bar{K}}^* \approx 220$ MeV

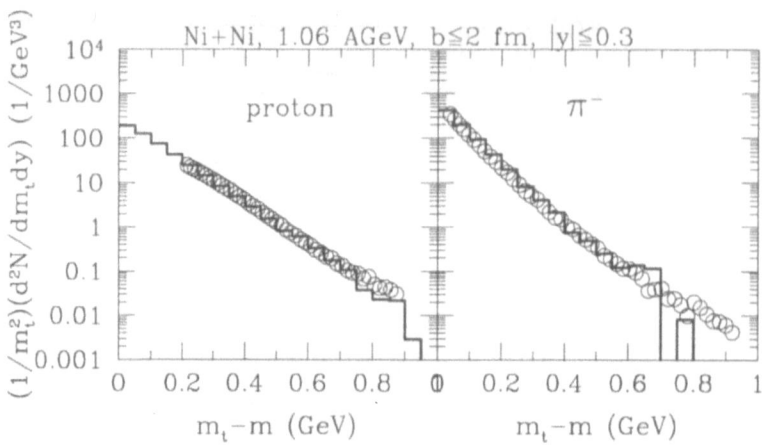

Figure 2. Proton and pion transverse momentum spectra in central Ni+Ni collisions at 1.06 AGeV. The open circles are experimental data from the FOPI collaboration.

[15]. With the electron chemical potential of $\mu_e(3\rho_0) = 214$ MeV [16], this implies that kaon condensation will take place at dnsity $\rho \sim 3\rho_0$.

From the Lagrangian we can also derive equations of motion for nucleons [8]

$$\frac{d\mathbf{x}}{dt} = \frac{\mathbf{p}^*}{\sqrt{\mathbf{p}^{*2} + m_N^{*2}}}, \quad \frac{d\mathbf{p}}{dt} = -\nabla_x(\sqrt{\mathbf{p}^{*2} + m_N^{*2}} + U_V), \tag{6}$$

for hyperons

$$\frac{d\mathbf{x}}{dt} = \frac{\mathbf{p}^*}{\sqrt{\mathbf{p}^{*2} + m_Y^{*2}}}, \quad \frac{d\mathbf{p}}{dt} = -\nabla_x(\sqrt{\mathbf{p}^{*2} + m_Y^{*2}} + (2/3)U_V), \tag{7}$$

and for kaons

$$\frac{d\mathbf{x}}{dt} = \frac{\mathbf{p}^*}{\omega_{K,\bar{K}} \mp b_k\rho_N}, \quad \frac{d\mathbf{p}}{dt} = -\nabla_x\omega_{K,\bar{K}}, \tag{8}$$

where $m_N^* = m_N - U_S$ and $m_Y^* = m_Y - (2/3)U_S$, with U_S and U_V being nucleon scalar and vector potentials. The minus and the plus sign in the last equation correspond to kaon and antikaon, respectively.

In addition to propagations in their mean field potentials, we include typical two-body scattering processes such as $BB \leftrightarrow BB$, $NN \leftrightarrow N\Delta$ and $\Delta \leftrightarrow N\pi$. Kaons and hyperons are mainly produced from the following baryon-baryon and pion-nucleon collisions, namely $BB \rightarrow BYK$, $\pi N \rightarrow YK$. The cross section for the former is taken to be the Randrup-Ko parameterization [17], and the cross section for the latter is taken from Cugnon *et al.* [18]. The antikaon production cross section in baryon-baryon collisions is taken from the recent work of Sibirtsev *et al.* [19], while that in pion-nucleon collisions is fitted to experimental data. Antikaons can also be produced from hyperon-pion collisons through the strangeness exchange process, namely, $\pi Y \rightarrow \bar{K}N$. The cross section for this process is obtained from the reverse one, $\bar{K}N \rightarrow \pi Y$, by the detailed-balance relation. The latter cross section, together with the $\bar{K}N$ elastic and

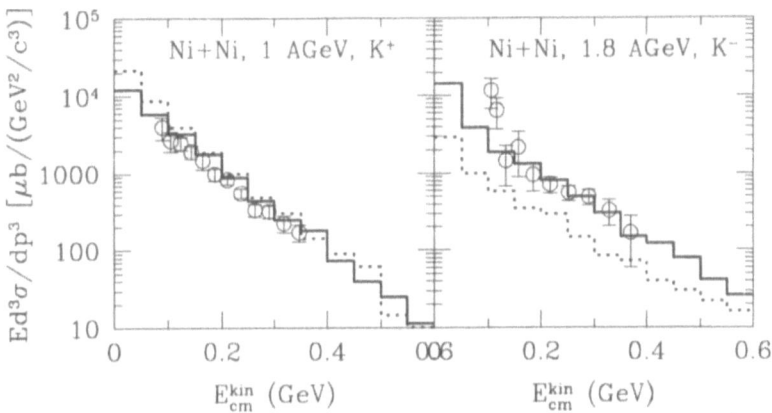

Figure 3. Left panel: K^+ kinetic energy spectra in Ni+Ni collsions at 1.0 AGeV. Right panel: K^- kinetic energy spectra in Ni+Ni collisions at 1.8 AGeV. The solid and dotted histograms are the results with and without kaon medium effects. The open circles are the experimental data from the KaoS collaboration.

absorption cross sections, are parameterized based on the available experimental data. We note that the antikaon absorption cross section is relatively large, and increases significantly with decreasing antikaon momentum, while kaons and hyperons undergo mainly elastic scattering with nucleons.

RESULTS AND DISCUSSIONS

We have studied strangeness production and flow mainly for Ni+Ni collisions at 1-2 GeV/nucleon. Our results for proton and pion transverse mass spectra in central Ni+Ni collisions at 1.06 AGeV are shown in Fig. 2. We determine our centrality to be $b \le 2$ fm in order to compare with the FOPI data [20] which correspond to a geometric cross section of 100 mb. To get free protons we applied a density cut of $\rho \le 0.15\rho_0$. It is seen that our model describes both the proton and pion spectra very well.

In the left panel of Fig. 3 we show our results for K^+ kinetic energy spectra in Ni+Ni collisions at 1.0 AGeV, with impact parameter $b \le 8$ fm. The solid histogram gives the results with kaon medium effects, while the dotted histogram is the results without kaon medium effects. The open circles are the experimental data from the KaoS collaboration [21]. It is seen that the results with kaon medium effects are in good agreement with the data, while those without kaon medium effects slightly overestimate the data. We note that kaon feels a slightly repulsive potential, thus the inclusion of the kaon medium effects reduces the kaon yield. The slopes of the kaon spectra in the two cases also differ. With a repulsive potential, kaons are accelerated during the propagation, leading to a larger slope parameter as compared to the case without kaon medium effects.

The results for K^- kinetic energy spectra in Ni+Ni collisions at 1.8 AGeV are shown in the right panel of Fig. 3. The solid and dotted histograms are the results with and without kaon medium effects. It is seen that without medium effects, our

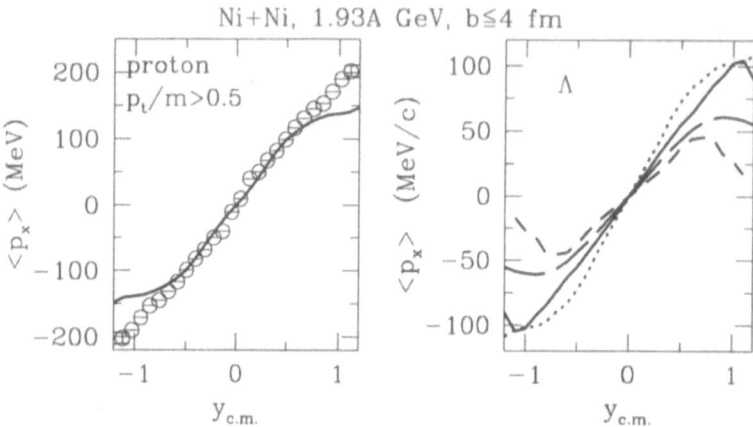

Figure 4. Left panel: Proton flow in Ni+Ni collisons at 1.93 AGeV and $b \leq 4$ fm, including a transverse momentum cut of $p_t/m > 0.5$. The open circles are the experimental data from the FOPI collaboration [20]. Right panel: Λ flow. The short-dashed curve is for primordial Λ's, the long-dashed curve includes elastic ΛN scattering, and the solid curve include both rescattering and propagation. The proton flow is shown by the dotted curve for comparison.

results are about a factor 3-4 below the experimental data. With the inclusion of the medium effects which reduces the antikaon production threshold, K^- yield increases by about a factor of 3 and our results are in good agreement with the data. This is similar to the findings of Cassing *et al.* [10].

The KaoS data show that the K^- yield at 1.8 AGeV agrees roughly with the K^+ yield at 1.0 AGeV. This is a nontrial observation. At these energies, the Q-values for $NN \rightarrow NK\Lambda$ and $NN \rightarrow NNK\bar{K}$ are both -230 MeV. Near their thresholds, the cross section for the K^- production is about one order of magnitude smaller than that for K^+ production. In addition, antikaons are strongly absorbed in heavy-ion collisions, which should further reduces the K^- yield. The KaoS results of $K^-/K^+ \sim 1$ indicate thus the importance of secondary processes such as $\pi Y \rightarrow \bar{K}N$ [22], and medium effects which acts oppositely on the kaon and antikaon production in medium.

In addition to particle yield, the collective flow of particles provide complimentary information about hadron properties in dense matter. In the left panel of Fig. 4 we compare our results for proton flow in Ni+Ni collisions at 1.93 AGeV with the experimental data from the FOPI collaboration [20]. Note that both the data and our results include a transverse momentum cut of $p_t/m > 0.5$. The agreement with the data is very good. The Λ flow in the same system is shown in the right panel of Fig. 4. The flow of the primodial Λ's, as shown in the figure by short-dashed curve, is considerably smaller than that of protons shown by dotted curve. Inclusions of ΛN rescattering (long-dashed curve) and propagation in potential (solid curve) increase the Λ flow in the direction of proton flow. Both the FOPI [20] and the EOS [23] collaborations found that the Λ flow is very similar to that of protons.

In the left panel of Fig. 5 we show our results for K^+ flow in Ni+Ni collisions at 1.93 AGeV, based on three scenarios for kaon potentials in nuclear medium. Without potential, kaons flow in the direction of nucleon as shown by the dotted curve. Without scalar potential, kaons feel a strong repulsive potential, and they flow in the opposite

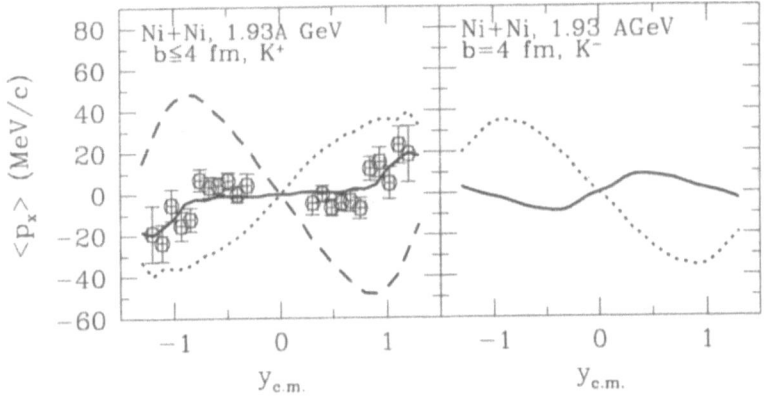

Figure 5. Left panel: K^+ flow in Ni+Ni colliisons at 1.93 AGeV and $b \leq 4$ fm, including a transverse momentum cut of $p_t/m > 0.5$. The open circles are the experimental data from the FOPI collaboration[20]. The dotted curve is the results without kaon potential, the dashed curve is the results without kaon scalar potential, and the solid curve gives the results with both the scalar and vector potential. Right panel: K^- flow in Ni+Ni collisions at 1.93 AGeV and $b = 4$ fm. The solid and dashed curves are the results with and without antikaon medium effects.

direction to nucleons as shown by the dashed curve. With both the scalar and vector potentials, kaons feel a weak repulsive potential, and one sees the disappearance of kaon flow, which is in good agreement with the experimental data from the FOPI collaboration [20]. In the right panel of Fig. 5 we show K^- flow in Ni+Ni collsions at 1.93 AGeV and $b = 4$ fm. Without medium effects one sees a clear antikaon antiflow signal, because of strong absorption of K^- by nucleons. Including the attractive antikaon potential, those antikaons that survive the absorption are pulled towards nucleon and thus show a weak antikaon flow signal. The experimental measurement of K^- in heavy-ion collisions will be very helpful in determining K^- properties in nuclear matter.

SUMMARY

In summary, we studied strangeness (K^+, K^-, and Λ) production and flow in Ni+Ni collisions at 1-2 AGeV. We based our study on the relativistic transport model including the strangeness degrees of freedom. We found that strangeness spectra and flow are sensitive to the properties of strange hadrons in nuclear medium. The predictions of the chiral perturbation theory that the K^+ feels a weak repulsive potential and K^- feels a strong attractive potential are in good agreement with recent experimental data from FOPI and KaoS collaborations.

The work of GQL, GEB and CHL were supported in part by Department of Energy under Grant No. DE-FG02-88ER40388, while that of CMK was supported in part by the National Science Foundation under Grant No. PHY-9509266. CHL was also partly supported by Korea Science and Engineering Foundation.

REFERENCES

1. C.-H. Lee, Kaon condensation in dense stellar matter, *Phys. Rep* 275:255 (1996).
2. N. Kaiser, T. Waas, and W. W.Weise, Low energy $\bar{K}N$ interaction in nuclear matter, *Phys. Lett.* B365:12 (1996).
3. R. Knorren, M. Prakash, and P. J. Ellis, Strangeness in hadronic stellar matter, *Phys. Rev* C52:3470 (1995).
4. X. S. Fang, C. M. Ko, G. Q. Li, and Y. M. Zheng, The relativistic transport model description of subthreshold kaon production in heavy-ion collisions, *Nucl. Phys* A575:766 (1994).
5. T. Maruyama, W. Cassing, U. Mosel, S. Teis, and K. Weber, Study of high-energy heavy-ion collisions in a relativistic BUU approach with momentum dependent mean fields *Nucl. Phys.* A573:653 (1994).
6. C. Hartnack, J. Jänicke, L. Sehn, H. Stöcker, and J. Aichelin, Kaon production at subthreshold energies, *Nucl. Phys.* A580:643 (1994).
7. G. Q. Li, C. M. Ko, and B. A. Li, Kaon flow as a probe of kaon potential in nuclear medium, *Phys. Rev. Lett.* 74:235 (1995).
8. G. Q. Li and C. M. Ko, Kaon flow in heavy-ion collisions, *Nucl. Phys.* A594:460 (1995).
9. G. Q. Li, C. M. Ko, and X. S. Fang, Subthreshold antikaon production in heavy-ion collisions, *Phys. Lett.* B329:149 (1994).
10. W. Cassing, E. L. Bratkovskaya, U. Mosel, S. Teis, and A. Sibirtsev, Kaon versus antikaon production at SIS energies, *Nucl. Phys.*, to be published.
11. C. M. Ko and G. Q. Li, Medium effects in heavy-ion collisions, *J. Phys.* G22:1673 (1996).
12. G. E. Brown and M. Rho, From chiral mean field to Walecka mean field and kaon condensation, *Nucl. Phys.* A596:503 (1996).
13. G. Q. Li, G. E. Brown, C.-H. Lee, and C. M. Ko, Nucleon flow and dilepton production in heavy-ion collisions, to be published.
14. E. Friedman, A. Gal, and C. J. Batty, Density dependent K^- nuclear optical potential from kaonic atoms, *Nucl. Phys.* A579:518 (1994).
15. G. E. Brown, Proc. Royal Dutch Academy Colloquium 'Pulsar Timing, General Relativity and the Internal Structure of Neutron Stars', to be published.
16. V. Thorsson, M. Prakash, and J. M. Lattimer, Composition, structure and evolution of neutron stars with kaon condensation, *Nucl. Phys.* A572:693 (1994).
17. J. Randrup and C. M. Ko, Kaon production in relativistic collisions, *Nucl. Phys.* A343: 519 (1980).
18. J. Cugnon and R. M. Lombard, K^+ production in a cascade model for high-energy nucleus-nucleus collisions, *Nucl. Phys.* A422:635 (1984).
19. A. Sibirtsev, W. Cassing, and C. M. Ko, Antikaon production in nucleon-nucleon reactions near threshold, *Z. Phys. A*, to be published.
20. N. Herrmann, FOPI collaboration, Particle production and flow at SIS energies, *Nucl. Phys.* A610:49c (1996).
21. P. Senger for the KaoS collaboration, kaon production in hadronic matter, *Heavy Ion Physics* 4:317 (1996).
22. C. M. Ko, Subthreshold K^- production in high energy heavy-ion collisions, *Phys. Lett.* B120:294 (1983).
23. M. Justice, EOS collaboration, Observation of collective effects in Λ production at 2 GeV/nucleon, *Nucl. Phys.* A590:549c (1995).

SEMIHARD PROCESSES IN NUCLEAR COLLISIONS

K. Werner

SUBATECH
Université de Nantes – IN2P3 – EMN
Nantes, France

INTRODUCTION

In the energy range 10 GeV $< \sqrt{s} <$ 50 GeV, hadronic interactions are well described in the framework of Gribov–Regge theory (GRT). Here, the elementary "exchange object" is the so–called Pomeron, and the theory is formulated entirely in terms of Pomeron exchanges.

At high energies, say $\sqrt{s} >$ 50 GeV, the soft Pomeron is not sufficient, the theory has to be generalized. It is well known that perturbative QCD (PQCD) comes into play, in particular, in case of large momentum transfer t, one may write inclusive cross sections as

$$\sigma(s) = \sum_{ij} \int dx_1 \, dx_2 \, dt \, f_i(x_1, t) \, f_j(x_2, t) \, \frac{d\sigma_{ij}}{dt}(x_1 x_2 s, t) \,, \tag{1}$$

with f_i, f_j representing the momentum distributions of partons in nucleons, and where $d\sigma_{ij}/dt$ represents parton–parton scattering according to an elementary QCD diagram.

Any formalism aiming at describing hadronic interactions in an energy range from 10 GeV up to several TeV should therefore have the following objectives:

- provide GRT for small energies ;

- reproduce PQCD results at high energies.

We are going to present such a formalism in this paper.

THE SOFT POMERON

In the following, we sketch the basic features of Gribov–Regge theory based on the soft Pomeron.

Advances in Nuclear Dynamics 3, edited by
Bauer and Mignerey, Plenum Press, New York, 1997

115

The elastic amplitude $A^{h_1 h_2}(s, t)$ for the scattering of hadron h_1 and hadron h_2 is given as

$$A^{h_1 h_2}(s, t) = \sum_n A_n^{h_1 h_2}(s, t), \tag{2}$$

where

$$A_n^{h_1 h_2}(s, t) = \frac{i^{n-1}\pi^{1-n}C^{n-1}}{n!} \int \prod_{i=1}^{n} d^2k_i \, \delta(k - \sum k_i) \prod_{j=1}^{n} G^{h_1 h_2}(s, k_j^2) \tag{3}$$

represents n Pomeron exchanges. As usual, the Mandelstam variables s and t are used. Actually, s is meant to be in units of some scale $s_0 = 1$ GeV and is therefore a dimensionless quantity. The function G is the Pomeron propagator, representing the exchange of a soft Pomeron, graphically expressed by a zigzag line:

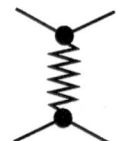

The Pomeron propagator is given as

$$G^{h_1 h_2}(s, k^2) = i \, \gamma_{h_1 h_2} \, e^{-(R_{h_1}^2 + R_{h_2}^2)k^2} \, s^{\Delta - \alpha' k^2}, \tag{4}$$

with

$$\Delta = \alpha(0) - 1, \tag{5}$$

with the so-called "intercept" $\alpha(0)$, the "slope" α', and the "Regge radii" R_h^2. Using the above parametrization for the Pomeron propagator, we obtain

$$A(s, t) = \frac{i}{4\pi C} \int d^2b \, \exp(i\vec{k}\vec{b}) \left\{ 1 - \exp\left[- C \, \chi_{\text{soft}}^{h_1 h_2}(s, b) \right] \right\}, \tag{6}$$

with the so-called "soft eikonal" $\chi_{\text{soft}}^{h_1 h_2}(s, b)$ being the Fourier transform of the Pomeron propagator,

$$\chi_{\text{soft}}^{h_1 h_2}(s, b) = \frac{1}{i\pi} \int d^2k \, G(s, k^2) \, \exp(-i\vec{k}\vec{b}), \tag{7}$$

which leads to

$$\chi_{\text{soft}}^{h_1 h_2}(s, b) = \frac{\gamma_{h_1 h_2} \, s^{\Delta}}{\lambda_{h_1 h_2}(s)} \exp\left\{ -\frac{b^2}{4\lambda_{h_1 h_2}(s)} \right\}. \tag{8}$$

λ is defined as

$$\lambda_{h_1 h_2}(s) = R_{h_1}^2 + R_{h_2}^2 + \alpha'(0) \ln s. \tag{9}$$

The total cross section is then

$$\sigma_{\text{tot}}^{h_1 h_2}(s) = \frac{2}{C_{h_1 h_2}} \int d^2b \, [1 - e^{-C_{h_1 h_2} \chi_{h_1 h_2}^{\text{soft}}(s,b)}], \tag{10}$$

We use essentially the same symbols as used in ref. [1], apart of using γ instead of N_0. We also consider here different hadron types and correspondingly different parameters for different hadron classes (this concerns γ, R^2, and C), see table .

symbol	in [1]	meaning
Δ	Δ	overcriticality
$\alpha'(0)$	$\alpha'(0)$	slope of Regge trajectory
$\gamma_{h_1 h_2}$	N_0	vertex constant
R_h^2	R^2	Regge radius
$C_{h_1 h_2}$	C	shower enhancement coeff.

THE SEMIHARD POMERON

We are now going to generalize the formalism discussed in the preceeding section to high energies. The basic idea [2] is to replace the soft Pomeron

by a so–called "semihard Pomeron", which is defined to be an ordinary soft Pomeron with the middle piece replaced by a "QCD parton ladder". In other words, we have a parton ladder sandwiched between two soft Pomerons:

$$(11)$$

Here, we use the symbol for the ladder diagram and zigzag symbol for the soft Pomeron. It is assumed that the ladder and the soft Pomeron may only be connected via gluons, i.e. the external legs of the ladder are necessarily gluons. Therefore the semihard Pomeron is said to be of gg–type (for gloun–gluon). Ladders with external quark lines can only couple directly to the hadron, on the projectile side, on the target side, or on both sides. Correspondingly we introduce qg–type, gq–type, and qq–type Pomerons (quark–gluon, gluon–quark, quark–quark):

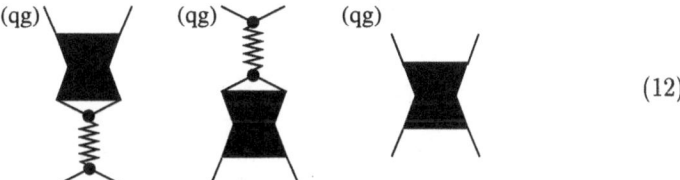

$$(12)$$

Here, the lines connecting ladder and soft Pomeron, are, as before, gluons, whereas the other external lines of the ladder are quarks. Associated with the four types of semihard Pomerons, we have four contributions to the semihard eikonal, which are discussed in the following, one after the other. In all cases, the eikonals are written as

$$\chi_t^{h_1 h_2}(s, b) = \int dx^+ dx^- \, \tilde{\chi}_t^{h_1 h_2}(s, b, x^+, x^-), \qquad (13)$$

where t refers to the type (gg or qg or gq or qq). The integrand of the semihard gg–type eikonal is defined to be

$$\tilde{\chi}_{gg}^{h_1 h_2}(s, b, x^+, x^-) = \frac{1}{2} f_e^{h_1}(x^+) f_e^{h_2}(x^-) \int \frac{dz^+}{z^+} \frac{dz^-}{z^-} r(z^+) r(z^-) \qquad (13)'$$

$$\chi^{\text{soft}}\left(\frac{1}{z^+ z^-}, b\right) \sigma_{\text{jet}}^{gg}(z^+ z^- x^+ x^- s) \, \Theta(z^+ z^- x^+ x^- s - 4q_0^2)$$

with $f_e^{h_1}(x)$ representing the momentum fraction distribution of the partons in the hadron representing the soft Pomeron end, parametrized as

$$f_e(x) = c_e \, x^{\alpha_e} \, (1 - x)^{\beta_e}. \tag{14}$$

The function

$$r(z) = r_0 \, (1 - z)^{\beta_{\text{cut}}} \tag{15}$$

provides a large z^+ (z^-) cutoff. The integrand of the semihard qg–type eikonal is

$$\tilde{\chi}_{qg}^{h_1 h_2}(s, b, x^+, x^-) = \frac{1}{2} f_q^{h_1}(x^+) \, f_e^{h_2}(x^-) \int \frac{dz^-}{z^-} r(z^-) \tag{15}$$
$$\chi^{\text{soft}}\left(\frac{1}{z^-}, b\right) \sigma_{\text{jet}}^{gg}(z^- x^+ x^- s) \, \Theta(z^- x^+ x^- s - 4q_0^2),$$

and the integrand of the semihard gq–type eikonal correspondingly as

$$\tilde{\chi}_{qg}^{h_1 h_2}(s, b, x^+, x^-) = \frac{1}{2} f_e^{h_1}(x^+) \, f_q^{h_2}(x^-) \int \frac{dz^-}{z^+} r(z^+) \tag{15}''$$
$$\chi^{\text{soft}}\left(\frac{1}{z^+}, b\right) \sigma_{\text{jet}}^{gg}(z^+ x^+ x^- s) \, \Theta(z^+ x^+ x^- s - 4q_0^2),$$

with $f_q^{h_1}(x)$ representing the momentum fraction distribution of quarks in the hadron, parametrized as

$$f_q(x) = c_q \, x^{\alpha_q} \, (1 - x)^{\beta_q}. \tag{16}$$

Finally, the integrand of the semihard qq–type eikonal may be written as

$$\tilde{\chi}_{qq}^{h_1 h_2}(s, b, x^+, x^-) = \frac{1}{2} f_q^{h_1}(x^+) \, f_q^{h_2}(x^-) \tag{16}'$$
$$\sigma_{\text{jet}}^{gg}(x^+ x^- s) \, \Theta(x^+ x^- s - 4q_0^2) \frac{1}{R_{qq}^2(0)} \exp\left\{ -\frac{b^2}{4 R_{qq}^2(0)} \right\}.$$

THE POMERON CONFIGURATION

Now we are going to discuss, how to determine the "Pomeron configuration", i.e. the precise specification of the type of interaction for all possible pairs of projectile and target nucleons. These specifications are based on the eikonals, determined earlier. We define the total eikonal to be

$$\chi_{\text{tot}}^{h_1 h_2}(s, b) = \chi_{\text{soft}}^{h_1 h_2}(s, b) + \chi_{\text{semi}}^{h_1 h_2}(s, b), \tag{17}$$

with the soft eikonal χ_{soft} given in eq. (8), and with the semihard eikonal χ_{semi} being the sum of the semihard eikonals of types gg, qg, gq, qq,

$$\chi_{\text{semi}}^{h_1 h_2}(s, b) = \chi_{gg}^{h_1 h_2}(s, b) + \chi_q^{h_1 h_2}(s, b), \tag{18}$$

with

$$\chi_q^{h_1 h_2}(s, b) = \chi_{qg}^{h_1 h_2}(s, b) + \chi_{gq}^{h_1 h_2}(s, b) + \chi_{qq}^{h_1 h_2}(s, b). \tag{19}$$

In case of nucleus–nucleus, nucleon–nucleus, or nucleon–nucleon scattering, we have obviously $h_i = p$ or n. However, the following considerations also apply to hadron–nucleus and hadron–hadron scattering in general, for arbitrary hadrons. We nevertheless refer to "nucleons of the projectile nucleus" or "nucleons of the target nucleus", which have to be understood as "projectile hadron" and/or "target hadron" in case.

The strategy to determine the configuration is as follows: for given projectile and target nucleus, incident energy, and impact parameter, the distance between some projectile nucleon i and some target nucleon j is

$$b_{ij} = b + b_i - b_j \tag{20}$$

with b_i and b_j being the transverse coordinates of the two nucleons. In the following, we use the abbreviations

$$\chi_{\text{tot}} = \chi_{\text{tot}}^{h_1 h_2}(s, b_{ij}) \tag{20}'$$

$$\chi_{\text{soft}} = \chi_{\text{soft}}^{h_1 h_2}(s, b_{ij}) \tag{21}$$

$$\chi_{\text{semi}} = \chi_{\text{semi}}^{h_1 h_2}(s, b_{ij}) \tag{22}$$

and corresponding abbreviations for χ_{gg}, χ_{qg}, χ_{gq}, χ_{qq}, and for χ_q.

The probability for an inelastic interaction (at least one cut Pomeron) is given as

$$\sigma_{\text{inel}} = \left\{ 1 - e^{-2\chi_{\text{tot}}} \right\}, \tag{23}$$

In case of an inelastic interaction, we have at least one cut Pomeron, and we have to determine the number of cut Pomerons and their types. We expand

$$\sigma_{\text{inel}} = 1 - e^{-2\chi_{\text{tot}}} \tag{24}$$

as

$$\sigma_{\text{inel}} = \sum_{m \geq 1} \sigma_{\text{inel}}^{(m)}, \tag{25}$$

with

$$\sigma_{\text{inel}}^{(m)} = \frac{(2\chi_{\text{tot}})^m}{m!} e^{-2\chi_{\text{tot}}}, \tag{26}$$

where $\sigma_{\text{inel}}^{(m)}$ represents the probability of m cut Pomerons. In the Monte Carlo procedure, we generate m according to the distribution $\sigma_{\text{inel}}^{(m)}$. A particular cut Pomeron is soft or hard with probabilites

$$\frac{\chi_{\text{soft}}}{\chi_{\text{tot}}} \tag{27}$$

and

$$\frac{\chi_{\text{semi}}}{\chi_{\text{tot}}} = 1 - \frac{\chi_{\text{soft}}}{\chi_{\text{tot}}}. \tag{28}$$

In case of a semihard Pomeron, a particular type t (gg, qg, gq, qq) occurs with probability

$$\frac{\chi_t}{\chi_{\text{semi}}}. \tag{29}$$

As a summary of this section, we list in the following the algorithm to determine the Pomeron configuration.

- determine impact parameter b

- loop over all projectile–target pairs i, j

- determine the distance $b_{ij} = b + b_i - b_j$ between the nucleons

- calculate the eikonals

$$\chi_{\text{tot}} = \chi_{\text{tot}}^{h_1 h_2}(s, b_{ij}) \qquad (29)$$

$$\chi_{\text{soft}} = \chi_{\text{soft}}^{h_1 h_2}(s, b_{ij}) \qquad (30)$$

$$\chi_{\text{semi}} = \chi_{\text{semi}}^{h_1 h_2}(s, b_{ij}) \qquad (31)$$

and

$$\chi_{gg} = \chi_{gg}^{h_1 h_2}(s, b_{ij}) \qquad (31)'$$

$$\chi_{qg} = \chi_{qg}^{h_1 h_2}(s, b_{ij}) \qquad (32)$$

$$\chi_{gq} = \chi_{gq}^{h_1 h_2}(s, b_{ij}) \qquad (33)$$

$$\chi_{qq} = \chi_{qq}^{h_1 h_2}(s, b_{ij}) \qquad (34)$$

$$(35)$$

- calculate the cross sections (probabilities)

$$\sigma_{\text{tot}} = \frac{2}{C} \left\{ 1 - e^{-C \chi_{\text{tot}}} \right\}, \qquad (35)'$$

$$\sigma_{\text{inel}} = \frac{1}{C} \left\{ 1 - e^{-2C \chi_{\text{tot}}} \right\}, \qquad (36)$$

$$\sigma_{\text{dif}} = \left(1 - \frac{1}{C} \right) (\sigma_{\text{tot}} - \sigma_{\text{inel}}) \qquad (37)$$

- realize an interaction with probability

$$\sigma_{\text{inel}} + \sigma_{\text{dif}} \qquad (38)$$

- in case of an interaction, consider it to be diffractive or nondiffractive with probabilities

$$\frac{\sigma_{\text{dif}}}{\sigma_{\text{inel}} + \sigma_{\text{dif}}} \qquad (39)$$

and

$$\frac{\sigma_{\text{inel}}}{\sigma_{\text{inel}} + \sigma_{\text{dif}}} \qquad (40)$$

- in case of a nondiffractive interaction, determine number m of cut Pomerons according to the distribution $\sigma_{\text{inel}}^{(m)}$ see eq. (26)

- loop over cut Pomerons m

- determine the nature of the Pomeron. Take it to be soft with probability

$$\frac{\chi_{\text{soft}}}{\chi_{\text{tot}}} \qquad (41)$$

and to be semihard with probability

$$\frac{\chi_{\text{semi}}}{\chi_{\text{tot}}} \qquad (42)$$

- In case of a semihard Pomeron, determine its type t (gg, qg, gq, qq) according to the probabilities

$$\frac{\chi_t}{\chi_{\text{semi}}}. \qquad (43)$$

OUTLOOK

Having determined the configuration, one calculates the energy sharing among the Pomerons, for the semihard Pomerons also the energy sharing among the soft and the hard pieces. In case of semihard Pomerons, the parton ladders have to be generated explicitely. Finally, partons and remnants constitute kinky strings. The details of these different procedures will be explained in a future publication.

REFERENCES

1.   K. Werner, Physics Reports 232 (1993) 87–299
2.   N.N. Kalmykov, S.S. Ostapchenko, Phys. At. Nucl. (USA), 56 (1993) N3 346

STRANGENESS PRODUCTION IN Pb+Pb AT 158 GeV/NUCLEON

G. Cooper[1] for the NA49 Collaboration

[1]Nuclear Science Division
Ernest Orlando Lawrence Berkeley National Laboratory
University of California
Berkeley, California 94720

INTRODUCTION

At sufficiently high energy density and volume, strongly interacting matter is predicted to exist in a state, called the quark-gluon plasma (QGP), in which the color degrees of freedom are deconfined. In a QGP, the rate of production of strange quark-antiquark pairs should be higher than in the case of hadronic matter, leading to enhanced strangeness in the observed final state.

Strongly interacting matter at high energy density may be studied in collisions of relativistic nuclei. By comparing collisions of increasing energy and system size, one hopes to observe a transition to the QGP. A possible signature of such a transition is an increased yield of strange hadrons relative to non-strange hadrons [1]. However, production of strangeness may also be enhanced with increasing system size as rescattering effects become more important.

Lead ion beams of 158 GeV/nucleon were provided by the CERN SPS beginning in 1994. Central collisions with stationary lead nuclei produce matter of the highest energy density in the largest volume available to experiments at present. Detailed measurement of the strange particle yields in these reactions are important to help detect the formation of the QGP in these reactions.

THE NA49 EXPERIMENT

NA49 is a large acceptance, charged hadron spectrometer designed to identify as many produced particles as possible. The detector is shown in Figure 1.

In the standard configuration, the lead target is placed just upstream of two large superconducting magnets with a total bending power of 9 T·m. Two time projection chambers (TPC) are situated inside the magnets. They track charged particles with momenta below about 5 GeV/c and identify neutral strange particles by reconstructing

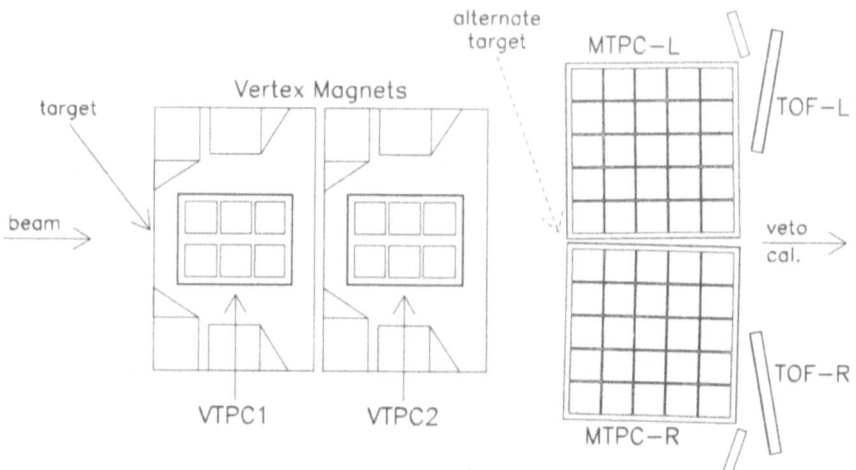

Figure 1. Layout of the NA49 experiment showing the momentum analyzing vertex magnets, the time projection chamber tracking detectors VTP1/2 and MTPC-L/R, and the time-of-flight detectors TOF-L/R. Triggering on spectator energy is accomplished with a veto calorimeter downstream. The standard target position just upstream of the magnets is indicated. Also shown is the target position in the alternate configuration.

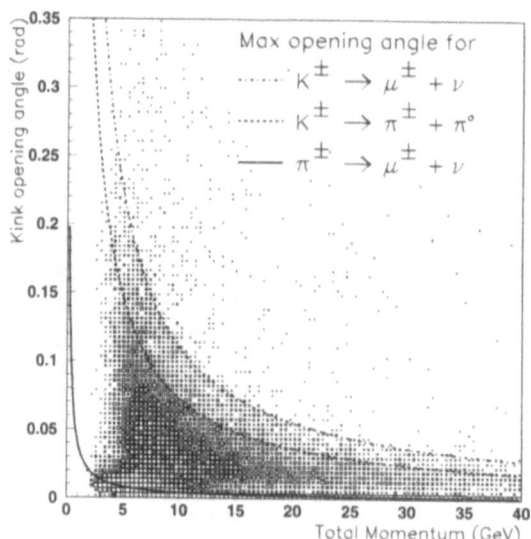

Figure 2. Reconstructed kink decay opening angle as a function of measured primary track momentum. The curves indicate maximum opening angle allowed by kinematics for the three most frequent decay modes. The data clearly show these limits.

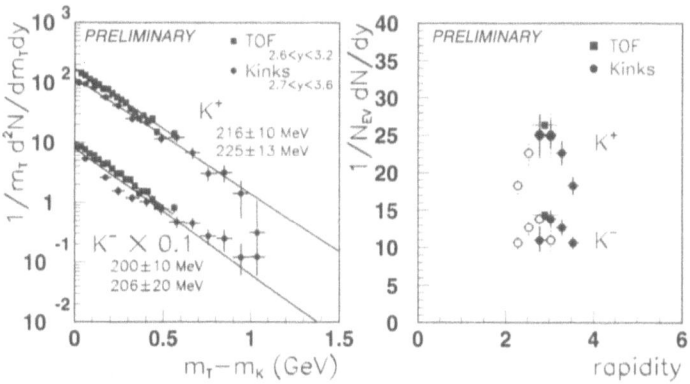

Figure 3. Corrected spectra for charged kaons obtained from the TOF system and the kink analysis. Extracted slopes from the transverse mass spectra (left) are consistent between the two measurements. Integrated and extrapolated rapidity density spectra are shown (right).

their decay. Two large main-TPCs downstream of the magnets track charged particles with momenta above about 3 GeV/c. In a limited momentum range of about 4 to 7 GeV/c, charged particles can be identifed with a combined measurement of velocity from the time-of-flight (TOF) detectors and track ionization in the main-TPCs. In addition, the large volume main-TPCs can be used to identify charged kaons by reconstructing their decays.

In a special configuration, the magnetic field is turned off and the target is placed just upstream of the main-TPCs. This configuration extends the acceptance for reconstruction of neutral strange particle decays.

Triggering is accomplished with a forward calorimeter, which characterizes the centrality of an event by measuring the energy of spectator nucleons. The trigger in the standard configuration selects the five percent most central events. The trigger in the special configuration is slightly less central, selecting the seven percent most central events.

KAON MEASUREMENTS

Approximately two percent of the produced charged kaons decay within the main-TPCs. The majority of these decay in a channel which produces a single charged particle and one or more neutral particles. Such a decay will be reconstructed as a kinked track. The distribution of kink opening angle as a function of reconstructed track momentum is shown in Figure 2. Kinematic constraints limit the opening angle for each kaon decay mode. A large number of pions also decay within the main-TPCs. However, as is also shown in Figure 2, pion decays are kinematically limited to smaller opening angles. Thus, the pion background can be excluded by selecting kinks with a sufficiently large opening angle.

The remaining background, which consists primarily of large-angle scattering and inelastic reactions of primary particles in the TPC gas, is estimated from the reconstructed kinks with opening angles larger than the highest kinematic limit and from simulation. Acceptance and reconstruction efficiency are estimated by reconstructing artificial kaon decays embedded in raw data.

The corrected transverse mass spectra for the charged kaons are shown in Figure 3. Also shown are the spectra obtained from the TOF measurement. Fits to

$$\frac{1}{m_\perp}\frac{dN}{dm_\perp} = A\exp(-m_\perp/T)$$

are shown and are consistent for the two measurements. The transverse mass spectra are numerically integrated and extrapolated to high transverse mass to obtain rapidity density spectra shown in Figure 3. For the decay measurement, the extrapolation is only 1.5 percent. For the TOF measurement, the extrapolation is about 20 percent.

A large K^+/K^- ratio of 1.83 ± 0.05 is measured at mid-rapidity. This ratio is consistent with that reported by the NA44 collaboration [2]. This ratio is indicative of a large baryon density [3], which is consistent with separate measurement of the net proton spectrum made by NA49 [4].

In the special field-off configuration, K^0_S are reconstructed by their decay into two charged pions, which form a V with a vertex displaced from and coplanar with the beam-target interaction point. Reconstructed Vs include decays of K^0_S, Λ, $\bar{\Lambda}$, and photon conversions. The opening angles for conversions are limited to small angles. In this field-off configuration, momenta of the decay tracks are not measured, so no invariant mass calculation is made. Instead, for pairs of tracks which form a displaced V, a neutral primary particle track is assumed to join the beam-target interaction point and the V-vertex, giving opening angles for each leg of the V. With these two angles and an assumed neutral particle mass, all momenta can be calculated.

Lambda decays into a proton and a pion are a background for the K^0_S measurement. However, in the laboratory frame, the proton trajectory is confined to small opening angles from the lambda trajectory due to kinematic constraints. Therefore, a region of the opening angle space can be selected to exclude lambda decays. The remaining background is combinatorial and is estimated from simulation and event mixing. Acceptance and efficiency are estimated from reconstructing simulated events and artificial K^0_S decays embedded in raw data.

The corrected transverse mass spectra are shown in Figure 4. The slope of the exponential fit is consistent with those obtained for the charged kaon measurements. An extrapolation to both low and high transverse mass must be made to obtain rapidity density spectra, shown in Figure 4. About 60 percent of the yield comes from the extrapolation.

Also shown in Figure 4 is the comparison of the rapidity density of K^0_S to the average of the K^+ and K^- rapidity densities. For an isospin symmetric system, one expects these two yields to be approximately equal. For systems with net isospin, such as Pb+Pb, deviations from equality are expected, but a deviation as large as measured here is puzzling.

The kaon measurements can be compared to results obtained previously for lighter systems. The NA35 collaboration measured the neutral kaon spectrum over large acceptance in S+S at the slightly higher energy of 200 GeV/nucleon [5]. The Pb+Pb and S+S results can be compared directly by scaling the S+S results by the ratio of the number of participants in the two systems. A parameterization of the scaled S+S result is shown in Figure 4. NA35 reported an enhanced K/π ratio in S+S and S+Ag relative to nucleon-nucleon and nucleon-nucleus reactions at the same energy. On the basis of the comparison to S+S, this enhancement is seen to persist in Pb+Pb with approximately the same magnitude.

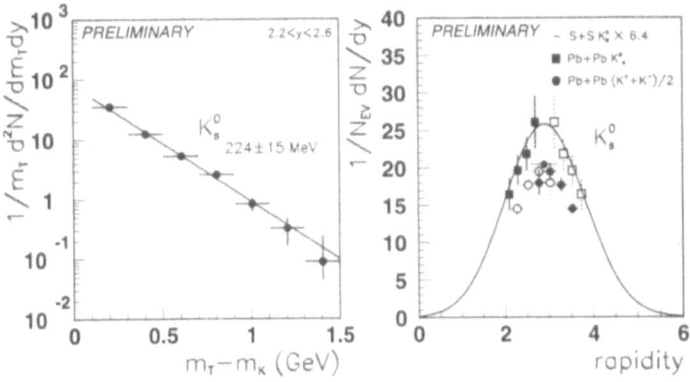

Figure 4. Corrected spectra for K_S^0 obtained from MTPC-L/R in the alternate configuration. The slope of the transverse mass spectrum (left) is consistent with that of the charged kaons (Figure 3). The rapidity density is shown and compared to charged kaons (right). Also shown is a parameterization of the K_S^0 yield from S+S at 200 GeV/nucleon scaled by the ratio of the number of participating nucleons in the two systems.

STRANGE BARYON MEASUREMENTS

Strange baryon spectra are measured in both the standard and special configurations. Preliminary results from the standard configuration for lambda and anti-lambda were reported in Ref. [6]. In the special configuration, a region in the space of V opening angles is selected for lambda decays. In contrast to the K_S^0 measurement, it is not possible to find a region of this space where only lambda decays occur, and therefore K_S^0 decays are a source of background. This background is subtracted based on the K_S^0 measurement. Also, since the charges of the decay tracks are not known, it is not possible to determine whether the decaying particle is a lambda or anti-lambda, and the measurements yields a combined lambda plus anti-lambda measurement. The remaining backgrounds, acceptance, and efficiency are estimated as in the K_S^0 measurement.

The corrected transverse mass spectra for the lambda measurements in both the standard and special configurations are shown in Figure 5. Fitted inverse slopes of

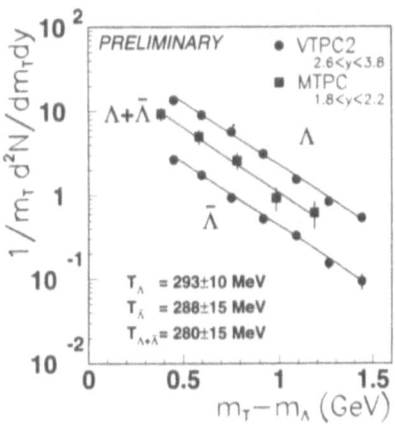

Figure 5. Corrected transverse mass spectra for Λ and $\bar{\Lambda}$ obtained from VTPC2, and $\Lambda + \bar{\Lambda}$ obtained from MTPC-L/R in the alternate configuration. The extracted slopes are consistent among the measurements, which were made in different rapidity intervals.

280-293 MeV are observed. The measured kaon and lambda inverse slopes show an increase with increasing particle mass, which was observed in Pb+Pb [6], and which indicates significant transverse expansion of the reaction volume.

Ref. [6] reported an integrated and extrapolated lambda rapidity density of 23 ± 7 in the interval $2.6 < y < 3.8$. Compared to the mid-rapidity result for S+S of 2.2 ± 0.4 measured by NA35 [5], the Pb+Pb result scales by a factor of 10.5 ± 3.7. This factor is somewhat larger than the scaling seen for kaons, but still consistent with ratio of the participant number in the two systems.

CONCLUSION

Preliminary inclusive spectra of strange particles produced in Pb+Pb collisions at 158 GeV/nucleon were presented. The results show no additional enhancement of strangeness production relative to S+S. A large ratio of K^+/K^- at mid-rapidity indicates a baryon-rich central rapidity region. An unexplained deficit of charged to neutral kaons is also observed.

ACKNOWLEDGEMENT

This work was supported by the Director, Office of Energy Research, Division of Nuclear Physics of the Office of High Energy and Nuclear Physics of the U.S. Department of Energy under Contract DE-AC03-76SF00098, the Bundesministerium für Bildung und Forschung, Germany, the Alexander von Humboldt Foundation, the Polish State Committee for the Scientific Research, and the Polish-German Foundation. I am grateful to the NA49 Collaboration for producing the presented results and for their continuing support and encouragement.

REFERENCES

1. P. Koch, B. Müller, J. Rafelski, Phys. Rep. **142** (1986) 167.
2. N. Xu et al., NA44 collaboration, Nucl. Phys. **A610** (1996) 175c.
3. P. Koch, J. Rafelski, W. Greiner, Phys. Lett. B **123** (1932) 151.
4. M. Toy et al., contribution to this conference
5. T. Alber et al., NA35 collaboration, Z. Phys. C **64** (1994) 195.
6. P. Jones et al., NA49 collaboration, Nucl. Phys. **A610** (1996) 188c.

THE ISOSPIN DEPENDENCE OF TRANSVERSE FLOW

Gary D. Westfall

National Superconducting Cyclotron Laboratory and
Department of Physics and Astronomy
Michigan State University
East Lansing, Michigan 49924-1321

INTRODUCTION

Collective transverse flow can provide information about nuclear reactions at intermediate energy and details of the nuclear equation of state (EOS)[1, 2]. One of the most compelling observables related to transverse flow is the incident energy at which the attractive mean field balances the repulsive nucleon-nucleon scattering. This energy is called the balance energy, E_{bal},[3] and can be related to various properties of the EOS through transport models such as the Boltzmann-Uehling-Uehlenbeck (BUU) model. In this paper, we will relate E_{bal} to the in-medium nucleon-nucleon scattering cross sections[4, 5].

The beams were produced by the K1200 Superconducting Cyclotron at the National Superconducting Cyclotron Laboratory (NSCL) at Michigan State University (MSU). The MSU 4π Array was used to carry out all the measurements presented here[6].

The main ball of the MSU 4π Array consists of 170 phoswich detectors (arranged in 20 hexagonal and 10 pentagonal subarrays) covering $18° \leq \theta_{lab} \leq 162°$. The 30 Bragg curve counters (BCCs) installed in front of the hexagonal and pentagonal subarrays were operated in ion chamber mode with a pressure of 125 Torr of C_2F_6 gas. The hexagonal anodes of the five most forward BCCs are segmented, resulting in a total of 55 separate ΔE detectors (the BCCs served as ΔE detectors for charged particles that stopped in the fast plastic scintillator of the main ball). Consequently, the main ball was capable of detecting charged fragments from $Z = 1$ to $Z = 16$, with mass resolution for the hydrogen isotopes in the phoswiches. Low energy thresholds were approximately 18, 3.5, and 7 MeV/nucleon for fragments with $Z = 1$, 3, and 12, respectively. In addition to the main ball of the 4π Array, a high rate array (HRA) was also used. The HRA is a close-packed pentagonal configuration of 45 phoswich detectors spanning laboratory polar angles $3° \leq \theta \leq 18°$. With the HRA, Z resolution up to Z=20 and mass resolution for the hydrogen isotopes were obtained.

Data were taken with a minimum bias trigger that required at least one hit in the HRA (HRA-1 data), and a more central trigger where at least two hits in the main ball (Ball-2 data) were required.

ISOSPIN DEPENDENCE OF TRANSVERSE FLOW

In this section we demonstrate experimentally that directed transverse flow depends on the isotopic ratio of the system by measurement of flow in three $A_{proj} = 58 + A_{targ} = 58$ systems with different N/Z at one bombarding energy of 55 MeV/nucleon. Measured flow is stronger for the more neutron-rich system in agreement with BUU predictions [7], which is due mainly to fact that the free neutron-proton cross section is approximately three times higher than the neutron-neutron and proton-proton cross sections at this incident energy. Because most of the experimental conditions (kinematics, available excitation energy, detector configuration, trigger, etc.) were held constant, the change in flow is most likely due to the different N/Z of the three systems.

In Fig. 1 we display impact-parameter-inclusive distributions for three global observables from ^{58}Ni+^{58}Ni (solid histograms) and ^{58}Fe+^{58}Fe (dashed histograms) reactions at 55 MeV/nucleon. The left panel shows the total charged-particle multiplicity M_{chgd}; the center panel shows the total midrapidity charge Z_{mr} [8]; and the right panel shows the reduced total transverse kinetic energy $\hat{E}_t$ [9]. These distributions demonstrate that we are comparing similar data sets for these isotopic systems. The impact parameter distributions in the simple geometric picture resulting from the $\hat{E}_t$ spectra in the right panel are nearly identical.

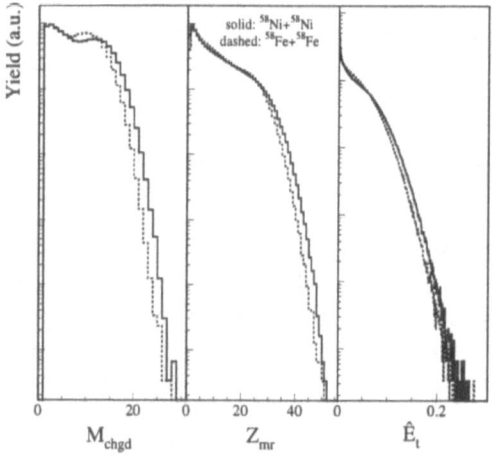

Figure 1. *Impact-parameter-inclusive spectra for the total charged-particle multiplicity M_{chgd}, the total midrapidity charge Z_{mr}, and the reduced total transverse kinetic energy $\hat{E}_t$. Solid (dashed) histograms are for $^{58}Ni+^{58}Ni$ ($^{58}Fe+^{58}Fe$) collisions at 55 MeV/nucleon.*

The enhancement at higher values of M_{chgd} and Z_{mr} for central ^{58}Ni+^{58}Ni reactions ($Z_{total} = 56$) is expected because a larger number of charged particles should be detected than in ^{58}Fe+^{58}Fe reactions ($Z_{total} = 52$). These fragments would account for a greater amount of participant charge, particularly when the system breaks into many pieces. Presumably these charges would carry off more transverse energy due to greater Coulomb repulsion tending to weaken the isospin effect for the most central collisions. These global variables could not be compared directly for the ^{58}Mn+^{58}Fe system because of poor statistics and some minor contamination (mainly ^{56}Cr) in the secondary beam.

Fig. 2 shows the extracted values of the directed transverse flow for three different fragment types from three different entrance channels plotted as a function of the isotopic ratio of the composite projectile plus target system where $(N/Z)_{sys} = (N_{proj} + N_{targ})/(Z_{proj} + Z_{targ})$. Using this definition we have: $(N/Z)_{sys}$ =1.07 for ^{58}Ni+^{58}Ni; $(N/Z)_{sys}$ =1.23 for ^{58}Fe+^{58}Fe; and $(N/Z)_{sys}$ =1.27 for ^{58}Mn+^{58}Fe.

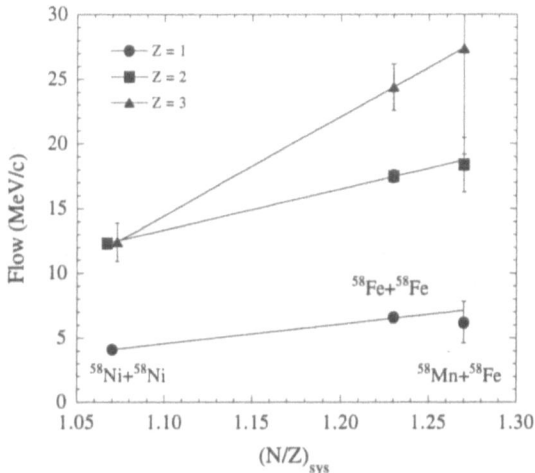

Figure 2. Directed transverse flow as a function of the isotopic ratio of the composite projectile plus target system for three different fragment types from three isotopic entrance channels. The extracted values of the flow are for impact-parameter-inclusive event sets at 55 MeV/nucleon. The lines are included only to guide the eye.

The data for fragments with $Z = 2$ and $Z = 3$ represent the slopes of linear fits over the reduced c.m. midrapidity region $-0.5 \le (y/y_{proj})_{c.m.} \le 0.5$. The fit range was reduced to $-0.5 \le (y/y_{proj})_{c.m.} \le 0.4$ for all three systems for the fits for fragments with $Z = 1$ because of the presence of a broad peak at projectile rapidity for ^{58}Mn+^{58}Fe not observed in the other systems. The points for $Z = 2$ from ^{58}Ni+^{58}Ni and ^{58}Mn+^{58}Fe have been offset in value of $(N/Z)_{sys}$ for clarity.

The results shown in Fig. 2 demonstrate clearly there is an isospin dependence for directed transverse flow even for the impact-parameter-inclusive data. The neutron-rich system ^{58}Fe+^{58}Fe exhibits larger flow values than ^{58}Ni+^{58}Ni for all three particle types. Although the difference between the flow values extracted for ^{58}Mn+^{58}Fe and ^{58}Fe+^{58}Fe is not statistically significant, the trends are consistent with the reactions involving the two stable beams.

Additional experimental evidence for the isospin dependence of directed transverse flow is shown in Fig. 3. The extracted values of the collective transverse flow in the reaction plane are displayed as a function of the reduced impact parameter for three different fragment types from ^{58}Fe+^{58}Fe and ^{58}Ni+^{58}Ni collisions at 55 MeV/nucleon. The extracted values of the flow are plotted at the upper limit of each $\hat{b} = (b/b_{max})$ bin. The errors shown are the statistical errors on the slopes of the linear fits.

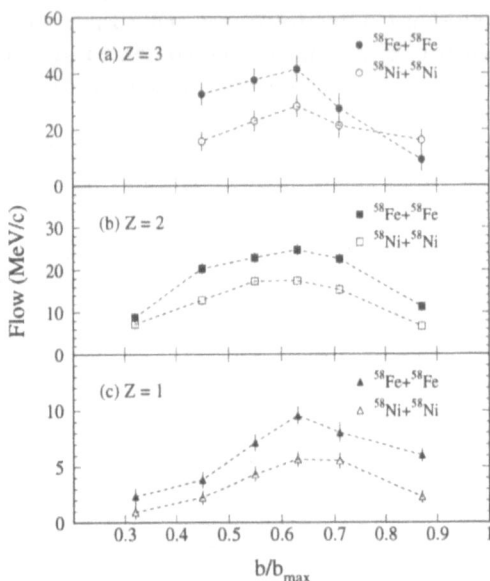

Figure 3. Directed transverse flow as a function of the reduced impact parameter for three different fragment types from $^{58}Fe+^{58}Fe$ and $^{58}Ni+^{58}Ni$ collisions at 55 MeV/nucleon. The extracted values of the flow are plotted at the upper limit of each $\hat{b}$ bin. The lines are included only to guide the eye.

The neutron-rich system $^{58}Fe+^{58}Fe$ systematically exhibits larger flow values than $^{58}Ni+^{58}Ni$ for all three particle types at all reduced impact parameter bins displayed (except for $Z = 3$ in the most peripheral bin). The largest difference in the magnitude of the flow between the isotopic entrance channels occurs for heavier mass fragments in semi-central collisions. The impact parameter dependence of the directed transverse flow shown is in qualitative agreement with previous work [2, 10, 11, 12], because the flow is maximal for semi-central events. The mass dependence of the directed transverse flow shown in Fig. 3 also demonstrates the well known increase in magnitude for heavier fragments [2, 10, 11, 13], (note the difference in vertical scale for each panel).

ISOSPIN DEPENDENCE OF THE BALANCE ENERGY

In this section we show that E_{bal} depends on the ratio of neutrons to protons (N/Z) of the system by measuring the disappearance of directed transverse flow in two different isotopic systems with different N/Z ratios[4]. Balance energies are larger for the more neutron-rich system at all measured impact parameters in agreement with BUU predictions. Because most of the experimental conditions (kinematics, available excitation energy, detector configuration, trigger, etc.) were held constant, the change in balance energies is due to the different N/Z of the two systems.

Fig. 4 shows the mean transverse momentum in the reaction plane $\langle p_x \rangle$ plotted versus the reduced c.m. rapidity $(y/y_{proj})_{c.m.}$ for the two isotopic entrance channels. The solid (open) squares are for fragments with $Z = 2$ from semi-central $^{58}Fe+^{58}Fe$ $(^{58}Ni+^{58}Ni)$ collisions at 105 MeV/nucleon. The upper limit of the reduced impact parameter bin for these events is $\hat{b} = 0.48$. The errors shown are statistical. The only difference between the two data sets is the N/Z ratio of the interacting system. The kinks in the spectra at $(y/y_{proj})_{c.m.} \approx -0.6$ are attributed to detector acceptance, but the transverse momentum analysis allows extraction of the flow with as little detector bias as possible [2] by not including affected regions of the spectrum. The vertical offsets from the origin occur because no recoil correction was applied in the reaction plane calculation. This does not affect the final values of the flow observables (balance energies) in this analysis [11]. Each spectrum shown in Fig. 4 is fit with a straight line over the midrapidity region $-0.5 \leq (y/y_{proj})_{c.m.} \leq 0.5$, and the slopes of these lines are defined as the directed transverse flow for each isotopic system. As expected, the directed transverse flow is similar for both isotopic entrance channels, but in what follows the difference is shown to be systematically significant in the data.

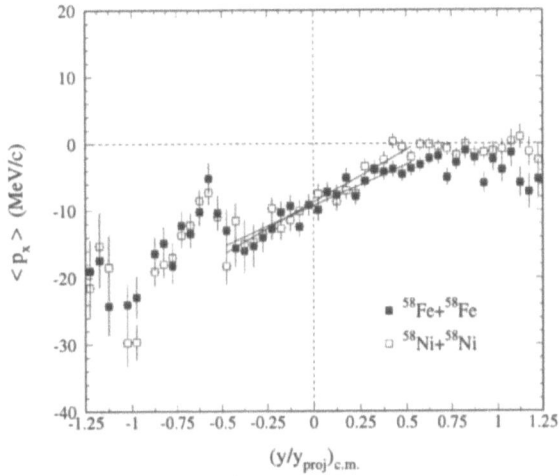

Figure 4. Mean transverse momentum in the reaction plane versus the reduced c.m. rapidity for $Z = 2$ fragments from semi-central collisions ($b/b_{max} = 0.48$) at 105 MeV/nucleon. The solid (open) squares are for $^{58}Fe+^{58}Fe$ ($^{58}Ni+^{58}Ni$). The straight lines are fits over the midrapidity region $-0.5 \leq (y/y_{proj})_{c.m.} \leq 0.5$.

The extracted values of the directed transverse flow plotted versus the incident beam energy are shown in Fig. 5. The solid (open) squares are for fragments with $Z = 2$ from semi-central $^{58}Fe+^{58}Fe$ ($^{58}Ni+^{58}Ni$) collisions. The errors shown are the statistical errors on the slopes of the linear fits (the systematic error associated with the range of the fitting region is +3 MeV/c and -1 MeV/c). The curves are included only to guide the eye. To extract the balance energy E_{bal}, the data were fit with a second-order polynomial allowing the fitting range to vary until χ^2 per degree of freedom was a minimum [11]. The second-order fits pass through minima for which the value of the abscissa corresponds to the balance energy for that particular entrance channel and $\hat{b}$ bin. The curves do not pass through zero at E_{bal} because no recoil correction was used in the reaction plane determination, as was done elsewhere [10, 11]. Collective transverse flow is assumed to be symmetric in the vicinity of the balance energy, and our measurements are unable to distinguish the sign (+ or -) of the flow, so that a local parabolic fit is the lowest order symmetric function that can be used without *a priori* knowledge of E_{bal}.

The horizontal displacement of the minima of the curves in Fig. 5 clearly indicates that E_{bal} is higher for ^{58}Fe+^{58}Fe than ^{58}Ni+^{58}Ni at this $\hat{b}$ bin. That the balance energy is larger for the more neutron-rich system is primarily attributed to the difference in nucleon-nucleon cross sections [7]. Directed transverse flow has already been shown to be sensitive to in-medium nucleon-nucleon cross sections [13, 14]. The neutron-proton cross section is approximately a factor of three higher than the neutron-neutron and proton-proton cross sections over the range of beam energies measured here [7]. This results in less repulsive collective flow from nucleon-nucleon scattering for the more neutron-rich system, pushing the balance energy higher in value.

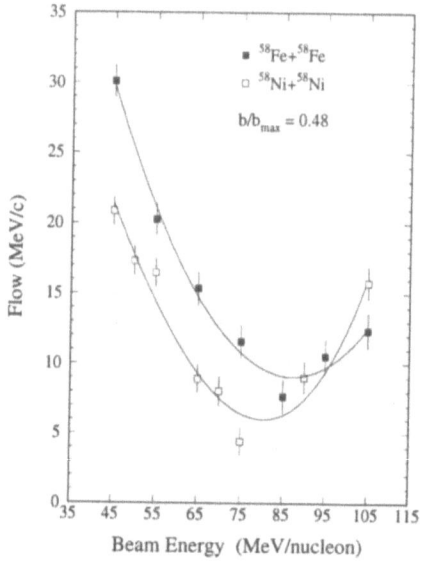

Figure 5. Excitation functions of the measured transverse flow in the reaction plane for $Z = 2$ fragments from semi-central collisions ($b/b_{max} = 0.48$). The solid (open) squares are for $^{58}Fe+^{58}Fe$ ($^{58}Ni+^{58}Ni$). The curves are included only to guide the eye.

Below the balance energy the attractive mean field has an even more dominant effect, resulting in higher flow values for the neutron-rich system [4, 5]. Above the balance energy where repulsive nucleon-nucleon scattering dominates, the converse is true, resulting in smaller values of the directed transverse flow for the neutron-rich system.

Additional experimental evidence for the isospin dependence of the balance energy is presented in Fig. 6. The solid (open) squares are the measured values of the balance energies for ^{58}Fe+^{58}Fe (^{58}Ni+^{58}Ni) extracted for four reduced impact parameter bins. These experimental values of $E_{bal}(b)$ are plotted at the upper limit of each $\hat{b}$ bin, and the values for ^{58}Ni+^{58}Ni have been slightly offset in the horizontal direction to show the error bars more clearly. The errors shown on the measured values of the balance energies are statistical. The balance energy increases as a function of impact parameter for both isotopic systems in agreement with previous work [10, 12, 16], and $E_{bal}(b)$ is systematically higher for the more neutron-rich system at all measured $\hat{b}$ bins.

The predictions of BUU model [7, 15] calculations which incorporate an isospin dependent potential and isospin dependent nucleon-nucleon scattering cross sections for ^{58}Fe+^{58}Fe (^{58}Ni+^{58}Ni) are shown as solid (open) circles in Fig. 6 for five $\hat{b}$ bins. The errors shown on the calculated points are statistical, and the values for ^{58}Ni+^{58}Ni have been slightly offset in the horizontal direction to show the error bars more clearly. That the balance energy is the same value for all fragment types [11, 13] facilitates comparison of the measured values of $E_{bal}(b)$ to predictions of transport models calculations which involve only nucleons. The balance energy has been shown to exhibit little sensitivity to the acceptance effects of our detector array [3], allowing direct comparison between experimental values and unfiltered theoretical results.

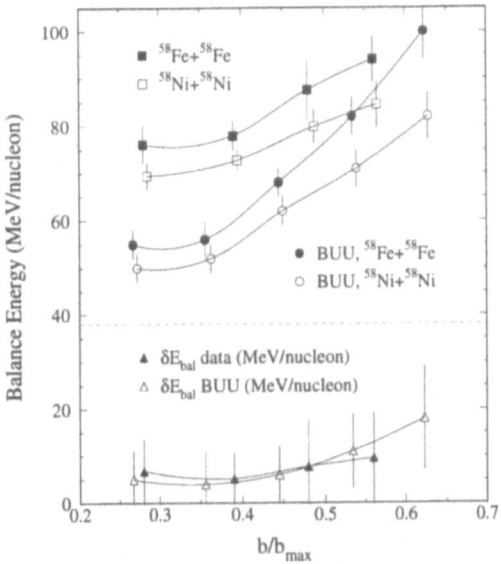

Figure 6. Measured balance energies as a function of impact parameter compared to the predictions of BUU model calculations with an isospin dependent mean field and isospin dependent in-medium nucleon-nucleon cross sections. The solid (open) squares are measured for $^{58}Fe+^{58}Fe$ ($^{58}Ni+^{58}Ni$) while solid (open) circles are BUU predictions for $^{58}Fe+^{58}Fe$ ($^{58}Ni+^{58}Ni$). The values for $^{58}Ni+^{58}Ni$ have been offset in the horizontal direction for clarity. The solid (open) triangles correspond to the difference in the balance energies between the isotopic systems at each reduced impact parameter bin for the data (BUU predictions). All curves are included only to guide the eye.

The trends in the values of $E_{bal}(b)$ predicted by the BUU model with isospin dependence are consistent with those for the measured values. The balance energy increases as a function of impact parameter for both isotopic systems, and $E_{bal}(b)$ is systematically higher for the more neutron-rich system at all impact parameter bins. That the overall magnitude of the values for the balance energies is underpredicted for central collisions by the BUU model has been attributed to a density dependent reduction of the in-medium nucleon-nucleon cross sections [13, 14]. This effect is stronger at smaller impact parameters where the interaction volume is larger than in peripheral collisions.

More importantly here, there is agreement between the data and the BUU model predictions for the magnitude of the isospin effect, which is demonstrated explicitly with the lower set of points in Fig. 6. The solid (open) triangles are the difference between the balance energies δE_{bal} for the data (BUU predictions) for the isotopic systems at each corresponding $\hat{b}$ bin. These δE_{bal} values are given in MeV per nucleon and are plotted on the same scale as the values of $E_{bal}(b)$. The errors shown are statistical.

There is good agreement between the data and the BUU model predictions for the overall magnitude of δE_{bal}, which is due mainly to the different N/Z ratios of the two isotopic systems. The difference in balance energies between isotopic systems was found to persist for BUU calculations made without Coulomb repulsion, indicating that the isospin effect is mainly due to the difference in the elementary nucleon-nucleon cross sections. The magnitude of δE_{bal} increases for more peripheral collisions where two extended neutron distributions overlap in the reaction of two neutron-rich nuclei [7].

CONCLUSIONS

We have demonstrated experimentally that collective transverse flow in nucleus-nucleus collisions depends on the isospin of the system. We used three symmetric systems, ^{58}Fe + ^{58}Fe, ^{58}Ni + ^{58}Ni, and ^{58}Mn+^{58}Fe, at one incident energy to show that flow depends on the isospin of the system. At all impact parmeters and for all observed particle types, flow is larger for the neutron-rich system. In addition, we have measured the balance energy for ^{58}Fe + ^{58}Fe and ^{58}Ni + ^{58}Ni and find that the balance energy is always higher for the neutron-rich system at all impact parameters. BUU calculation reproduce the difference between the two systems. However, BUU underpredicts the observed balance energy in central collisions as was observed previously.

This dependence of the flow on the isospin of the system supports our view of transverse flow as a balance between mostly repulsive nucleon-nucleon scattering and mostly attractive nuclear mean field interactions.

REFERENCES

1. H. Stöcker and W. Greiner, Phys. Rep. **137**, 277 (1986).
2. H.H. Gutbrod, A.M. Poskanzer, and H.G. Ritter, Rep. Prog. Phys. **52**, 1267 (1989).
3. C.A. Ogilvie *et al.*, Phys. Rev. C **42**, R10 (1990).
4. R. Pak *et al.*, Phys. Rev. Lett. **78**, 1022 (1997).
5. R. Pak *et al.*, Phys. Rev. Lett. **78**, 1026 (1997).
6. G.D. Westfall *et al.*, Nucl. Instr. and Methods **A238**, 347 (1985).
7. Bao-An Li *et al.*, Phys. Rev. Lett. **76**, 4492 (1996).
8. C.A. Ogilvie *et al.*, Phys. Rev. C **40**, 654 (1989).
9. L. Phair *et al.*, Nucl. Phys. **A548**, 489 (1992).
10. R. Pak *et al.*, Phys. Rev. C **53**, R1469 (1996).
11. R. Pak *et al.*, Phys. Rev. C **54**, 2457 (1996).
12. J.P. Sullivan *et al.*, Phys. Lett. B **249**, 8 (1990).
13. G.D. Westfall *et al.*, Phys. Rev. Lett. **71**, 1986 (1993).
14. D. Klakow, G. Welke, and W. Bauer, Phys. Rev. C **48**, 1982 (1993).
15. Bao-An Li and S.J. Yennello, Phys. Rev. C **52**, R1746 (1995).
16. A. Buta *et al.*, Nucl. Phys. **A584**, 397 (1995).

ACKNOWLEDGMENTS

Most of the work presented in this paper is derived from the Ph.D. dissertation of Robert Pak. The data were taken and analyzed in collaboration with the MSU 4π Group. This work supported by grant number PHY-95-28844 from the National Science Foundation.

APEX: WEAK POSITIVE EVIDENCE FOR 800 keV SHARP PAIRS

James J. Griffin

Department of Physics
University of Maryland
College Park, MD 20742

INTRODUCTION

The APEX experiment was mounted to test EPOS' earlier observations of sharp (e^+e^-) pairs from high-Z heavy ion collisions. In a recent letter the APEX collaboration[1] presents its results, and considers the earlier EPOS U+Th results[2, 3, 4, 5] in the light of their own new data. It asserts that their data offers no statistically significant evidence of sharp pairs and concludes "that the results of the present experiment represent a real disagreement with the previous observations".

Here this published APEX U+Th data is re-analyzed. We find that the best chi-squared fit to this data describes a smooth background plus a sharp pair line consisting of 123 pairs of width 23 keV at 793 keV and is significantly better than APEX' background-only description. The data even impose positive *lower* bounds of 23 sharp pairs per 20 keV near 800 keV at 99.0% Confidence Level. It is therefore untenable to argue from the APEX data against the existence of sharp pairs.

Ratios of various comparable pair counts from the APEX and EPOS experiments are cited to show empirically that both their background and their sharp pair distributions are mutually consistent; furthermore the two experiments are shown to be of comparable statistical potency, so that conflicts, if any, between them can be resolved only by other independent evidence.

Since our analysis implies a seemingly contradictory conclusion to that which APEX drewlfrom the from the very same data, it is essential to stipulate at the outset that the APEX (U+Th) data does unquestionably exclude an excess sharp pair count near 800 keV as large as the ≈ 2560 counts which they were expecting. In fact the APEX report establishes a specific (99%CL) upper bound* of 292 excess counts near 800 keV. That discrepancy leads the APEX collaboration to question the existence of

---

*APEX presents its upper bounds as cross sections. We use the conversion factor, 2560/5=512 excess counts/(μb/sr), to convert from APEX' sharp pair production cross section to sharp pair counts.

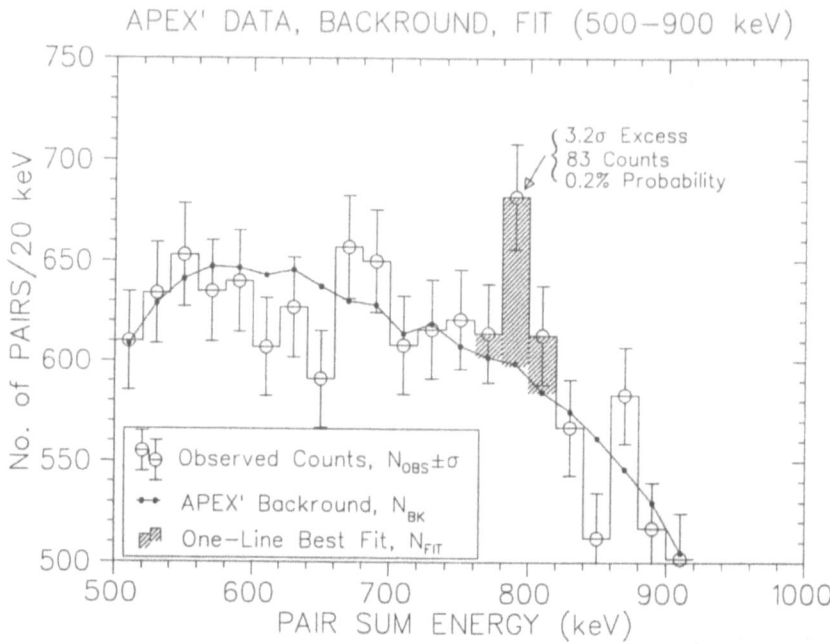

Figure 1. This figure presents the data from Figure 2 of Ref.[1] in the range from 500 to 920 keV where evidence for sharp pairs was previously reported. The largest excess (3.2σ) above the background occurs at 790 keV. The best gaussian fit produces the cross-hatched addition to the background, as described in the text. The data near 800 keV are tabulated in Table I.

the previously reported sharp pairs, and in particular of the EPOS line near 800 keV, and to draw the conclusion quoted above, of a "real disagreement" with the EPOS results.

The present analysis confirms these results, and prescribes a range (from 217 to 251 counts) of even smaller upper bounds upon the excess pair counts near 800 keV. Nevertheless, the present analysis also provides statistically significant positive evidence that APEX actually counted 123 ± 46 sharp pairs near 800 keV. This inference contradicts none of APEX emprical evidence. In particular it honors the quantitative upper bound of the APEX report.

In the end, we find that the APEX U+Th experiment recorded (at 793 ± 7 keV) 123 ± 46 sharp pairs, above a background of 1480 pairs/20keV, among a total of 40.8K EPOS-type[†] pairs. The EPOS experiment recorded (at 809 ± 8 keV) 97 ± 38 sharp pairs[5] above a background of 1280 pairs/20keV among a total of 50K pairs. Thus under direct comparison, the APEX and the EPOS pair data sets are of comparable size, and yield comparable numbers of background pairs and of sharp pairs above background near 800 keV. APEX has observed about the same proportion of sharp pairs to backround pairs as EPOS, and There is no experimental evidence of contradiction between them. Since their total pair counts are so similar, it follows also that neither experiment can claim clear statistical superiority for resolving differences in their pair distributions.

---

† "EPOS'-type" pairs are the RL(1,1) pairs accepted by the EPOS apparatus.

Since the APEX data even impose positive *lower* bounds of 23 sharp pairs per 20 keV near 800 keV at 99.0% Confidence Level, it is untenable to argue from them against the existence of sharp pairs.

APEX' Published Pair Sum Energy Data

Figure 1 summarizes the APEX' published[‡] data on 22K (e^+e^-) RL/wedgecut pairs selected by APEX[§] from its U+Th data to resemble those of the EPOS experiment. It presents the number of pairs counted in each 20 keV bin, APEX' "event-mixed" background,and the standard deviations $\sigma \propto N_{OBS}^{0.5}$, plotted as error bars. All of the data of Figure 1 were presented graphically in Figure 2a of the APEX report, and in Figs. VI.1.2, and VI.1.3 of the thesis of M.R. Wolanski[10] on the APEX U+Th experiment.

Table I presents the relevant APEX U+Th data from 700 to 920 keV in numerical form. It lists for these eleven bins the sum energies, the observed pair counts, N_{OBS}, the APEX' event-mixed background distribution, N_{BK}, and the number of "Excess" (above the background) sharp pairs, N_{EXC}. Column five lists APEX' upper bounds upon the pair cross section and column six lists the values of our best fitting "One Sharp Line" distribution, N_{FIT}. columns seven and eight list the 99% CL lower and upper bounds upon the mean sharp pair count in each bin, as calculated below.

Chi-Squared and Confidence Level Analysis of APEX' Pair Data

It is evident in Figure 1, that the largest excess pair count occurs in the bin centered at 790 keV, and that its magnitude (83.0 counts) is about 3.2σ. Therefore we compare with APEX' "Background-Only" hypothesis the alternative "One-Sharp-Line" hypothesis that the excess pair count is well explained by the sum, $N_{FIT}^{TOT} = N_{BK} + N_{FIT}$, of the event-mixed background and a distribution, N_{FIT}, arising from a single sharp pair line of gaussian form centered at energy E_S, with FWHM of Γ, and with an integrated total number of pairs equal to S,chosen to minimize the chi-squared sum over the 21 bins from 500 to 920 keV.

The results of the chi-squared analysis are summarized in Table II. The minimum value of chi-squared occurs at the values (S= 123.4 counts, $\Gamma = 23.1$ keV, $E_S = 792.8$ keV), and yields $\chi^2 = 16.65$, and $\tilde{\chi}^2 = 0.98$, as compared with the Background-Only values of 28.30 and 1.41. Clearly the One-Sharp-Line description is statistically superior to the Background-Only description presented in the APEX report. The statistical significance of that superiority is further quantified by means of Confidence Level analysis for upper and lower bounds, as discussed below.

We next apply Confidence Level analysis to the APEX data of Figure 1, to compute the bounds which the APEX data impose (at the 99% Confidence Level) upon the greatest mean value, ν_U, and upon the least mean value, ν_L, of excess sharp pairs for 20-, 40-, and 60- keV intervals, corresponding to the treatment of the APEX data bin by bin, and as sums of two and of three adjacent bins. The resulting bounds for the eleven 20 keV bins are isted in columns ten and eleven of Table I.

[‡]The author is grateful to the APEX collaboration for supplying the data published in Figure 2 of Ref[1] in numerical form. We here consider only the data from 500 to 920 keV, where earlier studies[2, 5, 6, 7, 8] have indicated the occurrence of pairs of sharp sum energy, and tabulate only the data from 700 to 920 KeV. See Ref.[9] for more complete and detailed information.

[§]These however are not "Qualitatively Similar" EPOS-type pairs, as we discuss further below.

TABLE I. U+Th PAIR DATA, FITS and BOUNDS							
APEX' RESULTS:					One Line	99%CL	
Measured, Simulated and Inferred					Best Fit	Bounds	
E_{SUM}	N_{OBS}	N_{BK}	N_{EXC}	$(d\sigma/d\Omega)^{UB}$	N_{FIT}	ν_L	ν_U
710	608	613.8	-5.8	0.28	0.0	0.3	61.8
730	616	618.8	-2.8	0.28	0.0	0.3	64.2
750	621	607.7	+13.3	0.32	0.1	0.5	76.2
770	614	602.1	+11.9	0.30	11.8	0.5	74.8
790	682	599.0	+83.0	0.57	83.1	24.1	145.4
810	613	584.5	+28.5	0.33	28.2	1.0	89.1
830	567	574.8	-7.8	0.21	0.3	0.2	58.2
850	512	561.5	-49.5	0.12	0.00	0.1	35.9
870	583	546.2.	+36.8	0.32	0.00	1.7	95.2
890	517	529.8	-12.8	0.17	0.00	0.2	52.4
910	502	504.9	-2.9	0.18	0.00	0.3	57.9

Table I. The table lists the APEX pair data in the range 700-920 keV, from Figure 2a of Ref.[1], as follows. (1)The bin energies, E_{SUM}; (2) APEX' pair counts, N_{OBS}; (3) APEX' event-mixed pair background, N_{BK}; (4) the excess pair count above the background, N_{EXC}; (5) the APEX upper bound (in μb/sr) upon the pair cross cross section; (6) our best Gaussian fit, N_{FIT}, to the excess in column four; and (7) ν_L; and (8) ν_U, the 99%CL lower and upper bounds upon the mean excess counts in each bin.

The greatest interest for the present discussion lies in the results for the *lower* bounds imposed by the data. For all of the energy bins except that at 790 keV, the lower limits have a value (≤ 2 counts per 20 keV), indicating that the APEX data fails to provide 99% CL "statistically significant" (evidence to support even a positive mean excess pair count even as small as two. But for the 20 keV bin centered at 790 KeV, the lower bound becomes substantial, with a valueof 24.1 counts per 20 keV (about 0.9 σ). The APEX data therefore provides statistically significant evidence that a non-zero positive excess of sharp pairs above the background occurs in a 20 keV bin hear 800 keV.

Analogous 99% CL upper and lower bounds have been established by considering also the counts/40keV and the counts/60 keV which arise by combining two and three adjacent 20 keV bins. For each bin grouping a total line strength also has been inferred by assuming our best fitting line shape. Each of these three 99%CL analyses agree with the chi-squared result that the APEX' evidence for a mean total excess sharp pair count near 800 keV which is greater than 23 counts and less than 251 counts is statistically significant.

We emphasize that the quantitative results of the present analysis contradict in no way the quantitive empirical results of the APEX report[1]. In particular, their 99%CL upper limit of 292 total sharp pair counts (0.572μb/sr in their Figure 2b) near 800 keV is consistent with, and in fact less restrictive than any of the present upper bounds upon the total counts. Moreover, the APEX upper bounds of Ref.[1] exhibit a maximum near 800 keV, like that which Table I presents for the present upper bounds, due to the large excess of pair counts measured in the 790 keV bin. Presumably the APEX analysis would also have produced a positive 99% CL *lower* bound, also greater than that presented here, if one had been extracted.

TABLE II. BEST χ^2 FIT to APEX' U+Th PAIR DATA					
	No.of	Fit Quality:		Probability:	99% Sh.Pair Range:
	D.of F.	χ^2,	$\tilde{\chi}^2$;	$P_{\chi \geq}$	$N_{SP}^L \leq N_{SP} \leq N_{SP}^U$
APEX' Background-Only:	20	28.30	1.41	10.2%	No Sharp Pairs
Background+Sharp Line:	17	16.65	0.98	47.8%	$24 \leq 123 \leq 251$

Table II. The Table compares two fits to twenty one of the $(e^+ e^-)$ pair counts reported by the APEX collaboration in Ref.[1]. The three parameter Background+One-Sharp-Line fit is a better fit than APEX' one parameter Background-Only description, as discussed further in the text.

Discussion of APEX Positive Evidence for Sharp Pairs

APEX reported its data and concluded that ".....the results of the present experiment represent a real disagreement with the previous observations". Yet the present analysis of the same data shows that the APEX data corroborates the existence of sharp pairs at a confidence level exceeding 99%. How can the same data support both conclusions?

Because of space limitations, we argue here simply that there is no *empirical* basis for the APEX' claim of real disagreement with EPOS, omitting the discussion of how APEX' excessive (by $\approx 9X$) expectation of ~ 2500 sharp pair counts arose from their assuming that a $5.0\mu b/sr$ cross section, later revised[11] to the value of $1.4\mu b/sr$, was a correct average value over their 0.17 MeV/U beam spread, instead of EPOS' 0.07 MeV/U thinner target beam spread. (See Ref.[9] for a more extended discussion.)

Thus we ask whether APEX' observed 123 count strength is consistent with the EPOS' sharp pair count (97±38 counts, FWHM= 40 keV) at 809 keV. We here present a purely empirical positive answer in terms of data-only ratios of "Qualitatively Similar" pair counts. These ratios indicate that for "Qualitatively Similar" pair counts the APEX and the EPOS U+Th pair data bases are quite comparable both in size and quality, and they suggest no contradictions between them.

We first emphasize that it in any *purely empirical* quantitative comparison of the APEX and EPOS pair experiments, the APEX pairs must be of the type accepted in the EPOS experiment. We refer to this as the requirement of "Qualitative Similarity". Without Qualitative Similarity, the expression for a ratio of comparable measured pair counts from the two experiments involves a ratio of the underlying cross sections for the pair production processes which is not a measured quantity, and either precludes the comparison or, when non-measured information is intrroduced to resolve the matter, renders it non-empirical in the strict sense.

For example, one cannot know empirically whether the APEX' comparison[1] of their own (1,n) pair set (in which any number n of electrons are accepted in coincidence with one positron) with EPOS' (1,1)-only pair set (in which pairs only from events in which one and only one electron is in coincidence with the positron are accepted) ought to provide evidence on the existence of sharp pairs or simply on the differences between sharp pair production from (1,1) events and from from (1,n) events. If one *assumes*, implicitly or explicitly, that the probability of sharp pairs is the same in the (1,n) events as it is in the (1,1) events, then a result can be obtained, but its reliability is contingent

TABLE III. COMPARATIVE MEASURES OF APEX and EPOS PAIR DATA BASES					
	$L^{TOT}(\mu b^{-1})$	$N_{TOT}^{LR(1,1)}$	$\Delta N_{800}^{LR(1,1)}$	N_{SP}^{800};	$(N_{SP}^{800})/(\Delta N_{800}^{LR(1,1)})$
APEX:	$7000^{(a)}$	$(40.8K)^{(c)}$	$(1480)^{(c)}$	$\leq 123\pm36^{(d)}$	≤ 0.083
EPOS:	$2196^{(b)}$	50K	1280	97 ± 38	0.076
(APX/EPS):	3.2	0.8	1.2	≤ 1.3	≤ 1.09

Table III. For the RL(1,1) pairs of the EPOS-type, the APEX pair data base is somewhat smaller overall and somewhat larger near 800 keV, but not 3X larger as it luminosity would suggest. Furthermore, the ratio of the observed APEX and EPOS *sharp* pair counts near 800 keV is quite consistent with the ratio of background pairs in the same energy range. This table offers no evidence of any contradiction between APEX' and EPOS' sharp pair results. To the contrary it presents two very similar backround pair distributions which quite consistently provide two very similar sharp pair signals near 800 keV. Notes to the table entries follow: (a) APEX' measured luminosity[1]; (b) EPOS' luminosity inferred in Table V of Ref.[9] by comparison of its positron data with APEX', following the method of Cowan and Greenberg[12]; (c) To guarantee Qualitatively Similar APEX/EPOS comparisons (see text) we consider only APEX pairs of EPOS' RL(1,1) type. These RL(1,1) data are taken from Ref.[10], Figures VI.2.2, and VI.2.2(c); (d) See Note No.18.

upon the correctness of the assumption, and the analysis can no longer be considered as purely empirical.

We therefore consider in Table III only APEX' counts of pairs which are of the "EPOS-type": the leptons must be observed in opposite arms of the experiment ("RL" pairs), and the pairs must arise only from events in which one positron and one and only one electron were emitted, ("(1,1) pairs")[¶].

EPOS', APEX' Qualitatively Similar Pair Databases Are Nearly Equivalent

Table III summarizes some quantitative characteristics of the APEX and EPOS U+Th experiments, including their total luminosities and selected characteristics of their Qualitatively Similar RL(1,1) pair sets: their pair count totals and their pair counts per 20 keV near 800 keV. Also presented in the last two columns are the observed counts, N_{SP}^{800}, of excess sharp pairs and the ratio of these to background counts per 20 keV near 800 keV.

Table III shows that despite the APEX' ~3X larger luminosity, it collects only about the same number of coincident EPOS-type pairs as EPOS both overall and in 20 keV intervals near 800 keV. These ratios indicate that the overall size and the general shape of the APEX background pair distribution is similar to that of EPOS. It follows that in direct comparisons of APEX' and EPOS' measured pair data, APEX can claim no clear statistical superiority over EPOS.

Besides the counts of their background distributions, the numbers of excess sharp pairs near 800 keV and their ratios to the background pair counts near 800 keV in columns five and six are also comparable for the two experiments. Thus Table III shows that as regards both their coincident pair background distributions and their excess sharp pair distributions[||], the APEX and EPOS data bases are comparable within

---

[¶] Wedge cut pairs are not considered in this comparison because the APEX' wedge differed somewhat from EPOS'.

[||] We note that the APEX sharp pairs include (1,n) pairs not "Qualitatively Similar" to the EPOS-type. Nevertheless since they include the EPOS-type pairs as a subset, their count can provide upper bounds upon the corresponding Qualitatively Similar ratios.

~30%; Table III therefore not only speaks against any substantial contradiction between them, but it supports their overall mutual consistency. It also precludes any claim of statistical superiority for the APEX' pair data over EPOS', and raises the question why APEX counted so few pairs (both of the smooth background and sharp 800 keV types) compared with EPOS despite its larger luminosity, a question to be discussed further below.

The APEX/EPOS total luminosity (L^{TOT}) ratio inferred[9] from the positron yields and efficiencies is substantially ($\approx 3X$) larger than the corresponding ratios (≈ 1) of comparable pairs counted overall and per 20 keV near 800 keV. However,

Summary and Conclusions

The published APEX data exhibit a sharp pair line (123±46 counts) of width 23.4 keV at a sum energy of 793±7 keV, according to a chi-squared analysis. Confidence Level analysis of the APEX data also implies at the 99% confidence level a mean sharp pair value greater than 23 and fewer than 146 sharp pairs in the 20 keV bin at 790 keV. For our best fitting line shape, these bounds correspond to total sharp pair counts greater than 24 and less than 217. These results, and others based upon the combining of two and three bins, all agree that some sharp pair count greater than 23 and less than 251 should be expected in any repetition of the APEX experiment. All of these implications of the APEX data also honor the sharp pair upper bound (292 counts near 800 keV) inferred by APEX from the same data.

Regarding the question of conflict between the APEX and EPOS experiments, we have presented quantitative purely empirical indices of the two data bases which show that, although APEX' measured luminosity (as inferred from the total positron counts and efficiencies of the two experiments) is $\approx$3X larger than EPOS' , nevertheless APEX' actually counted altogether 20% fewer (40.8K/50K) "Qualitatively Similar" EPOS-type pairs than were counted by the EPOS experiment, and some 20% more (1480/1280) pairs per 20 keV near 800 keV. These ratios show that for direct APEX/EPOS comparisons of coincident pairs, the APEX data base is rather comparable to the EPOS', and certainly not substantially larger. Furthermore, the ratios of sharp pairs to background pairs near 800 keV agree within 10% for the two experiments, suggesting mutual consistency between APEX and EPOS also with regard to their *sharp* pairs near 800 keV.

Thus, under purely empirical criteria, the APEX data must be judged to be at least weakly corroborative of the EPOS' 800 keV sharp line. On the other hand, the APEX evidence is very comparable with EPOS: no one who doubted the EPOS results needs to be convinced by those of APEX.

Consideration elsewhere[9] of the APEX collaboration's *expectations* for sharp pairs vis a vis EPOS' data indicates that those expectations inflate the implications of the EPOS data arbitrarily and should therefore be disregarded in assessing the experimental situation.

In short, the APEX experiment provides independent corroborative evidence for a sharp (e^+e^-) pair line near 800 keV. Furthermore, the background RL(1,1) pair distributions of the APEX and EPOS experiments are roughly comparable in size and shape. In that context, the rough similarity also of their sharp pair counts can be considered as merely another aspect of an overall consistency between their distributions. Thus by purely empirical measures, APEX' and EPOS' pair results agree, although APEX' RL(1,1) pair counts are $\sim$3X fewer than expected from EPOS' and the luminosities for reasons not yet understood. In the end there is no purely empirical basis in the APEX experiment to support the claim that it contradicts the EPOS' sharp pair results.

Acknowledgement

The author is grateful to Drs. Thomas Cowan, Mark Wolanski, K.C. Felix Chan, Frank Wolfs, and Thomas Trainor for their help in the course of this work. The support of the U. S. Department of Energy under grant no. DE-FG02-93ER-40762 is gratefully acknowledged.

REFERENCES

1.   I. Ahmad *et al.*, *Phys.Rev.Lett.* **75**, 2658 (1995).
2.   T. E. Cowan *et al.*, *Phys.Rev.Lett.* **56**, 444 (1986).
3.   T. E. Cowan *et al.*, in *Physics of Strong Fields*, edited by W. Greiner (Plenum Press, New York, 1987), p. 111.
4.   T. E. Cowan, Ph.D. thesis, Yale U., 1988.
5.   P. Salabura *et al.*, *Phys.Lett.* **B245**, 153 (1990).
6.   E. Berdermann *et al.*, *Nucl.Phys.* **A488**, 683c (1988).
7.   W. Koenig *et al.*, *Phys.Lett.* **B218**, 12 (1989).
8.   I. Koenig *et al.*, *Z.Phys.* **A346**, 153 (1993).
9.   J. J. Griffin, u. of MD. PP #97-080 (to be published).
10. M. R. Wolanski, Ph.D. thesis, University of Chicago, August, 1995.
11. R. Ganz *et al.*, *Phys.Lett.* **B389**, 4 (1996).
12. T. E. Cowan and J. Greenberg, *Phys.Rev.Lett.* **77**, 2838 (1996).

MULTIPLICITY DISTRIBUTIONS IN Au + Au COLLISIONS AT VARIOUS AGS BEAM ENERGIES

James Chang[1] for the E866[2]/E917[3] Collaborations

[1]Physics Department, University of California
Riverside, CA 92521
[2]BNL-UCBerkeley-UCRiverside-Columbia-INS(Tokyo)-Kyoto-LLNL-
Maryland-MIT-Tokyo-Tsukuba-Yonsei
[3]ANL-BNL-UCRiverside-Columbia-UIChicago-Maryland-MIT-Rochester-
Yonsei

INTRODUCTION

We report the multiplicity distributions from the collision of Au + Au at the AGS experiments E866 and E917. During the 1995 run, data was taken for beam energies of 2, 4, and 11 GeV/nucleon (GeV/n). The 1996 run was composed of 6, 8, and more of 11 GeV/n beam,* thus giving us an opportunity to study the nuclear dynamics of the same colliding system at various energies. We were able to obtain the multiplicity distributions from the E866/E917's global detector New Multiplicity Array (NMA) almost immediately, since it requires very little overhead to generate physics results. Here we share some of these results. Comparisons to models are not done here, mostly due to lack of time between data-taking and reporting of the results.

THE EXPERIMENT AND THE NMA DETECTOR

The experiments E866 and E917 are third and fourth generation continuations of the original E802 experiment. At the heart of E802/859/866/917 is the Henry Higgins Spectrometer. The E866 saw the introduction of a second spectrometer arm, called the Forward Spectrometer. This narrow-angle spectrometer allows particle identifications at much more forward angles than the Henry Higgins, thus allowing a wider rapidity coverage for particle spectra.

---

*The actual beam energies, according to AGS, are 2.94, 4.97, 6.91, 8.91, and 11.75 GeV/n. These beam energies are loosely referred to as 2, 4, 6, 8, and 11 GeV/n.

Advances in Nuclear Dynamics 3, edited by
Bauer and Mignerey, Plenum Press, New York, 1997

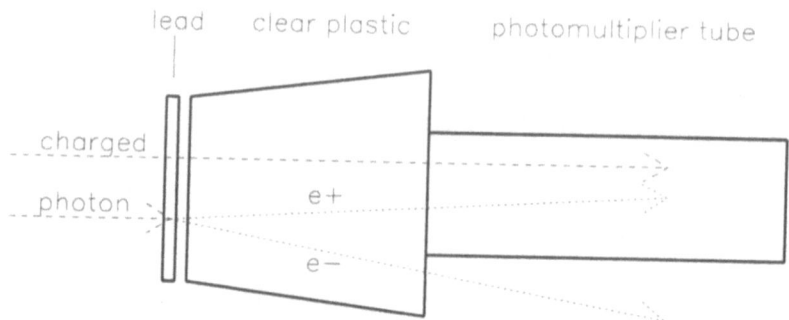

Figure 1. A typical NMA module. Charged hadrons traverse the 5 cm thick plastic, yielding total of about 400 photoelectrons from the Cerenkov radiation. The lead thickness is such that 25% of photons get converted to e^+e^- pairs. ADC is read out from the photomultiplier tube.

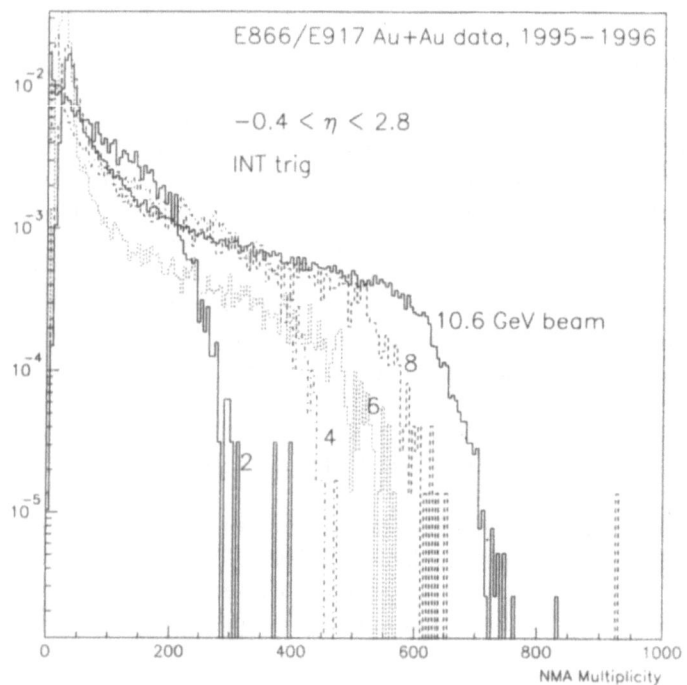

Figure 2. Detected multiplicity distributions for various beam energies.

The experiment has two major global detectors to characterize the collisions. First is the Zero-degree Calorimeter (ZCAL) that measrures the projectile fragment energy. The measured energy is translated into the projectile fragment size, thus giving information about the geometric centrality of the collision. The second detector measures the multiplicity of particles being emitted from the collision. The E802/859 used the Target Multiplicity Array (TMA) to detect any charged particles. The use of gold beam for E866, however, increased the multiplicity of particles to a level that TMA could not handle efficiently. Thus, the New Multiplicity Array (NMA) was designed and built for the Au + Au collisions being studied in E866.

The NMA is an array of 346 Cerenkov-light detecting modules. A typical module is shown in figure 1. Each module is composed of a clear plastic mated directly to a photomultiplier tube. Since the index of refraction of the plastic is about 1.5, the Cerenkov light begins to turn on at $\beta = v/c = 0.67$. This feature allows the detector to be sensitive mostly to fast charged particles, bulk of which are the produced charged pions. Furthermore, we wanted the detector to be sensitive to the produced neutral pions as well. We achived this by placing a sheet of lead in front of the clear plastic. The photons from the decayed π^os would then convert to e^+e^- pairs in the lead. The thickness of lead was chosen so that the conversion probability is 25%. This yields, on the average, one particle response for each π^o, thus putting π^o roughly on equal footing with the charged pion. The lead sheet does not affect the charged hadrons much, since the nuclear interaction length in lead is much longer. The added benefit of having a lead sheet in front of the module is that essentially all of the low-energy δ-electrons from the target are blocked out.

The modules are arranged in close-packed fashion to cover the polar angle θ from 7.0^o to 112.3^o, with the beam defining the Z-axis. Instead of using the variable θ, we usually use the variable *pseudorapidity* $\eta = -ln(tan(\theta/2))$. For relativistic particles ($\beta \approx 1$), $\eta \approx y$, the rapidity variable. In our case, particle identification is not possible, so we resort to using η The array is segmented in such a way that the η-bite $\Delta \eta$ is some convenient number, most of them being 0.20. The array covers the azimuth angle ϕ, with gaps left open for free passage of particles from the target to the spectrometers. The array is also segmented in ϕ, with most of the modules having a ϕ-bite $\Delta \phi$ of 10°.

The response of each module is such that it can distinguish a single hit, double hits, etc. One then adds up hits from all the modules to yield a raw multiplicity. The multiplicities reported in this paper are corrected to account for the gaps in ϕ. Since one cannot extrapolate beyond the η-coverage of the detector, the multiplicities reported are for the range of $-0.4 < \eta < 2.8$. The segmentations in η and ϕ allow us to obtain the $dN/d\eta$ and $dN/d\phi$ (reaction plane) on an event-by-event basis.

MULTIPLICITY DISTRIBUTIONS

Figure 2 shows the multiplicity distributions for minimum-bias interaction (hardware) triggers. Again, the multiplicities reported here are corrected for the gaps in ϕ and is restricted in the range $-0.4 < \eta < 2.8$. One can see that the shape of the distributions is not universal. This may be attributed to the slight variations in the trigger conditions during data-taking. For this report, however, we are only concerned about the high-multiplicity end of the spectrum, so these variations do not matter. In fact, one can look at the SPEC-trigger (at least one track in the spectrometer) or any other higher level triggers, and the high-multiplicity end of the distribution does not change.

Table 1. Maximum detected multiplicity values at various beam energies

Beam energy (GeV/n)	Maximum multiplicity	Detector resolution (%)
2.94	282	8.9
4.97	438	8.0
6.91	540	7.0
8.91	616	7.0
11.74	722	6.0

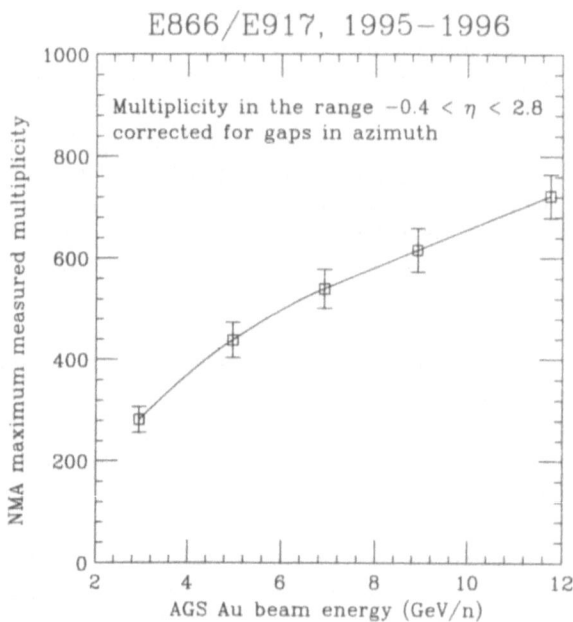

Figure 3. Detected maximum multiplicities at various beam energies.

Having established the stability of high end of the distribution, we judiciously decide what the maximum multiplicites are. These values are, for each of the beam energies, listed in table 1 and plotted in figure 3.

As one can see, the maximum measured multiplicity is basically linear as a function of beam energy. The error bars represent the overall detector resolution.

dN/dη

We now present the *dN/dη* distribution for *central* collisions. The term *central* simply means events which have multiplicities greater than 66% of the maximum multiplicity. The reason for selecting on NMA for centrality is that the ZCAL detector does not give an accurate response to beam with lower energies. The traditional geometric centrality of 4% $\sigma_{INTERACTION}$ for E866/917 full energy beam corresponds to ZCAL's measured energy of 350 GeV. This, in turn, corresponds to 66% of NMA's maximum value (for the full energy beam).

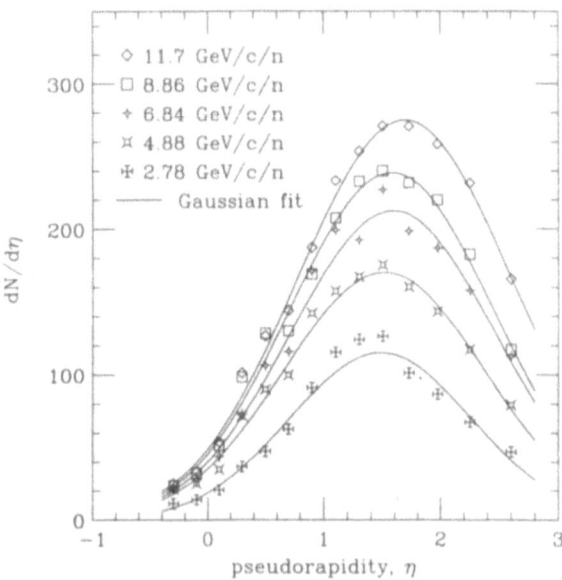

Figure 4. $dN/d\eta$ distributions for various beam momenta. Note that corresponding beam energies are 2.94, 4.97, 6.91, 8.91, and 11.75 GeV/n.

Figure 4 shows the $dN/d\eta$ distributions for various beam energies. A Gaussian fits are also included to get estimates of peak locations and relative widths. The error bars have been left out, since the statistical errors are negligible, and the systematic errors have not been worked out completely. Figure 5 shows a plot of fitted Gaussian peak location η_o as a function of beam energy. Also plotted are the mid-rapidity values of the symmetric colliding system at the given beam energies. Note that it is not strictly correct to compare η to rapidity. The divergence of η_o with mid-rapidity as beam energy decreases shows that more of the particles at lower beam energies have lower βs, thus skewing the η distribution from the rapidity distribution. Figure 6 shows a plot of relative widths of the $dN/d\eta$ as a function of beam energy. Not surprisingly, it is uniform.

$dN/d\phi$

While the multiplicity determination is the primary purpose of the NMA, its segmentation in ϕ gives us an opportunity to measure the reaction plane on an event-by-event basis. As stated in the detector description, most of the modules in the array had $\Delta\phi$ of 10°. There are two main problems with the NMA setup for reaction plane measurement. First is the presence of gaps in ϕ for the two spectrometers. This reduces the efficiency of reaction plane determination. Second is the NMAs sensitivity to the photons from the decaying π^os. These photons will, to some degree, wash out the reaction plane signal. Despite these two obstacles, simulations have shown us that we should be able to observe reaction planes, especially in non-central collisions.

The resolution of NMA for reaction plane is estimated to be around 40°, comparable to the HODOSCOPE detector in the same experiment. HODOSCOPE is a detector

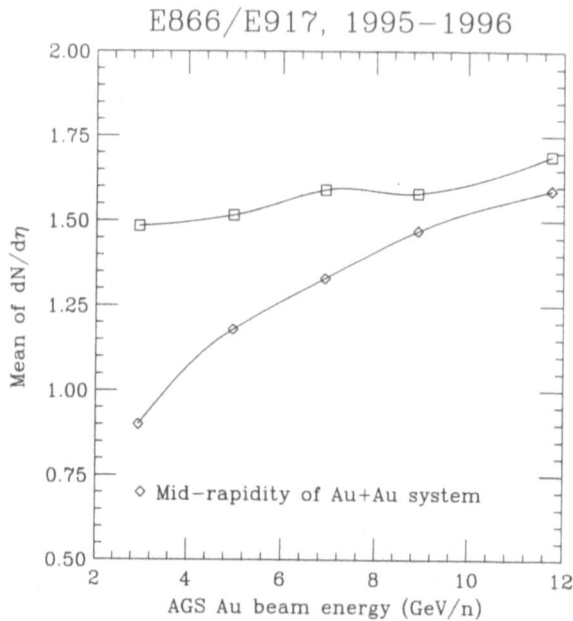

Figure 5. Mean of $dN/d\eta$ distribution as a function of beam energy.

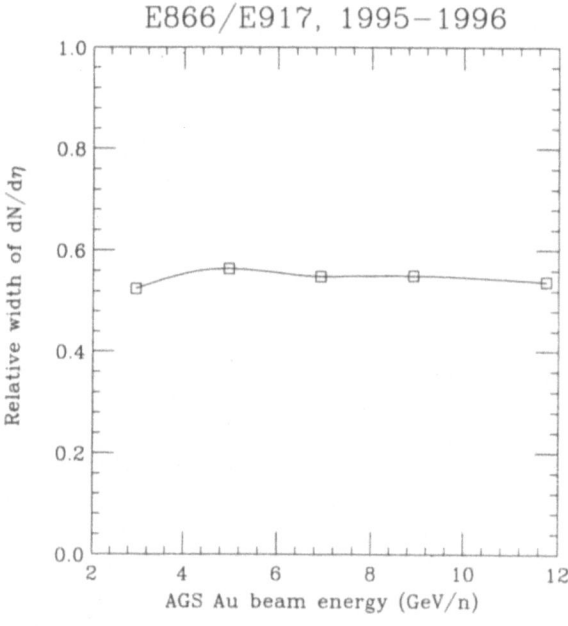

Figure 6. Relative width of $dN/d\eta$ distribution as a function of beam energy.

that measures the protons' horizontal and vertical positions at about 10 m downstream from the target. The HODOSCOPE *does* see a reaction plane, while the NMA does *not* see a reaction plane. Keeping in mind that the NMA is mostly sensitive to pions, one way to interpret this null result is to say that, at least at the level of NMA's resolution, pions are emitted isotropically.

SUMMARY

We presented results of the recent Au + Au data at the AGS, and summarize the report as:

1. The number of particles increases linearly as the beam energy increases.

2. The pseudorapidity distributions of particles show that the mean increases just slightly (linearly), and the relative width stays uniform, as the beam energy increases.

3. Reaction plane is not observed by the NMA, even though it has sufficient resolution Given the fact that NMA is mostly sensitive to pions, we interpret this result as isotropic emission of pions at the NMA's level of sensitivity.

FLOW MEASUREMENTS OF $Au + Au$ AT 10.8 GeV/c

Wen-Chen Chang and Stephen C. Johnson
for the E877 Collaboration

Physics Department
The State University of New York at Stony Brook
Stony Brook, NY, 11794

Measurements of collective flow effects in $Au + Au$ collisions at the AGS utilizing the E877 apparatus are presented. Both transverse energy and multiplicity distributions show anisotropies with respect to the reaction plane. Calorimetric measures of transverse energy coupled with charged particle multiplicity distributions lead to a separation of the nucleon and pion flow. Further, the azimuthal anisotropy of identified particles with respect to the reaction plane is measured utilizing the forward spectrometer. The resulting slope of the $\langle p_x \rangle$ for nucleons is comparable to those at Bevalac energies while the pions display a small flow signal opposite to the nucleons. The second Fourier moment is also measured and is found to be directed opposite to measurements at lower energies.

INTRODUCTION

Collective flow effects in heavy ion collisions have been an area of study and interest because they may provide insight into the integrated evolution of the source distribution. Theoretical studies have pointed to a minimum in the mean transverse momentum in the direction of the reaction plane ($\langle p_x \rangle$) as a measure of the softest point of the equation of state of nuclear matter, a signature of a phase-transition into a quark-gluon plasma. While general agreement abounds that a minimum in this variable could point to this softest point, the precise beam energy which should produce this minimum in the directed flow is theoretically unclear, ranging from AGS [1] to RHIC energies.

In what follows we will present measurements of the azimuthal distribution of global variables and indentified particles; these measurements include both nucleon and pion flow distributions and the triple differential cross-section. The second Fourier moment, refered to as elliptical flow, will also be shown.

FOURIER ANALYSIS

The E877 apparatus for the '93 and '94 runs of $Au + Au$ collisions has been described elsewhere [2]. The primary detectors for the measurement of flow are two calorimeters, the TCAL and PCAL, and a multiplicity detector directly in front of the PCAL. Together, these detectors have near 4π coverage. At Bevalac energies, most particles from the collision are identified and the reaction plane resolution is determined by dividing one event into two samples and determining the reaction plane from each sample. E877 does not have this method available, since we identify particles only within a small region of phase space. Instead, the method we adopt [7, 4] utilizes four windows in pseudorapidity, each with complete azimuthal coverage, to determine the reaction plane resolution: the PCAL is subdivided into three windows and the TCAL is used for a fourth window. Determining the reaction plane in these windows is mathematically sufficient for determining the resolution of each window. All data in this paper, unless otherwise noted, are corrected for this reaction plane resolution which would otherwise tend to decrease the true signal. The forward spectrometer includes two drift chambers, four multiwire proportional counters and a time-of-flight wall which collectively identify particles near beam rapidity and at low p_T.

Studies of flow progress via the mathematical tool of Fourier analysis. [6] In principle, any azimuthal global distribution G with respect to the reaction plane (ϕ) can be expanded as a Fourier series:

$$\frac{dG}{d\phi} = \frac{|G|}{2\pi} Re(\sum_{n=0}^{\infty} \nu_n e^{in\phi}) \tag{1}$$

where ϕ is the azimuthal angle, such that the Fourier coefficient can be determined from measurements on an event-by-event basis as:

$$\nu_n = |\nu_n| e^{-in\Psi_n} \tag{2}$$

where Ψ_n is the orientation of the multipole n and $|\nu_n|$ is its dimensionless strength. In practice this procedure may be utilized to extract Fourier moments from calorimetric or multiplicity measurements. For example, the transverse energy distribution may be expressed as

$$E_T = \sum_{i=1}^{N} E_i \sin(\theta_i) \tag{3}$$

such that

$$E_x = \sum_{i=1}^{N} E_i \sin(\theta_i) \cos(\phi_i) \tag{4}$$

$$E_y = \sum_{i=1}^{N} E_i \sin(\theta_i) \sin(\phi_i) \tag{5}$$

and

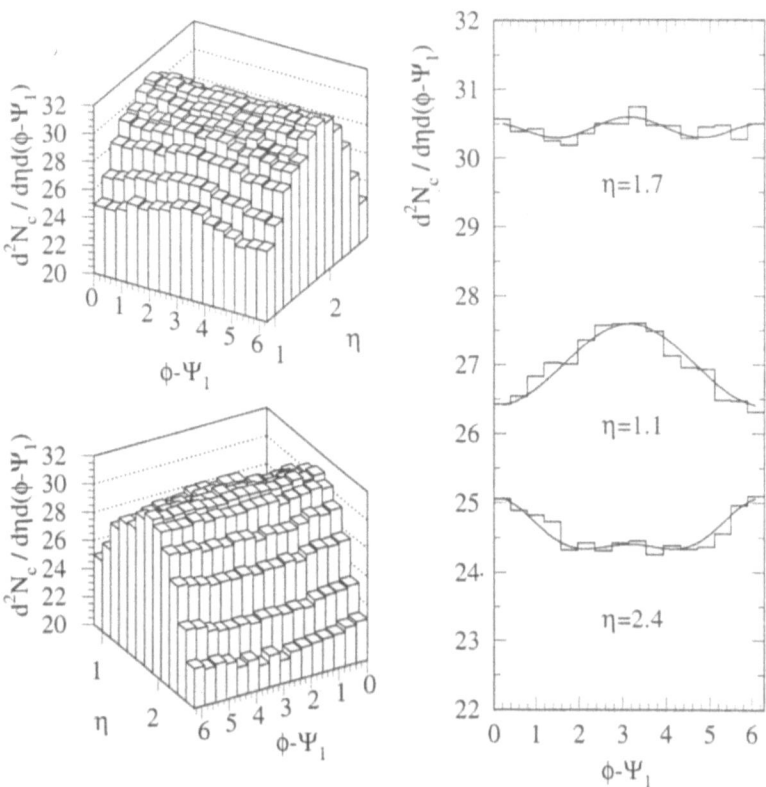

Figure 1. The charged particle multiplicity distribution as a function of pseudo-rapidity and azimuthal angle with respect to the reaction plane. The two plots on the left are identical but viewed from different directions. The inset is a selection of slices of these plots in pseudo-rapidity. Figure taken from [4,7].

$$\Psi_1 = \tan^{-1}\frac{E_y}{E_x} \qquad (6)$$

This procedure is performed by determining the reaction plane using one of the windows (3 in PCAL and 1 in TCAL) and quantifying the anisotropy of the global variables of transverse energy (E_T) and multiplicity (N_{ch}) in another detector in order to avoid auto-correlations [7]. As an example, the event shape of the N_{ch} distribution is shown in Fig. 1. Note that at forward rapidities the N_{ch} distribution is peaked at $\phi - \Psi_1 = 0$, that is in the reaction plane, while at backward rapidities the distribution is peaked at π. Representative slices of this distribution are displayed in the inset. At forward and backward pseudo-rapidities the distribution shows a clear first Fourier component, though of opposite sign, while at mid-pseudo-rapidities the first moment passes through zero; a clear second moment is seen at mid-pseudo-rapidity.

Shown in Fig. 2 are the results for the extraction of the first two Fourier coefficients from the transvere Energy E_T as a function of centrality. In the center plot one can clearly see the "S-shaped" distribution as a function of η. This first component disappears for central and peripheral events, as is required by symmetry, and reaches a maximum for mid-central events. The second moment reaches a constant positive

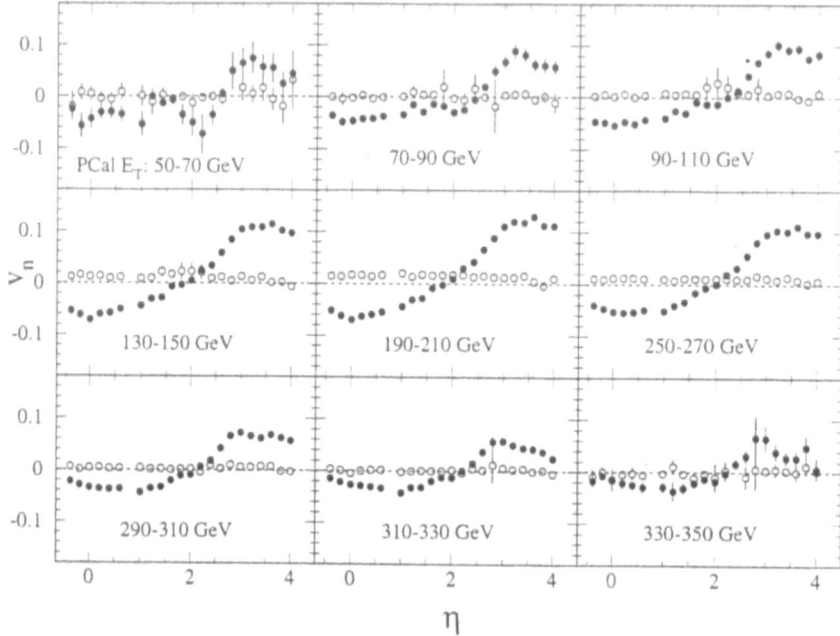

Figure 2. Fourier coefficients measured with the calorimeter (E_T) as a function of pseudo-rapidity and centrality. Plot from [4].

value for all η. The N_{ch} distribution shows the same behavior over a smaller range of η.

The components for the first Fourier coefficients as measured by N_{ch} and E_T are different due to the properties of our measuring devices: the calorimeter is sensitive to all nucleons and pions while the multiplicity detector is sensitive to only charged pions and less than 1/2 of all nucleons, the protons. This difference can be utilized to disentangle the nucleon flow from the pion flow as described in [4, 5, 7].

Triple Differential Multiplicity and $\langle p_x \rangle$

Utilizing the definition of the reaction plane from the calorimeters and the identification of charged particles from the forward spectrometer, triple differential multiplicity distributions can be determined. The preliminary measured triple differential cross-sections with respect to the reaction plane have been shown in [4] for protons and pions. The proton distribution was shown to be shifted towards positive p_x, into the reaction plane, while the pions display a barely noticeable shift to negative p_x.

From the triple differential cross section we may now glean the $\langle p_x \rangle$ from the distribution. Because E877 is not a full acceptance experiment, we cannot determine the mean transverse momentum simply from the distribution; in order to do so we need to extrapolate our results to larger p_t. Therefore, we determine the $\langle p_x \rangle$ by fitting our distribution to a reasonable parameterization and extrapolating $\langle p_x \rangle$ from the fit. In this analysis we used two parameterizations: (1) a distribution with a common temperature offset from zero

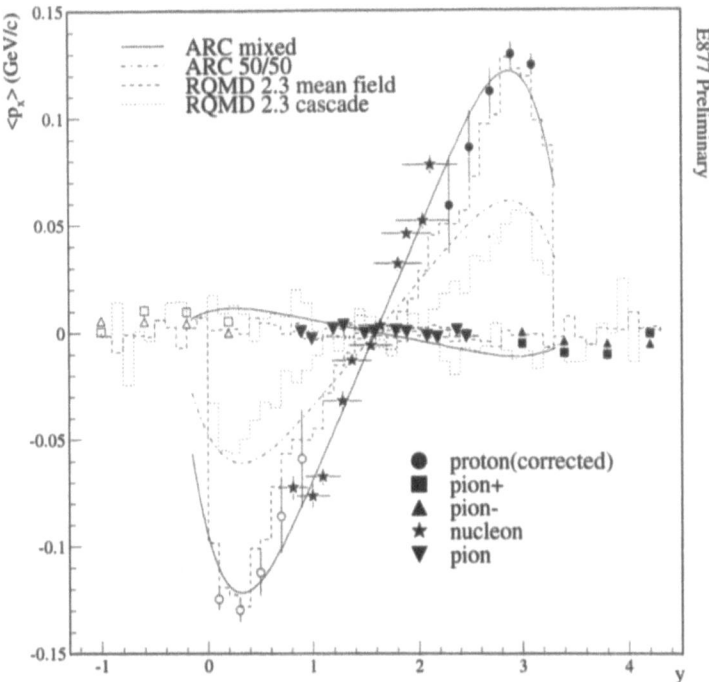

Figure 3. $\langle p_x \rangle$ as a function of rapidity for protons and pions as well as model predictions. See text for details. Data taken from [4].

$$d^2 N/dp_x dp_y \propto m_t \exp(-\frac{\sqrt{(p_x - \langle p_x \rangle)^2 + p_y^2 + m^2}}{T_B}) \qquad (7)$$

and (2) a transversely boosted thermal source withoug radial expansion

$$d^2 N/dp_x dp_y \propto E^* \exp(-E^*/T) \qquad (8)$$

where $E^* = \gamma E - p_t \beta_x \gamma \cos(\phi - \Psi_1)$ and β_x is the transverse flow velocity.

E877 has coverage to mid-rapidity for protons with only a small range in p_T; therefore, at some point our fits should fail due to insufficient coverage. This point is roughly determined at that point where the two parameterizations disagree on the $\langle p_x \rangle$; all points shown are those where the parametrizations agree on the extracted $\langle p_x \rangle$. These points for protons and pions are shown in Fig. 3 along with predictions of model calculations. Note that both ARC and RQMD codes in cascade mode cannot describe the data whereas with modifications they can both describe the data. RQMD chose a method of a mean-field [10] and ARC describes the data by utilizing repulsive trajectories below $E = 2m_\pi$, a 50% mix of repulsive and attractive trajectories above $E = 3m_\pi$ and a smooth variation in between [11].

Excitation Functions

Shown in Fig. 4 is a comparison of our measured flow signal [7] with those of Bevalac [9]. Here flow is defined as the slope of the $\langle p_x \rangle/A$ curve versus normalized rapidity:

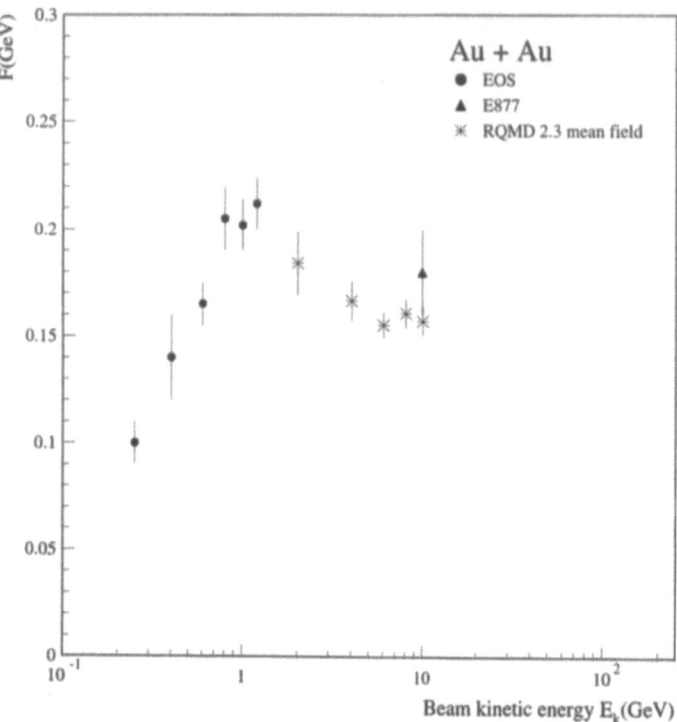

Figure 4. The slope of the $\langle p_x \rangle$ versus y/y_{beam} for a variety of beam energies and systems. Data from [7,9].

$$F \equiv \frac{d\langle p_x/A \rangle}{d(y/(y_{bm}))}\big|_{y=y_{cm}}. \qquad (9)$$

The curve shows a saturation at a beam energy of 1GeV with a possible weakening of the defined flow parameter as we approach AGS energies. This dropping of the measured flow signal may be related to a softening of the equation of state, possible evidence for the existence of a QGP. Further studies of collective motion around this region of beam energy should help clarify the possible existence of this softest point.

In Fig. 5 is a similar excitation function for the measured second Fourier moment for protons at mid-rapidity with a comparison to Plastic-Ball results at lower energies [12] and NA49 at higher energies [13]. Note that the measured value of the second moment at AGS energies is opposite in sign to that at lower energies. This effect has been described by the author of [3] as a consequence of the existence of spectators at lower energies which result in a "squeeze-out" effect while at higher energies the spectators move more quickly past the interaction zone and the source region is free to expand naturally. The NA49 collaboration has measured the second moment of the transverse energy at mid-pseudo-rapidity [13]; however, the orientation with respect to the reaction plane is currently unknown and therefore so is the sign of the second moment.

Conclusions

We have shown measurements of collective flow in $Au + Au$ collisions with the E877 apparatus. The $\langle p_x \rangle$ distribution is consistent with RQMD results with mean-

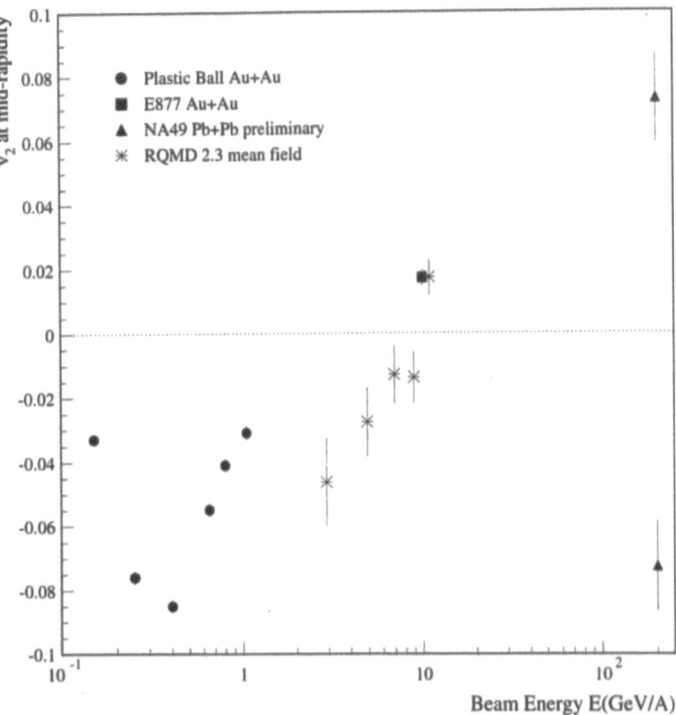

Figure 5. The second Fourier moment as a function of beam energy. NA49 [13] results do not include a determination of the reaction plane. Therefore the sign of the second moment is unknown.

fields and ARC with repulsive trajectories. The results at AGS are comparable with those at Bevalac, displaying a saturation or slight decrease of the measured $\langle p_x \rangle$, though the second moment changes sign compared to low energy measurements resulting in an in-plane expansion.

REFERENCES

1. D.H. Rischke, nucl-th/9608024; Heavy Ion Physics 1 (1995) 309.
2. J. Barrette *et al.*, E877 Collaboration, Nucl. Phys. **A590**,(1995) 259c and references therein.
3. H. Sorge, nucl-th/9610026 and these proceedings.
4. T.K. Hemmick, for the E877 collaboration, Nucl. Phys. A610, 63c (1996).
5. W.C. Chang, thesis, University at Stony Brook.
6. S. Voloshin, Y. Zhang, Z. Phys. **C70**, 665 (1996)
7. J. Barette et.al., Phys. Rev. **C 55**, in print.
8. S.A. Voloshin, nucl-th/9611038; Phys. Rev **C55** in print.
9. M.D. Pantlan, et al., Phys. Rev. Lett. **75**, 2100 (1995)
10. H. Stöcker, and W. Greiner, Phys. Rev. **137**, 278 (1986).
11. D.E. Kahana, HIPAGS '96 proceedings and references therein.
12. H.H. Gutbrod et.al., Phys. Rev. C **42**, 640 (1990).
13. T. Wienold, for the NA49 collaboration, Nucl. Phys. A610, 76c (1996).

Acknowledgments

Financial support from the US DoE, the NSF, the Canadian NSERC, and CNPq Brazil are gratefully acknowledged.

The Solenoidal Tracker at RHIC

W.J. Llope for the STAR Collaboration *

T.W. Bonner Nuclear Laboratory
Rice University, Houston, TX 77005-1892

INTRODUCTION

The Relativistic Heavy-Ion Collider[2] (RHIC) will provide glimpses of the universe some microseconds after the Big Bang in ^{197}Au+^{197}Au collisions at beam energies up to 100 GeV/nucleon/beam ($\sqrt{s}\sim$39 TeV). RHIC will also become the world's highest energy polarized proton collider when it delivers up to 250 GeV $\bar{p}$ beams. Detailed studies of parton distribution functions will be possible, including for the first time the evaluation of the gluon contribution to the nucleon spin in a region where the theory is perturbative. This more complete information on the parton distributions will be applied to understand the data from RHIC's p+A and A+A programs, where the ion mass, A, can be anything. Deviations of the character of RHIC's central ^{197}Au+^{197}Au collisions from purely hadronic interpretations will then be apparent. These deviations may signal the formation of the Quark-Gluon Plasma - a (predicted) state of matter that has never before been studied in the laboratory.

The construction of RHIC is well underway.[3] All of the RHIC dipoles and $\sim$3/4 of the quadrupoles are installed in the tunnel. The dipoles work beautifully, quenching at currents $\sim$30-50% larger than required. As the world watched this year's Super Bowl, ^{197}Au beams were successfully transported[4] all the way out to the end of the first sextant (i.e. to the RHIC 4 o'clock position), marking a major milestone in the RHIC construction project.

The first experiment approved for construction is the Solenoidal Tracker at RHIC[5] (STAR), which is shown in Figure (1). STAR is a large azimuthally symmetric apparatus designed for measurements of a variety of hadronic observables in a large acceptance. Central ^{197}Au+^{197}Au events at RHIC can result in several thousand particles in STAR's acceptance, allowing the measurement of quantities such as the source temperature, size, energy density, entropy, strangeness, and so on, on an *event by event* basis. This allows the selection of particular classes of interesting events for more detailed analyses, which may include searches for the Quark Gluon Plasma.

---

*An international collaboration of $\sim$400 members.[1]

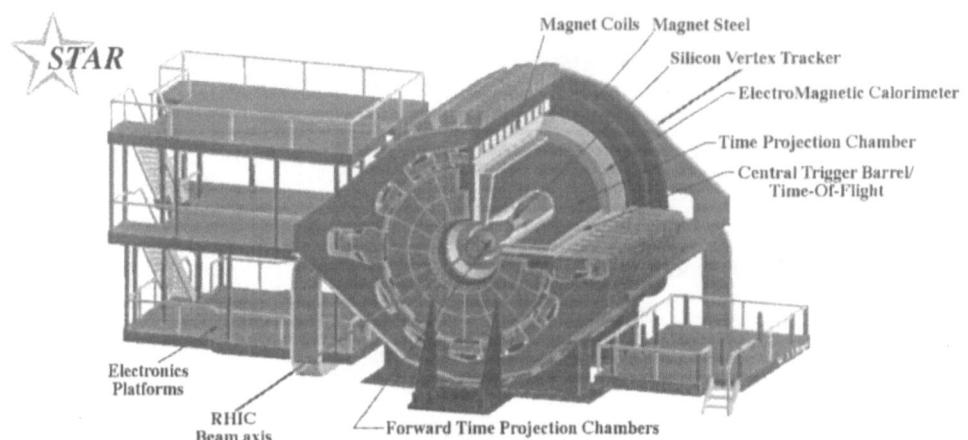

Figure 1. A cut-away view of the STAR apparatus showing both the Baseline and the AEE detector systems, as well as the various electronics platforms.

In this contribution, the status of the STAR project will be described. As talks on STAR have been presented[6] in several of the previous conferences in this series, this contribution will concentrate on the hardware progress made in the last year. This amounts to several hundred man-years of effort. The status described below is as of December, 1996, unless noted otherwise.

The STAR sub-systems are divided into two categories by the funding profile. The "Baseline" systems are presently under construction, while the Additional Experimental Equipment (AEE) systems are all in advanced stages of design and prototyping. The status and present schedule to completion for the Baseline systems are described in the following section. A discussion of the continually evolving System Test of the STAR Baseline systems is then presented, and followed by the status of the AEE efforts. The closing section describes the expected configuration of STAR and aspects of the beams available in the thirty-seven week first run of RHIC starting in the Fall of 1999.

STATUS OF THE STAR BASELINE

The Baseline components of STAR consist of the solenoidal magnet, the Time Projection Chamber (TPC), the TPC Front-End Electronics (FEE), the Central Trigger Barrel (CTB), as well as the Trigger and Data AcQuisition (DAQ) systems. These systems are under construction.

The STAR magnet is a 0.5 Tesla (maximum), 3.4 MW, warm coil solenoid in the shape of a 6.2 m long cylinder with an inner diameter of 5.26 m. The magnet endcaps will be removable, allowing the servicing of the inner detectors in the experimental hall. The floor of the experimental hall sags by 3/8 inches when STAR is in position, so 1/2" of travel in the vertical position of this 1200 ton magnet is possible by design. Space-trim and pole-tip coils will be used to correct for the distortions in the solenoidal field. The magnet is required to have radial components no larger than $\int [B_r / B_z] dz < 0.7$ cm, which corresponds to distortions of approximately one part in a thousand.

Figure 2. The STAR solenoidal magnet as of December, 1996. One of the two end rings is being test-fit onto the first nine backlegs, which have already been attached to the cradle. The next six backleg pieces to be installed are visible on the lower left.

Integrated radial distortions as low as 0.3 cm may be possible.

As shown in Figure 2, the first nine of the lower steel backlegs are in place. The installation of the two end rings and the next six backleg pieces is underway, and will complete the lower half of the magnet steel. The main and trim coils for the lower half will then be installed, and then the remaining 15 backlegs, coils, and the pole-tip steel will be installed. The magnet construction project is on schedule for acceptance testing in the middle of 1998.

The heart of STAR is the large cylindrical Time Projection Chamber. It is designed to track and momentum reconstruct up to 4000 particles with pseudo-rapidities in the range $|\eta| < 1.8$. To do this, each end of the TPC is segmented into 68,000 pads that are arranged in 24 sectors (12 inner and 12 outer). Each pad is sampled 512 times, which forms the equivalent of 70 million three-dimensional pixels in the $\sim 50 \text{m}^3$ active volume of the TPC. Detailed simulations imply that the momentum resolution, $\Delta p/p$, will be about 1.3% for momenta of a few GeV/c or less, rising to $\sim 3\%$ for momenta near 10 GeV/c. The device also provides direct particle identification information - the truncated-mean dE/dx resolution for full-length tracks is expected to be $\sim 6.7\%$.

At present, the twenty-four inner sectors plus two spares, and the twenty-four outer sectors plus two spares, have all been constructed and tested. The outer field cage has been inserted in the gas vessel. The central membrane and both of the end wheels have also been installed. The field cage high voltage system is complete. The construction of the inner field cage is underway. The TPC will be transported to Brookhaven in October, 1997, most probably in a large C-5A military transport aircraft.

The signals from the 136,000 TPC pads are amplified, shaped, and waveform-digitized by custom Front-End Electronics boards. This digital data is sent to the STAR data acquisition system over 1.5 Gbit/s optical fibers via so-called Readout Cards, which also contain the clock and trigger fan-outs, the event memory, and the slow controls interfaces. Four percent of the FEE boards are presently being fabricated. These will be used to instrument an entire sector in the System Test beginning in April, 1997. Full-scale production of the FEE boards will begin in June, 1997, and this effort

Figure 3. The outer field cage of the TPC (left frame), the interior of the gas vessel lined up for insertion onto the field cage (middle frame), and the two wheels (right frame). In this image, the central membrane is being stored on the nearer wheel.

will last for about 21 months. The first prototype of the readout card is being used in the System Test, while version two is in the final stages of layout. Tests of the final FEE and readout boards on the full TPC will begin in June, 1998.

The STAR Data AcQuisition system is sector-based. It receives digital data from the detectors over optical fibers at a rate of 100 Hz (7 GB/s). Custom integrated circuits and Intel i960 CPU chips are used to preprocess the TPC data,[†] calculate space points from the pixel information, and then apply fast tracking algorithms to these space points sector-by-sector. This track information is used in the third level trigger (see below) to select the most interesting events for archival to tape at a rate of 1 Hz (~12 MB/s). Other scenarios are possible within the DAQ architecture: instead of archiving waveform-digitized ADC data to tape, one could replace this information by calculated space points, or reconstructed track parameters. Physics programs other than central ^{197}Au$+^{197}$Au, such as $\vec{p}+\vec{p}$, may require the building of calorimeter-only events at rates of up to 1 kHz.

As noted in Figure 4, prototype versions of the DAQ Receiver and Mezzanine cards

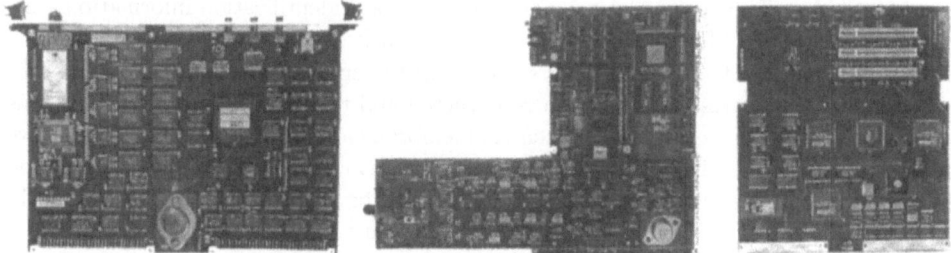

Figure 4. A few DAQ boards: "Rosebud", a programmable data source for fake raw data and sequencing (left image), "Rosie", a Receiver board plus Mezzanine board for use in the System Test (middle image), and "Cher", a prototype of the final receiver board (right image).

[†]This includes pedestal subtraction, zero suppression, ten bit to eight bit conversion, and the compilation of pointers to sequences on each TPC pad.

are being used in the System Test. The custom chip for the preprocessing of the TPC pixel data has been submitted to the foundry, and first versions have been returned for testing. Final production models of all of the DAQ hardware are expected to be completed and tested by late Summer, 1997.

Surrounding the entire active volume of the TPC are two segmented detector systems that provide fast information suitable for low-level triggering. The Central Trigger Barrel consists of 240 large scintillator slats mounted in 120 trays that are attached to the outside of the TPC. The anode wires covering the two ends of the TPC will also be read out, providing the equivalent of the signals from a standard Multi-Wire Chamber. The charged particle multiplicities in the range $|\eta|<1(1<|\eta|<2)$ are measured by the CTB(MWC), and used with various amounts of pixelization in the various levels of the STAR trigger. At present, ten CTB trays are under construction for use in the next round of the System Test beginning in May, 1997. As it's part of the TPC, the hardware for the MWC is already constructed, while the MWC electronics is being designed.

The STAR Trigger system consists of four detector systems and several custom electronics boards. The four trigger detectors are the CTB, the MWC (both described above), as well as the Vertex Position Detector (VPD), and the Veto Calorimeters (VTCs). The VPD consists of two small Pb/quartz ring detectors that surround the beam pipe in two locations at ± 3.6 m from the interaction region. These will be used to reject beam-gas events by locating the primary vertex to within ~ 1 cm. The two VTCs are positioned on the beam axis at about ± 15 m from the interaction region to measure the impact parameter via the number of spectator neutrons, and the location of the primary vertex to ~ 7 cm via timing. VTCs of exactly the same design will built for each RHIC experiment, providing a measure of the luminosities of ion beams in a consistent way, and hence allowing the comparison of the cross sections measured by the different RHIC experiments.

The trigger is arranged in several levels: Level-0 is an "accept" based on the coarsest CTB, MWC, and EMC information, while Levels-1 and -2 are "aborts" based on increasingly finer pixelizations of this information. The Level-3 trigger is performed in processors that are part of the DAQ system. The data processing and trigger decisions in Levels 0 , 1, and 2 are performed in custom electronics. Of these, the Data Storage and Manipulation board (DSM), the Trigger Control Unit (TCU), and the Trigger and Clock Distribution board (TCD), have been prototyped and are in use in the System Test. The production of the final Level-0 electronics will begin in October, 1997.

STAR SYSTEM TEST

The STAR System Test assembles STAR components at their current stage of development, and tests the functionality and interfaces in a reasonably realistic environment. The detectors and electronics used in this test will continue to evolve as the various components become more mature and/or more numerous. This effort is presently underway where the TPC is being constructed (Berkeley). It will move with the TPC to Brookhaven in October, 1997, and then continue until RHIC beams are available.

In the first round of tests,[7] the dip-angle response of an outer sector was measured using cosmic ray tracks and laser-induced tracks. The first fifteen available FEE cards

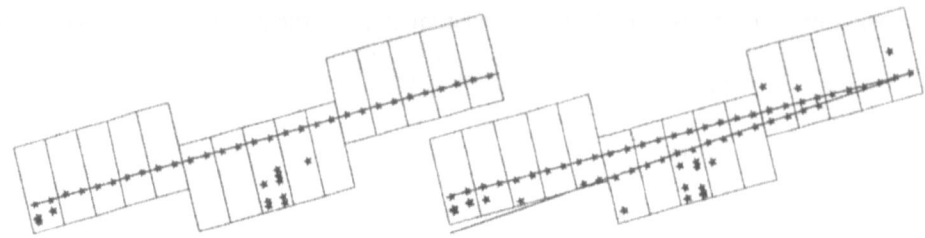

Figure 5. A cosmic ray (left image), and a reflected laser track (right image), as viewed by a TPC outer sector in the first round of the System Test.

and a prototype read-out board were attached to an outer sector. The triggers were provided by prototype components of the STAR Trigger, and system parameters are monitored by an EPICS-based slow controls system. Despite the fact the some of these first FEE boards were known to be flaky, the results obtained in the first round of tests were good. The hit residuals were measured[7] to be $\sim$390 and 780 μm in the ϕ and Z direction, respectively, which is close to the diffusion limits of $\sim$400 and 700 μm. By bouncing a laser pulse off of a mirror (*cf.* Figure 5), the two-track resolution was measured[7] to be $\sim$1cm, which is the value expected for outer sectors.

The components for the next round of this program are presently being assembled. This round will include one inner sector and one outer sector, 216 FEE boards, 6 read-out boards, 6 Rosie boards (*cf.* Figure 4), and 3 Cyclone boards (DAQ). Zero-suppression, gain corrections, and the interfaces to the Online monitoring system will be involved for the first time. Ten CTB trays (twenty scintillator slats) will be added to the set-up in May, 1997. The on-line system needed for the TPC field cage tests is to be available at this time as well.

STATUS OF THE AEE SYSTEMS

The "Additional Experimental Equipment" detectors include the Barrel Electro-Magnetic Calorimeter/Barrel Shower Maximum Detector (BEMC/BSMD), the Silicon Vertex Tracker (SVT), the Time-Of-Flight system (TOF), and the Forward Time Projection Chambers (FTPCs). These detectors are all in advanced stages of design and prototyping, and it is expected that at least patches of all of these detectors will be in place during the first year of RHIC running.

The EMC system consists a stack of 5 mm thick Pb plates and 5 mm thick plastic scintillator plates in the shape of an $\sim$18 X_0 deep barrel (the BEMC) and at least one $\sim$24 X_0 deep end-cap (the EEMCs). The barrel calorimeter is segmented into 4800 towers that subtend $(\Delta\eta, \Delta\phi)$=(0.05,0.05) and cover in total $|\eta|<1$ and 2π in azimuth. At a depth of approximately $\sim$4.5 X_0 in the stack, a wire/strip chamber Shower Maximum Detector (SMD) is embedded to provide the positions of electromagnetic showers to better than $\sim$1 cm, and information useful for electron/hadron discrimination. The EMC provides a number of fast triggers which can be obtained in no other way, and is crucial for the measurements of jets and direct-photons in both the $\vec{p}+\vec{p}$ and A+A programs.

In the last year, the design of the EMC and SMD has undergone a number of rather fundamental changes that significantly improve the performance of this system.

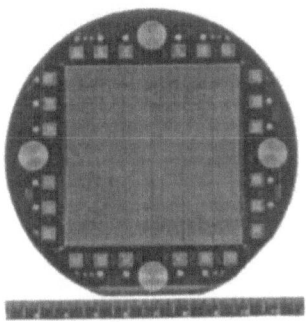

Figure 6. The final wafer of silicon drift detector (version 2.9) for the SVT (left image), and a full-scale wooden engineering model of the SVT (right image).

The optical read-out chain now involves either a "σ-groove" or an "α-groove", thicker scintillators, and more efficient multi-clad fiber, while the SMD segmentation in the η direction has been increased by a factor of five. The changes made to the design of the optical read-out chain improve the light output from the stack by roughly a factor of two, while the revised SMD design implies a SMD occupancy of only $\sim$23% even in central ^{197}Au$+^{197}$Au collisions. The EMC group is presently pushing to prototype the new aspects of the EMC and SMD designs for a in-beam test run at Brookhaven that is scheduled for May, 1997.

The SVT is a high resolution tracking and vertexing detector that immediately surrounds the beam-pipe and covers pseudo-rapidities $|\eta|<1$. It consists of 216 wafers of silicon drift detector arranged in three concentric barrels immediately surrounding the beam pipe with radii from 6 to 15 cm. Each drift detector has a two-dimensional position resolution on the order of 20×20 μm, and particle identification is possible via measurements of the particle energy loss. The SVT will be the major tool for the location of secondary vertices from the decay of strange and multi-strange particles such as the K_s, ϕ, Λ, Ω, and Ξ.

The performance of the final wafers of silicon drift detector (*cf.* Figure 6) is well understood, and the associated pre-amplifier plus shaping amplifier hybrid electronics have been prototyped. Presently underway are further tests of the hybrid and read-out electronics. The possibility for the commercial construction of the silicon wafers and the Beryllium support structures is also being explored.

The TOF system allows the direct particle identification of STAR global tracks by measuring the particle flight times with single-ended plastic scintillator slats. The TPC dE/dx information allows direct $\pi/K(K/p)$ separation for momenta less than $\sim$0.6(1.0) GeV/c. With the information of the TOF system, this range is greatly extended to $\sim$1.5(2.5) GeV/c, which significantly enhances the physics reach of STAR.

The TOF system is a barrel of 6000 scintillator slats which tile a cylinder immediately surrounding the TPC. The TOF system replaces the CTB, and many aspects of the two detectors are common, such as the trays, the cooling loops, and some parts of the electronics. Like the CTB, the Photo-Multiplier Tubes (PMTs) for the TOF system are inside the STAR solenoidal field, so Proximity Mesh Dynode PMTs are necessary. The cost of 6000 such PMTs from Hamamatsu is prohibitive - it is roughly equal to the *combined* cost of the two "small" RHIC experiments, PHOBOS and BRAHMS.

Figure 7. A prototype proximity mesh dynode photomultiplier tube.

However, with funds from the IPP[‡] and the CRDF[§], a cooperative effort was started to develop much lower-cost mesh dynode PMTs in Russia. This effort well underway, and approximately ten prototype mesh dynode PMTs have been delivered for testing. Of these, about half fully match the performance of the Hamamatsu PMTs at a fraction of the price. The focus at present concentrates on improving the acceptance to the level needed to produce thousands of these PMTs consistently.

The FTPCs are radial time projection chambers with microstrip gas chamber and pad read-out in the shape of two hollow barrels ($8 \leq R \leq 31$ cm and $150 \leq |Z| \leq 270$ cm) that surround the beam pipe. They are positioned inside the STAR magnet pole pieces. Simulations imply a momentum resolution of $\Delta p/p \sim 12\%$, and hence a good charge separation. The FTPCs cover pseudo-rapidities $2 \lesssim |\eta| \lesssim 4$ and transverse momenta below a few GeV/c - significantly increasing STAR's already large coverage to include more forward regions.

Two FTPC prototypes have been constructed, one uses a uniform drift field and microstrip read-out, while the other uses a radial drift field and microstrip gas chamber read-out. Tests using these prototypes have indicated that gas mixtures based on CO_2 or DME have a diffusions low enough to allow tracking in the FTPCs. The signal to noise ratio is on the order of 20:1, which makes feasible the use of the existing TPC electronics for the read-out of the FTPCs. The field cage for the FTPCs was designed and successfully tested.

TOWARDS DAY 1 OF RHIC.

After a RHIC test run and then a commissioning run, the first RHIC run for physics will begin in September 1999, and last until the summer shutdown 37 weeks later. The beams will be ^{197}Au at 100 GeV/c/nucleon each. The luminosity will increase during the course of the run, reaching $\sim 10\%$ of the design luminosity by week 37.[¶] Alternative beams and energies are possible in the first year, although lower energy ^{197}Au+^{197}Au collisions will be easier to tune (based on the initial 100 GeV ^{197}Au collisions) than proton, lighter A+A, and p+A collisions.[8]

[‡]The Institute for the Prevention of Proliferation, administered by the U.S. Department of Energy.
[§]The U.S. Civilian Research and Development Foundation for the Independent States of the Former Soviet Union, administered by the U.S. National Science Foundation.
[¶]The RHIC design luminosities are 2×10^{26} ions/cm^2/s for ^{197}Au beams and 1.4×10^{31} ions/cm^2/s for polarized proton beams.

On Day 1 of RHIC beam, STAR will consist of the solenoidal magnet, the fully instrumented TPC, the complete CTB, the MWC electronics, both VPDs, both VTCs, and patches of both the BEMC/BSMD and the TOF. The Trigger system will consist basically of just Levels 0 and 3; the more selective triggers possible in Levels 1 and 2 are not necessary given the relatively low luminosities that are expected in RHIC's first run. The SVT will be installed a few months into the first run, once the stability of the beam tune is assured.

At ~12 MB/s to tape, STAR will collect about one Terabyte of data per day of beam. A powerful computing infrastructure will therefore be necessary to store these data, and support the formation of both data summary tapes and physics results via analyses of these data. STAR recognizes the colossal amount of dependable software needed for these tasks. Recently, massive overhauls of both the STAR analysis shell and the STAR simulations framework were made. The new codes, called the STAR Analysis Shell (STAF) and AGI/GSTAR/G2T, respectively, are prepared to keep pace with STAR's growing needs for analyses of both experimental and simulated events.

The organization of the effort needed to be ready for the analysis of Day-1 data has started in STAR via the formation of seven "physics working groups" - entitled Event-by-Event, Strangeness, High-P_T, Spectra, Spin, Correlations, and Peripheral.[||] With the software developed by each of these groups complementing the precision hardware and electronics, STAR will be a powerful tool for studies of the wealth of RHIC physics on Day 1 and beyond.

ACKNOWLEDGEMENTS

Support from the U.S. Department of Energy under the Grant No. DE-FG03-93ER40772 is gratefully acknowledged.

REFERENCES

1. http://www.rhic.bnl.gov/star/starlib/doc/www/smd_1/organization.html
2. http://www.rhic.bnl.gov/
3. http://www.rhichome.bnl.gov/RHIC/
4. http://www.rhichome.bnl.gov/RHIC/Sextest/
5. http://www.rhic.bnl.gov/star/starlib/doc/www/star.html
6. J.N. Marx for the STAR Collaboration, Proceedings of the 12th Winter Workshop on Nuclear Dynamics, Snowbird, Utah, 1996, eds. W. Bauer and G.D. Westfall, Plenum Press (New York); J.N. Marx for the STAR Collaboration, Proceedings of the 10th Winter Workshop on Nuclear Dynamics, Snowbird, Utah, 1994, eds. J. Harris, A. Mignerey, and W. Bauer, World Scientific (Singapore); J.W. Harris, Proceedings of the 8th Winter Workshop on Nuclear Dynamics, Jackson Hole, Wyoming, 1992, eds. W. Bauer and B. Back, World Scientific (Singapore).
7. W. Betts *et al.*, presented at the IEEE 1996 Nuclear Science Symposium, Anaheim, California, Nov. 3-9, 1996, and submitted to the Conference Issue of the Transactions of Nuclear Science.
8. S. Peggs, RHIC Report RHIC/AP/115, Nov. 11, 1996.

[||]The convenors of these groups are I. Sakrejda, S. Margetis and K. Wilson, T. LeCompte, W.J. Llope, G. Eppley, J. Cramer, and S. Klein, respectively.

ISOSPIN DEPENDENCE OF NUCLEAR EQUATION OF STATE AND COLLISIONS OF NEUTRON-RICH NUCLEI

Bao-An Li[1], Che Ming Ko[1], Wolfgang Bauer[2] and Zhongzhou Ren[3]

[1]Cyclotron Institute and Department of Physics
Texas A&M University
College Station, TX 77843, USA
[2]National Superconducting Cyclotron Laboratory
and Department of Physics and Astronomy
Michigan State University
East Lansing, MI 48824, USA
[3]Department of Physics, Nanjing University
Nanjing 210008, P.R. China

INTRODUCTION

A new field of research in nuclear physics, i.e., the isospin physics, has recently emerged due to the rapid advance in radioactive ion beam experiments. These experiments have already given us novel information about the structure of unstable nuclei far from the stability valley[1, 2]. Since a transient state of nuclear matter with both appreciable isospin asymmetry and compression can be created in these collisions, they are thus also useful for extracting information about the isospin dependence of the nuclear equation of state[7, 8] and the nucleon-nucleon cross sections in medium[3, 4, 5, 6]. Knowledge on the EOS of asymmetric nuclear matter is essential for understanding both the structure of radioactive nuclei and neutron stars. For example, the cooling mechanism of neutron stars and the critical density for kaon condensation in dense stellar matter are both significantly influenced by the isospin dependence of the nuclear EOS[9, 10]. At present, theoretical predictions on the latter vary widely[11]. In this talk, we shall discuss the possibility of using the ratio of the number of fast neutrons to that of protons ($R_{n/p}(E_{kin}) \equiv dN_n/dN_p$) from collisions of neutron-rich nuclei at intermediate energies to study the isospin dependence of nuclear EOS, which we have recently proposed in ref.[8].

Advances in Nuclear Dynamics 3, edited by
Bauer and Mignerey, Plenum Press, New York, 1997

ISOSPIN DEPENDENCE OF NUCLEAR EQUATION OF STATE

Various theoretical studies[9, 12, 13, 14, 15, 16, 17, 18] have shown that the EOS of an asymmetric nuclear matter can be expressed as

$$e(\rho, \beta) = T_F(\rho, \beta) + V_0(\rho) + \beta^2 V_2(\rho), \tag{1}$$

where

$$T_F(\rho, \beta) = \frac{3\hbar^2}{20m} \left(\frac{3\pi^2 \rho}{2} \right)^{2/3} \left[\beta^{5/3} + (1 - \beta)^{5/3} \right] \tag{2}$$

is the Fermi-gas kinetic energy; $V_0(\rho)$ and $V_2(\rho)$ are the isospin-independent and -dependent potential energies. Higher-order terms in β are negligible[12, 14, 15]. Eq. (1) can be approximated by a parabolic form, i.e.,

$$e(\rho, \beta) = e(\rho, 0) + S_{sym}(\rho)\beta^2, \tag{3}$$

where $e(\rho, 0)$ is the EOS of symmetric nuclear matter and

$$S_{sym}(\rho) = \frac{1}{2} \frac{\partial^2 e(\rho, \beta)}{\partial \beta^2} \Big|_{\beta=0} = \frac{5}{9} T_F(\rho, 0) + V_2(\rho) \tag{4}$$

is the bulk symmetry energy. At normal nuclear matter density, the two terms in the above equation have about equal contributions to the symmetry energy, which is known to be in the range of 27-36 MeV[19].

The density dependence of the symmetry energy varies widely among theoretical studies. A $\rho^{2/3}$ dependence was obtained by Siemens using the Bethe-Goldstone theory[12], while the RMF theory predicts a linear dependence[20, 21]. Sophisticated calculations, such as the one by Wiringa et al. using the variational many-body theory[16], give different density dependences of $S_{sym}(\rho)$ depending on the nuclear interactions. Typical results from these studies are given by[17]

$$S_{sym}(\rho) = (2^{2/3} - 1)\frac{3}{5} E_F^0 u^{2/3} + \left(S_0 - (2^{2/3} - 1)\frac{3}{5} E_F^0 \right) F(u), \tag{5}$$

where E_F^0 is the Fermi energy and $F(u)$ has one of the following three forms

$$F_1(u) = \frac{2u^2}{1 + u}, \tag{5}$$
$$F_2(u) = u,$$
$$F_3(u) = u^{1/2},$$

with $u \equiv \rho/\rho_0$ being the reduced baryon density.

The mean-field potentials for neutrons and protons due to the symmetry energy are given by

$$V_{asy}^{n(p)}(\rho, \beta) = \partial w_a(\rho, \beta)/\partial \rho_{n(p)}, \tag{6}$$

where $w_a(\rho, \beta)$ is the symmetry energy density, i.e.,

$$w_a(\rho, \beta) = \left((S_0 - (2^{2/3} - 1)\frac{3}{5} E_F^0 \right) \rho F(u)\beta^2. \tag{7}$$

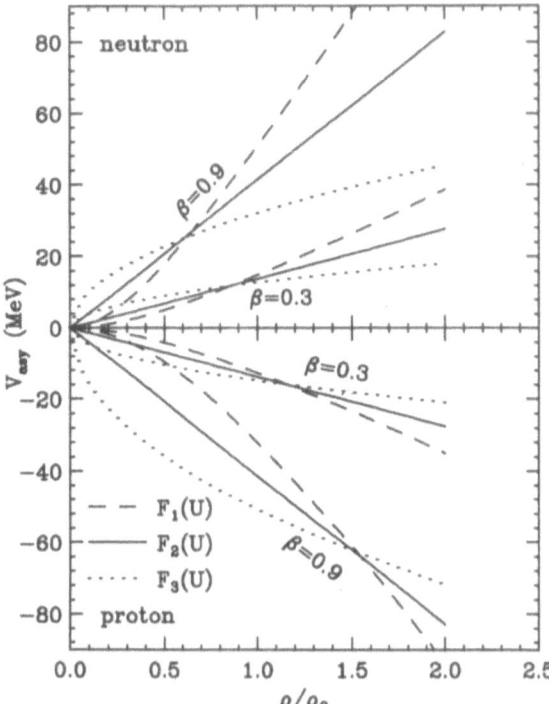

Figure 1. Neutron and proton symmetry potentials from the three forms of $F(u)$ (see text).

The nucleon symmetry potentials corresponding to $F_1(u)$, $F_2(u)$ and $F_3(u)$ are, respectively,

$$V_{\text{asy}}^{n(p)} = 2e_a \frac{u^2}{(1+u)^2}\beta^2 \pm 4e_a \frac{u^2}{1+u}\beta, \tag{7}'$$

$$V_{\text{asy}}^{n(p)} = \pm 2e_a u\beta, \tag{8}$$

$$V_{\text{asy}}^{n(p)} = -\frac{1}{2}e_a u^{1/2}\beta^2 \pm 2e_a u^{1/2}\beta. \tag{9}$$

The magnitude of the symmetry potential is shown in Fig. 1 using the three forms of $F(u)$ and $S_0 = S_{\text{sym}}(\rho_0) = 32$ MeV. It is seen that the repulsive (attractive) mean-field for neutrons (protons) depends sensitively on the form of $F(u)$, the neutron excess β, and the baryon density ρ. In collisions of neutron-rich nuclei at intermediate energies, both β and ρ can be appreciable in a large space-time region where the isospin-dependent mean fields, which are opposite in sign for neutrons and protons, are strong. This affects differently the reaction dynamics of neutrons and protons, leading to possible differences in their yields and energy spectra. Since the magnitude of the asymmetric part of the nuclear EOS is small compared to the symmetric part in eq. (3), to extract $S_{\text{sym}}(\rho)$ from experiments thus requires observables which are only sensitive to the asymmetric part but not the symmetric part of the nuclear EOS. Also, these observables should not depend strongly on other quantities that affect the reaction dynamics, such as the in-medium nucleon-nucleon cross sections. In the following, we shall demonstrate that the ratio $R_{n/p}(E_{\text{kin}})$ of pre-equilibrium neutrons to protons from collisions of neutron-rich nuclei meets these requirements.

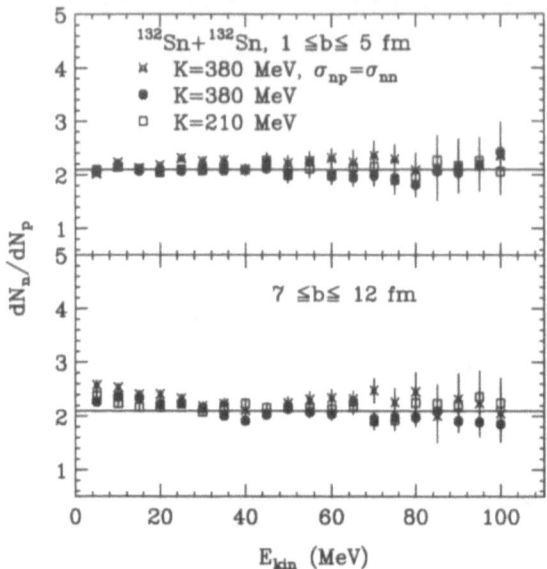

Figure 2. The ratio of preequilibrium neutrons to protons as a function of the nucleon kinetic energy for central (upper window) and peripheral (lower window) collisions calculated without Coulomb and symmetry potentials.

The n/p RATIO OF FAST NUCLEONS AS A PROBE OF THE ISOSPIN DEPENDENCE OF NUCLEAR EQUATION OF STATE

We shall use an isospin-dependent Boltzmann-Uehling-Uhlenbeck (BUU) transport model[3, 8, 22]. The proton and neutron densities calculated from the nonlinear relativistic mean-field (RMF) theory[23] are used as inputs to initialize the BUU model. The isospin dependence is included in the dynamics through nucleon-nucleon collisions by using isospin-dependent cross sections and Pauli blocking factors, the symmetry potential $V_{\mathrm{asy}}^{n(p)}(\rho, \beta)$, and the Coulomb potential V_c^p for protons. Furthermore, the nucleon mean field $V^{n(p)}(\rho, \beta)$ also includes a symmetric term for which we use the Skyrme parameterization, i.e.,

$$V^{n(p)}(\rho, \beta) = a(\rho/\rho_0) + b(\rho/\rho_0)^\sigma + V_c^p + V_{\mathrm{asy}}^{n(p)}(\rho, \beta). \tag{10}$$

In the above, the parameters a, b and σ are determined by the saturation properties and the compressibility K of symmetric nuclear matter. The symmetric term should also contain a momentum-dependent part, which is, however, not important for present study as we will show that the ratio of pre-equilibrium neutrons to protons from collisions of neutron-rich nuclei is essentially independent of the symmetric part of the nuclear EOS. The isospin-dependent BUU model has been used to explain successfully a number of isospin-dependent phenomena in heavy-ion collisions at intermediate energies[22]. More recently, the isospin dependence of collective flow and balance energy predicted in ref. [3] using this model was confirmed experimentally at NSCL/MSU by Pak *et al.*[4, 5, 6].

We have studied collisions of ^{112}Sn+^{112}Sn, ^{124}Sn+^{124}Sn and ^{132}Sn+^{132}Sn reactions at a beam energy of 40 MeV/nucleon. The first two reactions have been recently studied experimentally at NSCL/MSU by the MSU-Rochester-Washington-Wisconsin collaboration[24, 25]. Pre-equilibrium particles were measured in these experiments and are now being analyzed[26]. We note that the last reaction is included here only for the purpose of discussions and comparisons. To identify free nucleons, we use a phase-space coalescence method at 200 fm/c after the initial contact of the two nuclei, when the quadrupole moment of the nucleon momentum distribution in the heavy residue is almost zero, indicating the approach of thermal equilibrium. A nucleon is considered as free if it is not correlated with other nucleons within a spacial distance of $\Delta r = 3$ fm and a momentum distance of $\Delta p = 300$ MeV/c. Otherwise, they are bound in nuclear clusters. We have checked that the results are not sensitive to these parameters if they are varied by less than 30% around the above values. We first study effects of the compressibility K of symmetric nuclear matter and the in-medium nucleon-nucleon cross section on the ratio $R_{n/p}(E_{kin})$ by neglecting both the Coulomb and symmetry potentials in the BUU model. In Fig. 2 this ratio is shown as a function of nucleon kinetic energy for central (upper window) and peripheral (lower window) collisions of ^{132}Sn +^{132}Sn at a beam energy of 40 Mev/nucleon. Varying the compressibility K from 210 MeV (open squares) to 380 MeV (filled circles), we find that although the yields of both neutrons and protons increase, their ratio remains similar for all impact parameters. This is simply because the effects of symmetric EOS on both neutrons and protons are identical.

The experimental cross section for neutron-proton collisions is about three times that for neutron-neutron (proton-proton) collisions in the energy range studied here. Setting the two cross sections equal (fancy squares), we find that the yields and their ratios change by less than 10% even in peripheral collisions of ^{132}Sn +^{132}Sn. This result is also easy to understand since both colliding nucleons have the same probability to gain enough energy to become unbound. Thus, the in-medium, isospin-dependent nucleon-nucleon cross sections do not affect much the ratio $R_{n/p}(E_{kin})$. It is important to point out that in the absence of Coulomb and symmetry potentials the ratios are almost independent of the nucleon kinetic energy and have a constant value of about 2.1 ± 0.3 in both central and peripheral collisions of ^{132}Sn +^{132}Sn.

Including the Coulomb and the asymmetric term of the EOS in eq. (11), one can then study the effects of the symmetry energy $S_{sym}(\rho)$ since the Coulomb effect is well-known. The symmetry potential $V_{asy}^{n(p)}$ will have the following effects on pre-equilibrium nucleons. First, it tends to make more neutrons than protons unbound, so one expects that a stronger symmetry potential leads to a larger ratio of free neutrons to protons. Second, if both neutrons and protons are already unbound, the symmetry potential makes neutrons more energetic than protons. These effects are shown in Fig. 3 where we show the ratios calculated using the three forms of $F(u)$ for central (left windows) and peripheral (right windows) collisions of ^{112}Sn +^{112}Sn, ^{124}Sn +^{124}Sn, and ^{132}Sn +^{132}Sn, respectively. The ratios are seen to change smoothly from central to peripheral collisions. More detailed study on the impact parameter dependence of the ratio $R_{n/p}(E_{kin})$ will be published elsewhere[11]. The increase of the ratios at lower kinetic energies in all cases is due to the Coulomb repulsion which shifts protons from lower to higher kinetic energies. On the other hand, the different ratios calculated using different $F(u)$'s reflect clearly the effect mentioned above, i.e., with a stronger

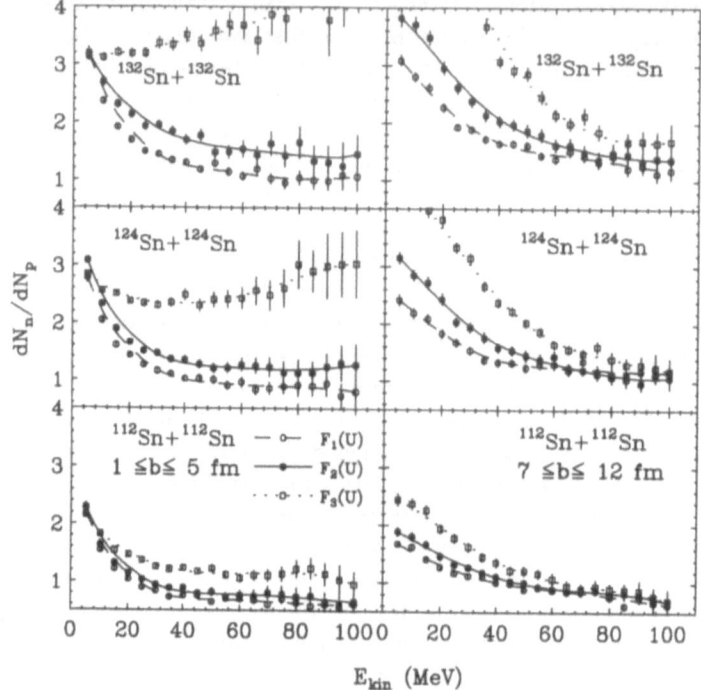

Figure 3. Same as in Fig. 2 but calculated with both Coulomb and symmetry potentials.

symmetry potential the ratio of pre-equilibrium neutrons to protons becomes larger for more neutron rich systems.

It is interesting to note that effects due to different symmetry potentials are seen in different kinetic energy regions for central and peripheral collisions. In central collisions, effects of the symmetry potential are most prominent at higher kinetic energies. This is because most of finally observed free neutrons and protons are already unbound in the early stage of the reaction as a result of violent nucleon-nucleon collisions. The symmetry potential thus mainly affects the nucleon energy spectra by shifting more neutrons to higher kinetic energies with respect to protons. In peripheral collisions, however, there are fewer nucleon-nucleon collisions; whether a nucleon can become unbound depends strongly on its potential energy. With a stronger symmetry potential more neutrons (protons) become unbound (bound) as a result of a stronger symmetry potential, but they generally have smaller kinetic energies. Therefore, in peripheral collisions effects of the symmetry potential show up chiefly at lower kinetic energies. For more neutron-rich systems the effects of the symmetry potential are so strong that in central (peripheral) collisions different forms of $F(u)$ can be clearly distinguished from the ratio of pre-equilibrium neutrons to protons at higher (lower) kinetic energies. However, because of energy thresholds in detectors, it is difficult to measure low energy nucleons, especially neutrons. Furthermore, the low energy spectrum also has appreciable contribution from equilibrium emissions during the later stage of the collisions. Therefore, the measurement of the ratio $R_{n/p}(E_{kin})$ in neutron-rich, central heavy-ion collisions for nucleons with energies higher than about 20 MeV is more suitable for extracting the EOS of asymmetric nuclear matter.

SUMMARY

In summary, we have shown that collisions of neutron-rich nuclei at intermediate energies reveal novel information about the EOS of asymmetric nuclear matter that is of interest to both nuclear- and astro-physics, such as the properties of radioactive nuclei, supernovae, and neutron stars. The ratio of the number of fast neutrons to that of protons is found to be sensitive to the asymmetric part, but not to the symmetric part of the nuclear EOS. It is also almost independent of the in-medium nucleon-nucleon cross sections. A large n/p ratio of fast nucleons has recently been observed in reactions induced by neutron-rich nuclei[26], and this is consistent with the findings of this work. A detailed comparison with the experimental data is underway and is expected to give useful information about the isospin-dependent nuclear EOS.

ACKNOWLEDGEMENT

The work of B.A.L and C.M.K was supported in part by the NSF Grant No. PHY-9509266, and that of W.B. was supported in part by the NSF Grant No. PHY-9403666 and NSF Presidential Fellow grant No. PHY-9253505. Z.Z.R was supported in part by grants from the Foundation of National Educational Commission of P.R. China.

REFERENCES

1. I. Tanihata, Prog. of Part. and Nucl. Phys., **35** (1995) 505.
2. P.G. Hansen, A.S. Jensen and B. Jonson, Ann. Rev. Nucl. Part. Sci. **45**, 591 (1995).
3. B.A. Li, Z.Z. Ren, C.M. Ko and S.J. Yennello, Phys. Rev. Lett. **76**, 4492 (1996).
4. R. Pak et al., Phys. Rev. Lett. **78**, 1022 (1997).
5. R. Pak et al., Phys. Rev. Lett. **78**, 1026 (1997).
6. G.D. Westfall *et al.*, this volume.
7. I. Tanihata, Preprint RIKEN-AF-NP-229, July, 1996.
8. B.A. Li, C.M. Ko and Z.Z. Ren, Phys. Rev. Lett. **78**, 1644 (1997).
9. J.M. Lattimer, C.J. Pethick, M. Prakash and P. Haensel, Phys. Rev. Lett. **66**, 2701 (1991).
10. C.-H. Lee, Phys. Rep. **275**, 255 (1996).
11. B.A. Li, C.M. Ko and W. Bauer, review article to be published.
12. P.J. Siemens, Nucl. Phys. **A141**, 225 (1970).
13. G. Baym, H.A. Bethe, and C.J. Pethick, Nucl. Phys. **A175**, 225 (1971).
14. O. Sjöberg, Nucl. Phys. **A222**, 161 (1974).
15. I.E. Lagaris and V.R. Pandharipande, Nucl. Phys. **A369**, 470 (1981)
16. R.B. Wiringa, V. Fiks and A. Fabrocini, Phys. Rev. **C38**, 1010 (1988).
17. M. Prakash, T.L. Ainsworth and J.M. Lattimer, Phys. Rev. Lett. **61**, 2518 (1988).
18. V. Thorsson, M. Prakash and J.M. Lattimer, Nucl. Phys. **A572**, 693 (1994).
19. P.E. Haustein, Atomic data and nuclear data tables, **39**, 185-395 (1988).
20. S.A. Chin, Ann. Phys. (N.Y.), **108**, 301 (1977).
21. C.J. Horowitz and B.D. Serot, Nucl. Phys. **A464**, 613 (1987).
22. B.A. Li and S.J. Yennello, Phys. Rev. **C52**, R1746 (1995).
23. Z.Z. Ren, W. Mittig, B.Q. Chen, and Z.Y. Ma, Phys. Rev. **C52**, R20 (1995).
24. G.J. Kunde et al., Phys. Rev. Lett. **77**, 2897 (1996).
25. J.F. Dempsey et al, Phys. Rev. **C54**, 1710 (1996).
26. D.K. Agnihotri *et al.*, this volume.

PHOBOS rising at Brookh(e)aven

Rudolf Ganz
Physics Department Univ. of Illinois at Chicago, Chicago, Il 60607-7059

The *PHOBOS* Collaboration:
M.D. Baker[6], D. Barton[2], R.R. Betts[1,9], A. Bialas[5], C. Britton[8], A. Budzanowski[4], W. Busza[6], A. Carroll[2], Y.H. Chang[7], A.E. Chen[7], Y.Y. Chu[2], T. Coghen[4], C. Conner[9], W .Czyz[5], P. Decowski[6], R. Ganz[9], E. Garcia-Solis[10] J. Godlewski[4], S. Gushue[2], C. Halliwell[9], R. Holynski[4], B. Holzman[9], K.T. Huang[7], U. Jagadish[8], J. Kotula[4], H.W. Kraner[2], P. Kulinich[6], M. Lemler[4], W. T. Lin[7], P. Maleki[4], S. Manly[12], D. McLeod[9], A. Mingerey[10], A. Olzewski[4], R. Pak[11], K. Pakonski[4], H. Palarczyk[4], M. Plesko[6], L.P. Remsberg[2], G. Roland[6], L. Rosenberg[6], A. Sanzgiri[12], S.G Steadman[6], G.S.F. Stephans[6], M. Stodulski[4], C. Taylor[3], A. Trzupek[4], G. Van Nieuwenhuizen[6], R. Verdier[6], B. Wadsworth[6], H. Wilczynski[4], F.L.H. Wolfs[11], D. Woodruff[6], B. Wosiek[4], K. Wozniak[4], B. Wysloch[6], K. Zalewski[5]

[1]Argonne National Laboratory, USA;[2]Brookhaven National Laboratory, USA;[3]Case Western Reserve University, USA;[4]Institute of Nuclear Physics Krakow, Poland;[5]Jagiellonian University Krakow, Poland;[6]Massachusetts Institute of Technology, USA;[7]National Central University, Taiwan;[8]Oak Ridge National Laboratory, USA;[9]University of Illinois at Chicago, USA; [10]University of Maryland, USA;[11]University of Rochester, USA;[12]Yale University, USA

INTRODUCTION

PHOBOS[1],[2],[3],[4],[5],[6] is the name of one of currently four experiments (*PHOBOS*, PHENIX, STAR, BRAHMS) being developed to explore the new regime of heavy ion collisions at the Relativistic Heavy Ion Collider (**RHIC**) at Brookhaven National Laboratory. At the RHIC facility, which will become operational in 1999, it will be possible to study hadronic matter at energy densities which are an order of magnitude higher than what has been produced in the laboratory before. Au+Au collisions at 200 AGeV center of mass energy may - on a small scale - reveal conditions similar to those which occured in the first few micro-seconds of the creation of the universe. In this regime of extremely high energy densities new phenomena such as the Quark-Gluon Plasma (QGP) are expected to manifest themselves.

Theory can give only guidelines for the processes which may occur in these collisions. However, several signatures such as strangeness enhancement or temporary restoration of chiral symmetry have been proposed (summarized in [7]) to appear in the QGP phase of the collision. Since the creation of this phase is expected to occur in an early stage of the collision this may or may not alter the final state particle composition. Such a change then could be utilized as an experimental probe to detect the transition between nuclear matter and the deconfined quarks in the QCD-plasma. To answer these types of questions, *PHOBOS* is equipped with a two arm spectrometer, which allows us to identify and study all charged hadrons emitted at mid-rapidity. The QGP might also become observable in changes of global event characteristics such as a rapid increase of the multiplicity of created particles. Therefore, the *PHOBOS* particle multiplicity measurement will cover all charged particles emitted in a pseudo-rapidity range spanning from -5.3 to +5.3, the largest coverage of all RHIC experiments.

To accomplish the broad range of physics motivated goals, the RHIC experiments will have to face new experimental challenges. They will have to deal with an enormously high number of particles ($\approx 12,000$) emitted simultaneously in one single central Au+Au collision. Additionally, these particles will then, when impinging on any material - the detector material itself - produce a correspondingly large number of secondary particles. Therefore, highly segmented detectors with multiple hit capability are required. Another difficulty for these experimentsis the data volume, which will be generated during the runs. When RHIC operates at its nominal luminosity of $L \approx 2 \times 10^{27}$ cm^{-2}s^{-1}, about 500 Hz minimum bias collisions and about 30 Hz central collisions will occur. The high multiplicity per event implies large event sizes of 10 – 20 kBytes in case of *PHOBOS* and even several Mbytes in case of STAR or PHENIX[8]. The data taking will be limited by the read-out rates (Bytes/s) the data acquisition can handle, but also - and even more important - the amount of storage space available. Since RHIC is envisioned to operate the whole year round, this may reach the order of hundreds of Tbytes each year* .

THE *PHOBOS* STRATEGY

Since the signatures of the new physics cannot be predicted unambiguously, *PHOBOS* initially is trying to study these collisions in an unbiased way. Therefore *PHOBOS* will study the production of all types of hadronic particles. The majority of the emitted charged hadrons will be detected by the multiplicity detector, which covers almost 4π of solid angle. This permits us to study their distribution in a pseudo-rapidity range of $|\eta| \leq 5.3$ on an event-by-event basis. 1% of all emitted particles will be studied in detail by the two-arm magnetic spectrometer in the mid-rapidity region, where the highest energy densities are expected. It will identify these particles and measure their momentum even down to very low transverse momentum (p_t). A time-of flight wall will extend the PID and momentum measurement to high momenta. A lead layer in the multiplicity detector setup will convert photons from the decay of π^0 and other neutral

---

*This amount is comparable for all four experiments since the larger event-sizes of PHENIX and STAR are balanced by the higher read-out rates of BRAHMS and *PHOBOS*.

resonances, which gives an experimental tool to observe changes in the ratio of charged to neutral particles. Strong fluctuations of this ratio have been predicted due to the formation of a Disoriented Chiral Condensate[9].

Compared to the other three RHIC projects *PHOBOS* is taking a rather unique approach in terms of the chosen detector technology. Instead of a composition of various detector types, *PHOBOS* will almost exclusively be based on a single technology, namely Silicon micro-strip and pad detectors. Each channel of these highly segmented detectors will be read out separately in order to deal with the high occupancy expected for central Au+Au collisions. The choice of this technology was driven by their commercial availability and reliability. Furthermore, these detectors have been used in many experiments, and specifically a type similar to the *PHOBOS* design[1],[5] was successfully operated in the CERN WA98 experiment (see P. Stankus in this proceedings).

A consequence of a unbiased data collection is that *PHOBOS* has to be capable of processing all detector signals at very high rates. For the silicon detector read-out this will be achieved by a custom designed pipeline chip, which establishes an almost continuous data flow between the 64 channel preamplifier chip and a dual-range 7 bit ADC which digitizes signals between $0 - 1$ MeV and $0 - 10$ MeV. This part of the electronics will be located directly on the silicon detectors and will allow us read-out rates of about 300 Hz for approximately 100,000 channels of the *PHOBOS* detectors.

THE *PHOBOS* DETECTOR

Figure 1 shows an overview of the *PHOBOS* setup. It can be separated into two distinct components. The first is the two-arm particle spectrometer, which consists of a multi-layer silicon-detector device partly located within the gaps of the two yoke magnet. It's extended by a time-of-flight wall of scintillation detectors[6]. The second is the multiplicity and vertex detector array. The center-piece of this assembly is the "Octagon" detector array. It surrounds the Beryllium beam pipe with eight rows of

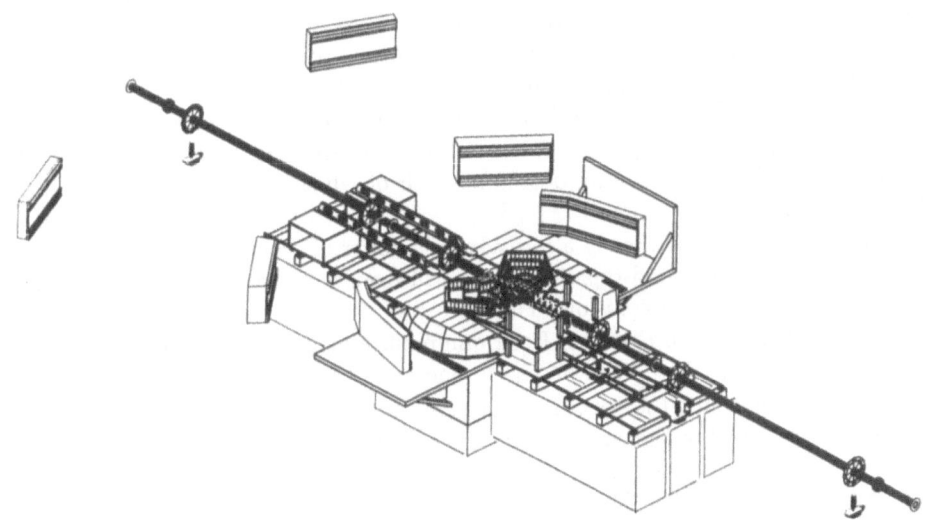

Figure 1. The *PHOBOS* experiment and its components shown in an AutoCad drawing. The upper part of the magnet's yoke structure has been removed for visibility.

silicon pad detectors positioned along the beam axis. The charged multiplicity detection is extended by two sets of three ring detectors matching the eight-fold symmetry in the Octagon.

Several high resolution strip detectors, which are embedded in the central part of the Octagon structure, will serve as the Vertex Finder. This two layer device allows us to project the tracks of emitted particles back to the interaction point/vertex. It also will provide a reference measurement and extension of the multiplicity measurement.

The four small rings surrounding the beam pipe in figure 1 indicate four sets of Cerenkov counters which will provide the Level 0 trigger. They will also be used as a timing reference for the time-of-flight wall. Two of them will placed between the first and the second closest multiplicity ring counter next to the center ("low η- counter"). The other two rings are placed behind the third multiplicity ring detector and cover an even higher pseudo-rapidity of $\eta \approx 5.5$ ("high η- counter"). These detectors are in the design phase and are not finalized yet.

The *PHOBOS* Multiplicity Array

The multiplicity detector (figure 1 and 2) covers a range of $-5.3 < \eta < 5.3$ in pseudo-rapidity. It measures the total charged multiplicity in sixteen bins of azimuthal angle $\Delta\Phi = 22.5^o$ and about 1000 bins of pseudo-rapidity ($\Delta\eta \approx 0.01$). The $\Delta\Phi$ separation corresponds to the eight separate rows of silicon strip-detectors surrounding the beam pipe in the center region and the corresponding segmentation of the ring counters into eight detector modules (see figure 2). Each of these modules is segmented by two rows (additional $\Delta\Phi$-seperation) of 32 pads ($\Delta\eta$ separation).

In each of the eight rows of the Octagon thirteen such wafers are oriented along the beam pipe each of which is segmented into 64 strips. Extended Monte Carlo studies have led to the chosen balance between accuracy of the measurement and a reasonable total number of channels. In the chosen design about 9,000 silicon channels of the Octagon multiplicity array have to be read out. This number takes into account the fact that four wafers have been removed in each of the two vertical planes as an aperture for the two spectrometer arms. A total of six modules on the top and bottom have been replaced by eight high resolution (384 channel) modules, which besides measuring the multiplicity, serve as a part of the Vertex Finder (see next section). Another 3,000 channels come from the six ring detectors, each of which is assembled with eight 64 channel modules.

Even though this is already a fairly large number of segments, the occupancy per detector channel in the mid-rapidity region still reaches values between one and three hits for a central Au+Au collision. Therefore we utilize the energy information which will be recorded for each channel. Each minimum ionizing particle (MIP) on average deposits about 100 keV $\times \cos\theta$ of its kinetic energy in the 300 μm thick silicon detectors (θ is the angle at which the particle penetrates the detector). By deriving θ from the vertex and the detector segment position, we can then resolve multiple hits on a single strip by the measured energy. Additionally we can extract from Monte Carlo calculations the average number of hits per strip (including those from secondary particles) per primary particle, emitted from the origin, within the corresponding pseudo-rapidity region. With these correction we then can reconstruct the multiplicity distribution of the emitted charged particles as shown for a full GEANT Monte Carlo simulation in figure 3. The reconstruction capability can be cross-checked experimentally by com-

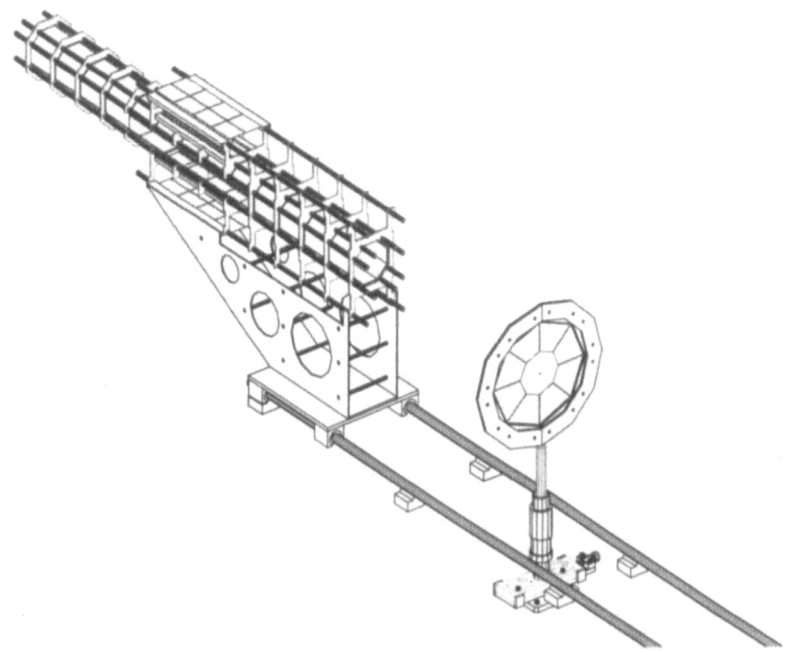

Figure 2. The *PHOBOS* Multiplicity array. The figure shows the center piece of the multiplicity detector array -the "Octagon"-, which also contains the Vertex detector. It is mounted on a rail system to facilitate the mounting procedure. On the right side one of the six ring counters is shown.

paring the distribution derived from three of the central 64 channel detectors with four of the 394 channel multiplicity/vertex finder modules. Both cover the same solid angle and the same pseudo-rapidity range, but the occupancy and thus the systematic error in the highly segmented detector is considerably lower.

There are a variety of questions which will be approached with the multiplicity measurement described above. The large coverage in η and ϕ allows us to study the multiplicity distributions on an event-by-event basis. Fluctuations in the multiplicity distributions might become a powerful probe to reveal the existence of the QGP. This can be understood by the following entropy argument[3]. As a result of the color deconfinement in the QGP phase, the entropy will increase significantly with the increasing number of degrees of freedom. In order not to lose entropy, the system should expand and generate a larger number of particles in the final state of the collision than a purely hadronic process. This should appear in our measurement.

Another aspect we can investigate when studying multiplicity distributions is the formation of a Disoriented Chiral Condensate[9], which is expected to produce fluctuations in the ratio of the number of charged to neutral particles emitted into a limited phase space region. These contributions should appear as fluctuations in our measured charged multiplicity distribution[3], which may be even more pronounced when we restrict the distribution to one of our sixteen ϕ bins.

The Vertex Finder

Besides serving as a part and a reference in the multiplicity measurement, the Vertex Finder has to reconstruct the position of the primary interaction vertex. This is

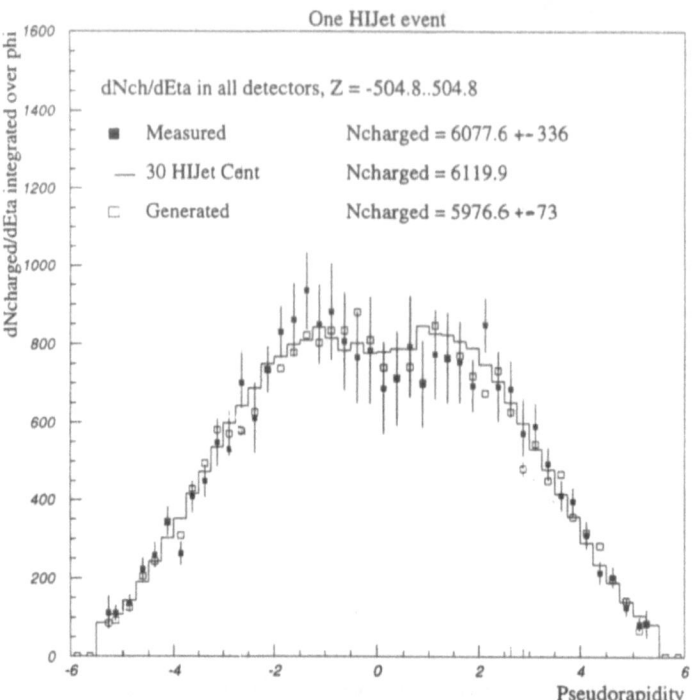

Figure 3. A reconstructed charged multiplicity distribution. The solid line shows an average of 30 generated HIJET central Au+Au collisions. The dashed curve is the distribution generated for one specific event and the "measured" points is the reconstructed distribution for the same event.

necessary since the region of the intersecting beam extends over 40 cm (1σ region) along the beam axis. For this purpose we have added another layer of silicon detectors (Outer Vertex Detector) at twice the distance (11 cm) on top of the four highly segmented modules (Inner Vertex Detector), which are embedded into the octagon (figure 2). This additional layer covers the same $\Delta\phi$ range of 45° and the same sector along the beam axis with eight modules of 192 channel strip detectors. In order to reconstruct the vertex position, all possible tracks out of the hits on the Outer Detector and the Inner Detector are projected onto the beam axis. All hit combinations produced by single particle tracks originating in the collision vertex will pile up as a narrow peak on top of the distribution of all the possible projections (figure 4). This method gives a vertex position resolution of $\delta z \approx 0.08\ mm$ within the covered range of $|z| \leq 20$ cm.

The design of *PHOBOS* contains two such Vertex Finder setups, one on the top and one on the bottom. As Monte Carlo studies have shown, one of these assemblies is sufficient to reconstruct the vertex. Therefore, by introducing a lead layer ($L_{rad}/2$), we utilize the lower part of the setup to measure the production of neutral particles by detecting secondary particles from conversion processes in the lead layer. Therefore this might also be used as a probe for a Disoriented Chiral Condensate, which is predicted to yield event-by-event fluctuations of the production of neutral particles ($\pi^o/\pi^\pm$ ratio).

The Two Arm Spectrometer and the TOF Wall

A detailed study of all hadronic particles will be carried out with the *PHOBOS* two arm spectrometer. The acceptance for each arm covers $\Delta\phi = 11°$ of azimuthal

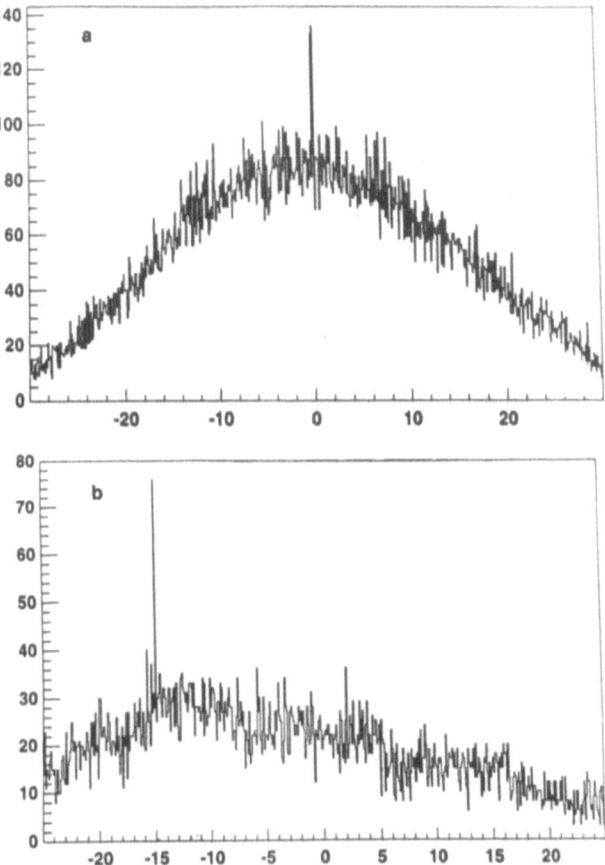

Figure 4. The *PHOBOS* Vertex reconstruction from a Monte Carlo simulation of a central Au+Au collision at a vertex at z= 0 cm (a) and z=-15 cm (b). The position of the narrow peak on top of the distribution indicates the vertex position.

angle and one unit of rapidity in the mid-rapidity region, e.g. $0.5 \leq \eta \leq 1.5$ for z=0 vertices. In each central collision we expect about 70 particles emitted into this phase space. These particles will be tracked in a series of eleven planes of silicon pad and strip detectors. The most distant six of the eleven planes are located in the 15.5 cm gap of a conventional magnet, which generates a field of about 2 Tesla. High momentum particles will leave the silicon detector device and will be detected by a time-of-flight wall composed of scintillation detectors.

The particle identification in this setup is done by three different methods, depending on the momentum of the particle. Pions and kaons with $p_t \leq 1.2$ GeV/c and $p_t \leq 1.9$ GeV/c respectively, will be identified by their time-of-flight, measured as a time difference ($\delta t \approx 100$ ps) between the Cerenkov trigger counters and the scintillation detector arrays posted in three different distances from the interaction region (figure 1). At lower momentum, $p_\pi \leq 0.6$ GeV/c and $p_K \leq 1.0$ GeV/c respectively, the average energy deposited in the silicon planes and the bending of the tracks in the last six detector planes will be used for particle and momentum identification. At very low momenta, the pions ($p_\pi \leq 60$ MeV/c) and the kaons ($p_K \leq 140$ MeV/c) will be stopped in the first seven silicon layers. The total of the deposited energy (equal to the total kinetic energy of the particle) and the energy deposited on average in these detectors will be used for the particle identification at these low energies.

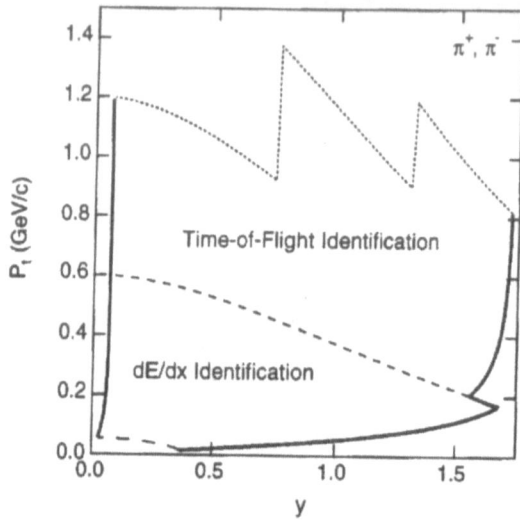

Figure 5. The *PHOBOS* spectrometer acceptance versus transverse momentum p_t and the rapidity of charged pions. The acceptance is shown separately for different methods of particle identification. The lower part of the diagram is identified by the energy deposited in silicon detectors, which is extended in the upper part of the diagram by the time of flight measurement of the scintillator wall (from [6]).

The acceptance for identified pions in the *PHOBOS* spectrometer is summarized in figure 5. *PHOBOS* covers lower transverse momenta than the other RHIC experiments. According to an anticipated exponential shape of the p_t-spectrum, this is the region where the highest yield of particles is expected. Moreover, at low-p_t in-medium effects are expected to be enhanced, such as a possible mass shift of the Φ-meson, which has been reported by the AGS E802/E859 collaboration[10]. Besides the measurements of production rates of charged particles, *PHOBOS* will also be capable of carrying out two particle correlation measurements, like HBT[4]. Due to the very fine segmentation of the silicon detectors, this will be done down to low relative momenta, where correlations should be more pronounced.

THE CURRENT STATUS OF *PHOBOS*

Since *PHOBOS* is now a construction project, several parts of the design have been finalized, pieces of the equipment have been prototyped, and will soon be manufactured.

The mechanical layout of *PHOBOS*, like the support structure of the detector and the magnet, as well as the water cooling system have been designed by the Krakow group. Several models have been built in order to study questions like the thermal behavior and mechanical stability of the setup, which is extremely crucial in the vicinity of the silicon detectors. Many details of the geometry have been included in the Monte Carlo simulation program to give realistic estimates of secondary particle production.

The silicon detector design, such as the layout of the etching masks, has converged and prototype detectors have been fabricated in Taiwan and are currently being tested thoroughly at MIT and UIC. For the summer, the production of a major part of the silicon detectors is planned. First versions of the pre-amp chip and a prototype silicon detector have been successfully used in a test- beam experiment at the AGS. The design and test of a pipeline chip will be done shortly. The layout of the TOF wall will be finalized contingent upon the response to a NSF proposal submitted by the Rochester group[6].

REFERENCES

1.   " *PHOBOS* Conceptual Design Report"; April 1994 *(unpublished)*
2.   B. Wyslouch, *Nucl. Phys. A566 (1994)305*
3.   M.D. Baker et al., " *PHOBOS* physics capabilities"; *Proceedings the Pre-Conf. Workshop 11th Int. Conf. on Ultra-Relativistic Nuclues-Nucleus collisions (1995)31*
4.   G. Roland et al., "Measuring two-particle Bose-Einstein correlations with *PHOBOS* at RHIC"; *Proceedings the Pre-Conf. Workshop 11th Int. Conf. on Ultra-Relativistic Nuclues-Nucleus collisions (1995)111*
5.   R.R. Betts et al. "Advances in Nuclear Dynamics 2"; *Plenum Press N.Y.(1996) 225*
6.   F.L.H. Wolfs "Construction of the *PHOBOS* time-of-flight wall"; *Internal Report NSRL Univ. of Rochester Nov. 1995*
7.   J.W. Harris and B.Mueller, "The Search for the Quark Gluon Plasma"; *Ann. Rev. Nucl. Part. Sci. 46(1996)71*
8.   C. Halliwell "Comparison of the RHIC experiments"; *Proceedings of the 36th Zakopane Theoretical Physics School 1996*
     `http::/sunhehi.phy.uic.edu/~phobos/phobos/public_info/presentations/`
9.   J.D. Bjorken *Acta Phys. Pol. B23(1992)637*
10. Y.Wang et al.; *Nucl. Phys A566(1994)379*
11. The transparencies of this talk can be found on the World Wide Web at:
     `http::/sunhehi.phy.uic.edu/~ganz/Marathon97/`
12. More information about the *PHOBOS* project are available at:
     `http::/sunhehi.phy.uic.edu/~phobos/`

FISSION HINDRANCE IN HOT NUCLEI

B. B. Back, D. J. Hofman, and V. Nanal

Argonne National Laboratory
Argonne, IL 60439, USA

INTRODUCTION

The role of dynamics in fission has attracted much interest since the discovery of this process over fifty years ago. However, the study of the dynamical aspects of fission was for many years hampered by the lack of suitable experimental observables against which theoretical calculations could be tested. For example, it was found that the total kinetic energy release in fission can be described equally well by very different dissipation mechanisms, namely the wall formula, that is based on the collisions of the nucleons with the moving wall of the system, as well as a bulk viscosity of the nuclear matter. Although early theoretical work [1] suggested that the fission process may be described as a diffusion process over the fission barrier, this was largely forgotten because of the success of a purely statistical model which instead of enumerating the ultimate final states of the process argues that the fission rate is determined at the 'transition state' as the system traverses the fission saddle point [2].

It was therefore significant when Gavron *et al* [3] showed that the transition state model was unable to describe the number of neutrons emitted prior to scission at high excitation energy in reactions of $^{16}O+^{142}Nd$. Subsequent experimental work using different methods to measure the fission dissipation/viscosity has confirmed these initial observations. It was therefore very surprising when Moretto *et al.* in recent publications [4, 5] concluded that their analysis of fission excitation functions obtained with α and ^3He induced projectiles was perfectly in accord with the transition state model and left no room for fission viscosity. In this paper we'll show that Moretto's analysis is flawed by assuming first chance fission only (in direct contradiction to the experimental observation of pre-scission neutron emission in heavy-ion induced fission), and reveal why the systematics presented by Moretto looked so convincing despite these flaws.

Advances in Nuclear Dynamics 3, edited by
Bauer and Mignerey, Plenum Press, New York, 1997

DISSIPATIVE FISSION

In measurements of the fission cross section from highly excited nuclei it is normally not possible to measure the mass and charge of the fission fragments to a precision that would allow for a unique identification of the fissioning system. In general, it is expected that the measured fission cross section has contributions from several steps of the evaporation cascade of the system as illustrated in Fig. 1. Only at low excitation energies where the removal of the first neutron lowers the excitation energy below the fission barrier is multi-chance fission effectively excluded. In hot nuclei one expects, however, that on average several neutrons are evaporated before the system traverses the saddle point on its way toward scission.

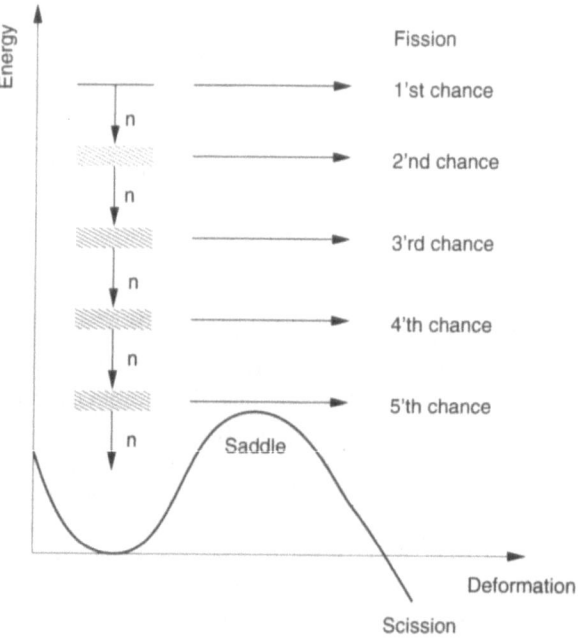

Figure 1. Schematic illustration of the neutron evaporation cascade and multichance fission .

The effect of dissipation or viscosity in the fission process is to lower and possibly delay the fission decay rate over the saddle point. This lowers the fission cross section, and increases the evaporation residue cross section, both of which can be measured. For the nuclei that fission despite this reduction one expects that a larger number of neutrons are emitted before the saddle point is traversed.

Within the diffusion model of fission [1, 6, 7] the fission decay width of the transition state model [2], Γ_{BW}, is modified in two ways: 1) it is reduced by a factor $(\sqrt{1+\gamma^2} - \gamma)$, which depends on the dissipation strength, γ, and 2) there is a buildup time of τ before the constant decay width is achieved:

$$\Gamma_f = \Gamma_{BW}(\sqrt{1+\gamma^2} - \gamma)[1 - \exp(-t/\tau)] \tag{1}$$

In addition, the descend from saddle to scission would be substantially slower in case of high viscosity, allowing for a larger number of neutrons to be emitted before the system scissions. Similar conclusions apply to the emission of GDR γ-rays prior to scission. With no dissipation, the saddle-to-scission time, t_{ss0} is of the order 2 - 5×10^{-21} sec. This descend time is prolonged by a factor $(\sqrt{1 + \gamma^2} + \gamma)$ in the presence of dissipation [8] resulting in

$$t_{ss} = t_{ss0}(\sqrt{1 + \gamma^2} + \gamma). \tag{2}$$

The value of the normalized dissipation strength, γ, is in most cases taken as a free parameter to be determined from comparisons to data. In most cases it is found that $\gamma \approx$ 5-10 is required to obtain agreement with data. It is of interest to note that the one-body dissipation mechanism [9, 10] predicts $\gamma \approx$ 5-6 for the heavy nuclei studied in these experiments, in quite good agreement with those obtained from experiment. Although most analyses do not discriminate between the fission motion inside and outside the saddle point, there is no á priori reason to expect that the strength of the dissipation be constant over the range of shaped needed to describe the fission process. In fact, it has been suggested that a rather strong shape dependence may be present [11].

The experimental observables for viscosity in the fission motion are therefore clear: 1) The combined effects of a reduced fission rate and an slowed down descent from saddle to scission are enhanced emission of pre-scission neutrons, γ-rays and charged particles, and 2) the reduced fission rate alone may be observed in a reduced fission cross section, which is counterbalanced by an increased cross section for evaporation residues. All of these effects have been observed in fission of heavy nuclei.

PRE-SCISSION EMISSION OF NEUTRONS AND γ-RAYS

The initial observation of an excess in the emission of pre-scission neutrons by Gavron et al. [3] spawned further studies of this effect in a large number of systems [12, 13, 14, 15, 16, 17], which firmly established the fact that the motion toward scission is highly viscous resulting in a reduced fission probability and a slowing down of the fission motion. As an example we show data for the reactions $^{16}O+^{208}Pb$ [16], and $^{19}F+^{232}Th$ [12], see Fig. 2a-b. We observe that the experimental data are in good agreement with model predictions assuming a dissipation strength of $\gamma = 5$ [18], whereas the purely statistical model ($\gamma = 0$) severely underestimate the multiplicity, ν_{pre} of pre-scission neutrons.

The effect of fission dissipation is also clearly seen in the emission of pre-scission γ-rays, especially in the Giant Dipole Resonance region. This has been studied for a number of reactions by the Stony Brook group [20, 19, 7, 21]. One example is shown in Fig. 2c-d for the system $^{32}S+^{208}Pb$ at $E_{beam} = 230$ MeV [20]. Again, this analysis shows a clear effect of the dissipation in the fission motion requiring a strength of $\gamma \approx$ 5-10 inside the barrier and a saddle-to-scission time of $\tau_{ss} = 30 \times 10^{-21}$ sec, corresponding to $\gamma = 5$ outside the saddle point as well.

A strong effect of fission viscosity is also observed in the cross sections for evaporation residues in heavy fusion reactions [22, 21, 23]. Thus it is found that the survival probability of the fused system does not decrease as additional decay steps are added to the evaporation cascade by increasing the excitation energy. It appears that the fission decay branch at high excitation energy is essentially hindered due to high viscosity.

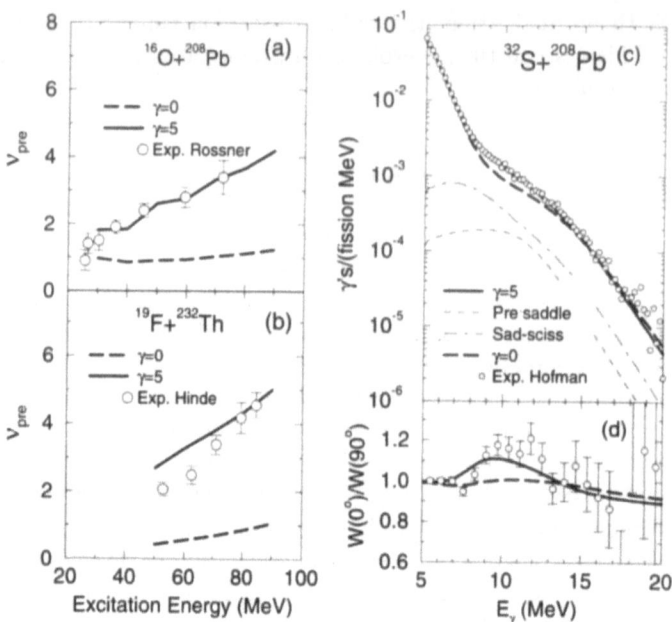

Figure 2. Pre-scission neutron multiplicity nu_{pre} (open circles) for the $^{16}O+^{208}Pb$ [16] (panel a) and $^{19}F+^{232}Th$ [12] (panel b) reactions are compared to statistical model calculations with ($\gamma=5$) and without ($\gamma=0$) fission dissipation as a function of excitation energy. Panel c: γ-ray spectrum in coincidence with fission fragments for $^{32}S+^{208}Pb$ at 230 MeV [20] is compared with statistical model calculations with ($\gamma=5$) and without ($\gamma=0$) fission dissipation. The contributions from pre-saddle and saddle-to-scission for $\gamma=5$ are also shown. Panel d: The γ-ray anisotropy relative to the fission axis is compared to calculations including fission dissipation (solid curve).

MORETTO'S ANALYSIS OF FISSION CROSS SECTIONS

In a recent publication Moretto *et al.*[4] have analyzed a large number of α induced fission excitation functions on targets ranging from ^{182}W to ^{209}Bi. More recent measurements of $^3He+^{197}Au, ^{208}Pb$, and ^{209}Bi [5] have been analyzed in the same manner. This analysis is based on the branching ratio between neutron emission and fission of the fused system (i.e. first chance fission *only*) using the transition state model with Fermi-gas expression for the respective level densities. The partial decay width for first chance fission, Γ_f, is

$$\Gamma_f \approx \frac{T_{sad}}{2\pi} \frac{\rho_{sad}(E - B_f - E_r^{sad})}{\rho_{gs}(E - E_r^{gs})} \qquad (3)$$

where T_{sad} and ρ_{sad} is the nuclear temperature and level density at the saddle point and ρ_{gs} is the level density at the ground state deformation, all taken at the excitation energy E and back-shifted for the rotational energy E_r and the fission barrier B_f. The cross section for *first chance* fission is

$$\sigma_f = \sigma_0 \frac{\Gamma_f}{\Gamma_{total}} \approx \sigma_0 \frac{1}{\Gamma_{total}} \frac{T_{sad}\rho(E - B_f - E_r^{sad})}{2\pi\rho_{gs}(E - E_r^{gs})}, \qquad (4)$$

where σ_0 is the fusion cross section, and Γ_{total} is the total decay width, which is dominated by the neutron decay branch. This expression may be re-written as

$$\frac{\sigma_f}{\sigma_0}\Gamma_{total}\frac{2\pi\rho_{gs}(E - E_r^{gs})}{T_{sad}} = \rho_{sad}(E - B_f - E_r^{sad}). \qquad (5)$$

Using the Fermi-gas expression for the level densities $\rho(E) \propto \exp(2\sqrt{aE})$ and taking the natural logarithm this expression reduces to

$$\ln(\frac{\sigma_f}{\sigma_0}\Gamma_{total}\frac{2\pi\rho(E - E_r^{gs})}{T_{sad}})/2\sqrt{a_n} = \ln(R_f)/2\sqrt{a_n} = 2\sqrt{\frac{a_f}{a_n}(E - B_f - E_r^{sad})}. \qquad (6)$$

Here a_n and a_f are the Fermi-gas level density parameters for the ground state and saddle point deformation, respectively. Using the measured fission cross section σ_f and the fusion cross section σ_0 obtained from an Optical Model calculation, Moretto plots the left hand side of this relation against the right hand side and finds a convincing linear relationship, when using reasonable choices of the fission barrier B_f, and the ground state shell correction Δ_{shell}. This is taken to prove that the fission decay of these hot systems is in perfect accord with the transition state model and under the assumption of *first chance fission* it is shown that this analysis excludes fission delay times of the order $\tau = 30\text{-}50\times10^{-21}$ sec, as obtained from the analysis of pre-scission neutron multiplicity data. This conclusion presents a serious puzzle and it is in direct conflict with the conclusions drawn from pre-scission neutron- and γ-emission data, as well as evaporation residue cross sections. However, in the following sections we'll show that this conflict is only *apparent* and that it is caused by the unjustified assumption of first chance fission and masked by an unfortunate way of plotting the data, that effectively eliminates the sensitivity to the cross sections.

MULTI-CHANCE FISSION

To determine whether the assumption of first chance fission is justified in these systems we have calculated the first chance fission cross sections for the ^{3}He+^{208}Pb reactions studied by Rubehn et al.[5], using the parameters given in this reference and a fusion cross section obtained from the modified proximity potential. This is shown as a dashed curve in Fig. 3a and is seen to under estimate the fission cross section by about a factor of three. By taking the multi-chance fission into account in a simplified statistical that uses the Sierk fission barriers [24] and ground state shell corrections from Möller & Nix [25] we find that the fission cross section is *over* predicted by a factor of three at the highest energies. The contributions to the fission cross sections from the various steps in the decay chain is illustrated in Fig. 3b. A similar result is found from a more complete statistical model calculation using the CASCADE code [26].

Since the true model calculation over-predicts the fission cross section, there is now room for the effects of fission hindrance or fission delay. Thus we find that introducing a fission delay time of $\tau=100\times10^{-21}$ sec gives a reasonable agreement with the

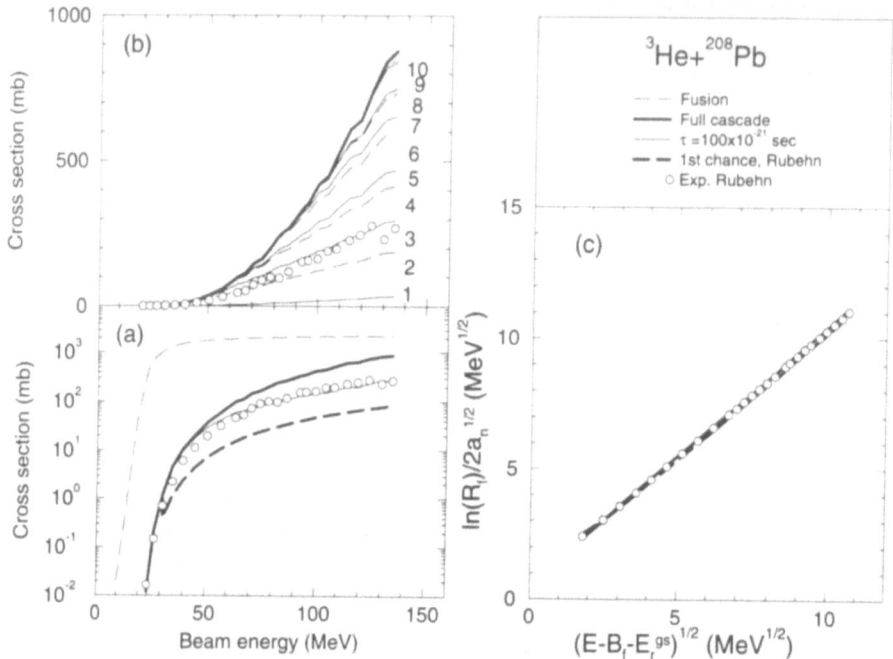

Figure 3. Panel a: experimental fission cross sections σ_f (open circles) are compared to the first chance fission cross section [5] (thick dashed curve), the full cascade (solid thick curve), and a calculation including a fission delay time of $\tau=100\times10^{-21}$ sec (thin solid curve). The fusion cross section is shown as a thin dashed curve. Panel b: The importance of multi-chance fission is shown in a cumulative fashion. Panel c: Moretto plot for experimental data (open circles), first chance fission (dashed line) and the full cascade (solid line).

data, within this simple model. We therefore disagree with the conclusion of Moretto et al.[4] and Rubehn et al. [5], who find that the fission cross section data are not compatible with the notion of fission decay times of the order obtained from pre-scission neutron- and γ-emission data. As demonstrated here, this conclusion was reached on the erroneous premise that only first chance fission occurs in these reactions, an assumption that is directly contradicted by the observation of substantial pre-scission neutron multiplicities in similar systems [12].

LARGE NUMBERS - SMALL NUMBERS

We are therefore left with the question of how the plots presented by Moretto et al.[4] and Rubehn et al.[5] could look so convincing when the calculated cross sections deviated by a factor of three from the measurements. This is illustrated in Fig. 3c where the quantity $\ln(R_f)/2\sqrt{a_n}$ is plotted as a function of $\sqrt{E-B_f-E_r^{gs}}$ for the experimental cross sections (solid circles), the first chance fission cross section (same as the dashed line in panel a), and the total fission cross section from the full cascade (same as the wide solid curve in panel a). We see that these are essentially indistinguishable in this plot, despite the fact that they are based on cross sections that are up to an order of magnitude apart. At the highest energies, an order of magnitude difference in the cross sections translates to a 2.3% difference in the plotted quantity, which, of course, is barely noticeable. This arises because the term $\ln(\sigma_f/\sigma_0)$ is completely overwhelmed by the remaining term $\ln(\Gamma_{total}2\pi\rho(E - E_r^{gs})/T_{sad})$ containing mainly the logarithm of a level density. The Moretto plot is thus dominated by the level densities that appear in both the abscissa and the ordinate. The sensitivity of this plot to the measured quantity is very small, and it is therefore a very unfortunate way of presenting the data. As we have demonstrated, it easily leads to unwarranted conclusions.

SUMMARY

In this paper we have discussed the concept of viscosity in the fission process and its observable effects in pre-scission neutron, and γ-emission. Other observables related to the competition between fission and neutron emission in the decay cascade of an excited nuclear system have also provided additional evidence for dissipation in the fission motion. In direct conflict with these observations, Moretto et al.[4] and Rubehn et al.[5] have recently concluded that their analysis of a large number of fission excitation functions leave no room for fission delay or hindrance. In this paper we have shown that this analysis is incorrect because, 1) only first chance fission is considered and 2) the method of analysis effectively eliminated the sensitivity to the measured quantities. We therefore conclude that there is no ambiguity between the different experimental observables and we find that a detailed analysis of all the data within a single statistical model, including the effects of fission viscosity is called for.

This work was carried out under the auspecies of the U. S. Department of Energy under Contract No. W-31-109-Eng-38.

REFERENCES

1. H. A. Kramers, Physica **7**, 284 (1940)
2. N. Bohr and J. A. Wheeler, Phys. Rev. **56**, 426 (1939)
3. A. Gavron et al. Phys. Lett. **176B**, 312 (1986)
4. L. G. Moretto, K. X. Jing, R. Gatti, and G. J. Wozniak, Phys. Rev. Lett. **75**, 4186 (1995)
5. Th. Rubehn et al. Phys. Rev. C54, 3062 (1996)
6. P. Grangé, Li Jun-Qing, and H. A. Weidenmüller, Phys. Rev. C27, 2063 (1983)
7. R. Butsch, D. J. Hofman, C. Montoya, and P. Paul, Phys. Rev. C44, 1515 (1991)
8. H. Hofmann and J. R. Nix, Phys. Lett. **122B**, 117 (1983)
9. J. Błocki et al. Ann. Phys. (NY) **113**, 330 (1978)
10. J. R. Nix and A. J. Sierk, Phys. Rev. C21, 396 (1980)
11. P. Fröbrich and I. I. Gontchar, Nucl. Phys. **A556**, 281 (1993)
12. D. J. Hinde et al. Nucl. Phys. **A452**, 550 (1986)
13. J. O. Newton et al, Nucl. Phys. **A452**, 550 (1986)
14. D. J. Hinde et al. Phys. Rev. C39, 2268 (1989)
15. D. J. Hinde et al. Phys. Rev. C45, 1229 (1992)
16. H.Rossner et al. Phys.Rev. C45, 719 (1992)
17. D. Hilscher, Nucl. Phys. **A471**, 77 (1987)
18. B. B. Back in Proceedings of the International School-Seminar on Heavy Ion Physics, Dubna, Russia, May 10-15, 1993, ed. Y. Oganessyan, Dubna Press, 1993
19. R. Butsch et al. Phys. Rev C41, 1530 (1990)
20. D. J. Hofman et al. Phys. Rev. Lett. **72**, 470 (1994)
21. D. J. Hofman, B. B. Back, and P. Paul, Phys. Rev. C51, 2597 (1995)

22. K.-T. Brinkmann, *et al.* Phys. Rev. **C50**, 309 (1994)
23. B. B. Back *et al.* In proceedings of the 'International Workshop on Physics with Recoil Separators and Detector Arrays", New Delhi, Jan 30 - Feb 2, 1995, Allied Publishers LTD, New Delhi, India, 1995
24. A. J. Sierk, Phys. Rev. **C33**, 20239 (1986)
25. P. Möller and J. R. Nix, ADNDT **59**, 185 (1995)
26. F. Pühlhofer, Nucl. Phys. **A280**, 267 (1977)

HADRON-INDUCED MULTIFRAGMENTATION

W.-C. Hsi,[1] K. Kwiatkowski,[1] G. Wang,[1] D.S. Bracken,[1] H. Breuer,[2] Y.Y. Chu,[3] E.Cornell,[1] F. Gimeno-Nogues,[4] D.S. Ginger,[1] S. Gushue,[3] M.J. Huang,[5] R.G. Korteling,[6] W.G. Lynch,[5] K.B. Morley,[7] E. Ramakrishnan,[4] L.P. Remsberg,[3] D.Rowland,[4] M.B. Tsang,[5] V.E. Viola,[1] S.J. Yennello,[4] N.R. Yoder,[1] and H. Xi,[5]

[1] Indiana University, Bloomington, IN 47405
[2] University of Maryland, College Park, MD 20742
[3] Brookhaven National Laboratory, Upton, NY 11973
[4] Texas A & M University, College Station, TX 77843
[5] Michigan State University, East Lansing, MI 48824
[6] Simon Fraser University, Burnaby, BC V5A 1S6 Canada
[7] Los Alamos National Laboratory, Los Alamos, NM 87545

INTRODUCTION

For light-ion-induced reactions in the bombarding energy regime 1 - 10 GeV, dramatic changes occur for inclusive observables associated with highly excited heavy residues.[1] For example, the excitation functions for IMF production (IMF: $3 \leq Z \leq 15$) increase by more than two orders of magnitude over this energy range–while the N-N total scattering cross section changes only gradually. This has been interpreted in terms of large deposition energies in the residue, giving rise to multifragmentation events. Subsequent 4π studies[2] have verified this interpretation and further shown that the most highly excited residues are produced with very low velocities ($v/c \sim 0.01$). This latter fact suggests the importance of baryonic resonances in the excitation process; the former demonstrates that light-ion-induced reactions create hot, multifragmenting systems that can be observed in the laboratory with minimal kinematic distortion.

Above beam energies of about 5 GeV, the IMF cross sections become nearly constant, giving rise to the concept of limiting fragmentation. Another striking feature is the evolution of the IMF angular distributions, which evolve from gently forward peaked below 5 GeV to sideways peaking at higher energies.[3] This suggests that dynamical processes play at least some role in the IMF production process. Finally, the character of the IMF spectra–which are thermal in character, but skewed to very low energies[2] –gives a strong hint that multifragmentation occurs from an expanded/dilute source.

In order to investigate these and related phenomena associated with the dynamics of the multifragmentation process, E900 was conducted at the Brookhaven AGS accelerator during the 1996 proton running cycle. Here we report preliminary results of these studies, which were the first measurements of GeV hadron-induced reactions on a heavy target nucleus in which fragments were Z-identified with large solid-angle acceptance, good granularity and low energy thresholds.

MEASUREMENTS

Spectra of light-charged particles (LCP = H and He isotopes) and IMFs were measured with the Indiana Silicon Sphere (ISiS) 4π detector array[4] at the Brookhaven AGS accelerator (E900). Secondary positive beams of momentum 6.0, 10.0, 12.8 and 14.6 GeV/c and negative beams at 5.0, 8.2 and 9.2 GeV/c were incident on a ^{197}Au target. Here we associate the positive beam with protons, although at the lowest momentum a significant π^+ component is present in the beam. The negative beam is predominantly composed of π^- (> 90%). The event focus was defined by a 1 x 1 cm^2 (2.0 mg/cm^2) and 2 x 2 cm^2 (1.8 mg/cm^2) ^{197}Au foil suspended on two 50 μm tungsten wires attached to a 5 x 5 cm^2 Al frame. Average beam intensities were approximately 4 x 10^6 particle/s and blank runs were performed periodically to monitor possible effects of beam halo or muon background.

The ISiS array consists of 162 gas-ionization chamber/500 μm silicon/28 mm CsI detector telescopes, each having an energy acceptance of $0.7 \leq E/A \leq 95$ MeV/fragment nucleon. The angular coverage was $\Theta = 14°$ - 86.5° in the forward hemisphere and $\Theta = 93.5°$ - 166° in the backward direction, with nearly full ϕ coverage over this range.

Data acceptance was set by a multiplicity trigger that required fast signals from three or more silicon detectors in the array. In order to detect energetic hadrons with energy-loss too low to trigger the corresponding silicon discriminator, all CsI linear signals corresponding to E > 16 MeV were accepted for each event that passed the multiplicity trigger, along with its coincident silicon linear signal. These events provided information on the multiplicity of fast-cascade LCPs (and/or pions). In order to minimize beam-halo effects, the ISiS array was complemented by a 15x15 cm^2 total beam counter (TB), an annular ring veto scintillator (RV), a 28 mm diameter scintillator of beam definition counter (BC) and a segmented inner/outer scintillator array (UV) upstream from the target. Thus, the acceptance logic was:
TB * $\overline{RV}$ * BC * $\overline{UV}$ * ISIS
A second segmented inner scintillator array was also placed downstream from the target for beam alignment purposes.

MULTIPLICITY DISTRIBUTIONS

One important goal of these studies was to look for evidence of deposition-energy saturation for hadron momenta above $\sim$ 5 GeV/c. This is suggested by the observation of limiting fragmentation in inclusive studies,[5] 4π studies of multiplicity and total energy distributions from ^{3}He-induced reactions,[2] and both INC and BUU simulations[2,6] of the reaction dynamics.

In Fig.1, we show the probability distributions, $N_c/\sum N_c(i)$ as a function of total charged-particle multiplicity, $N_c = N(LCP) + N(IMF)$, where at least one IMF must be present. Fast cascade particles (as described above) are not included in these distributions. In Fig. 1, the left-hand frame gives the data for proton beams and the right-hand frame presents that for π^- beams. The center frame compares the results for p and π^- projectiles at the same total kinetic energy of 9.1 GeV (9.2 GeV/c π^- and 10.0 GeV/c p). Charged-particle multiplicities up to $N_c \sim 30$ are measured in each case, with a most probable value of $N_c \sim 10$. The measured IMF probability distributions are shown in Fig. 2 in the same format as those for total charged particles in Fig. 1. Measured IMF multiplicities up to $N_{IMF} = 10$ are observed, comparable to results for the 4.8 GeV ^{3}He + ^{197}Au reaction,[2] peripheral ^{197}Au + ^{197}Au reactions[7] and heavy-ion-induced reactions on similar systems.[8]

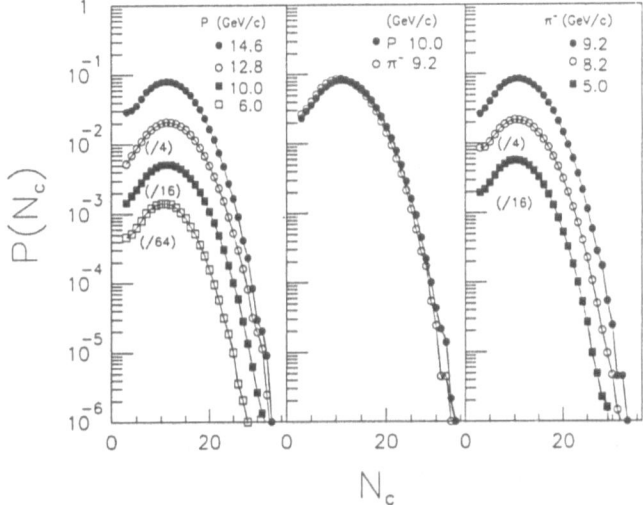

Figure 1. Probability distributions as a function of beam momentum for measured charged-particle multiplicities from ^{197}Au for events with at least one IMF. Left: proton beams; right: π^- beams, and center: p and π^- beams at 9.1 GeV total energy. Distributions are scaled according to legend on figure.

Two common features characterize Figs. 1 and 2. First, the observed multiplicity distributions in each figure are remarkably insensitive to beam momentum for both hadron types. This result indicates that the deposition energy is nearly constant from 5 - 15 GeV/c incident momentum and can be understood as a tradeoff between the higher available energy of the projectile and the increased transparency of the heavy target nucleus to fast cascade hadrons.[6] Thus, these data substantiate previous conclusions concerning a similar observation in the 1.8 - 4.8 GeV ^{3}He + ^{nat}Ag system,[2] and also provide a more complete understanding of the origin of limiting fragmentation.[5]

A second observation is that both projectile types yield nearly identical charged-particle and IMF multiplicity distributions, as illustrated by comparison of the p/π^- results at 9.1 GeV total energy in the center frames of Figs. 1 and 2. Previous emulsion studies suggested a similar conclusion.[9] This independence of hadron type indicates that the initial collision step is similar in both cases and that, on average, the subsequent hard scatterings and resonance excitations evolve similarly during the fast cascade–at least as far as deposition energy in the heavy residue is concerned. Both the

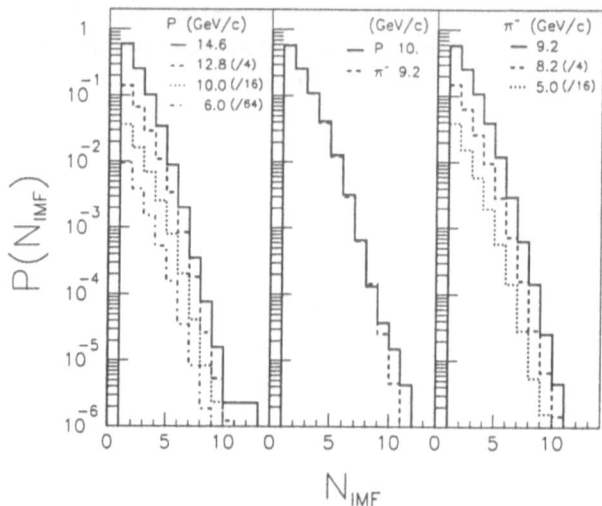

Figure 2. Probability distributions as a function of beam momentum for measured IMF multiplicities from ^{197}Au for events with at least one IMF. Left: proton beams; right: π^- beams, and center: p and π^- beams at 9.1 GeV total energy. Distributions are scaled according to legend on figure.

bombarding-energy and hadron-type independence are in qualitative agreement with intranuclear cascade calculations of Toneev[10] over this range of beam momentum.

In order to illustrate the dependence of the observed IMF multiplicity on collision violence, in Fig. 3 we have plotted the average number of IMFs, $\langle N_{IMF} \rangle$, versus the total detected charge in an event, Z_{obs}, for all systems. These values do not include corrections for geometric acceptance ($\sim 70\%$) or energy thresholds (E/A ~ 0.7 MeV/A), which are nearly identical for all systems. The parameter Z_{obs} has previously been shown[2] to be a more sensitive gauge of excitation energy deposition than LCP or total-charged-particle multiplicity in light-ion-induced reactions; it scales nearly linearly with total transverse energy and total thermalized energy. Again, little variation is observed as a function of beam momentum or projectile type. The Z_{obs} dependence is very similar to the 4.8 GeV ^{3}He + ^{197}Au reaction[2] and the 6 GeV ^{86}Kr + ^{197}Au reaction[8] for events with the same value of Z_{obs}. Thus, the dependence of N_{IMF} on deposition-energy-related variables appears to be similar for all systems, including heavy ions. The average observed IMF multiplicity for all events with one or more IMFs is $\langle N_{IMF} \rangle \approx$ 1.6 - 1.7, also nearly independent of beam momentum or hadron type.

Although the multiplicities for ejectiles that appear to originate in a thermal-like source show little variation with beam momentum or hadron type, when the fast cascade component of the data is added in, there is a systematic evolution with beam momentum. In Fig. 4, the transverse energy distributions, $E_T = E^*\sin\theta$, are plotted for each system for events with one or more IMFs. Here one observes the expected increase in E_T with beam momentum.

ANGULAR AND CHARGE DISTRIBUTIONS

The extent to which dynamical versus thermalized mechanisms contribute to multifragmentation remains an important question. At energies below ~ 5 GeV the IMF

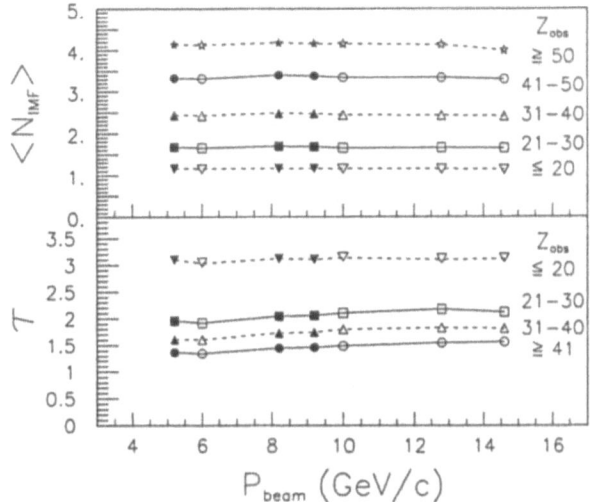

Figure 3. Beam momentum dependence of average IMF multiplicity (upper) and power-law exponent (lower) gated on Z_{obs} intervals, as indicated on figure.

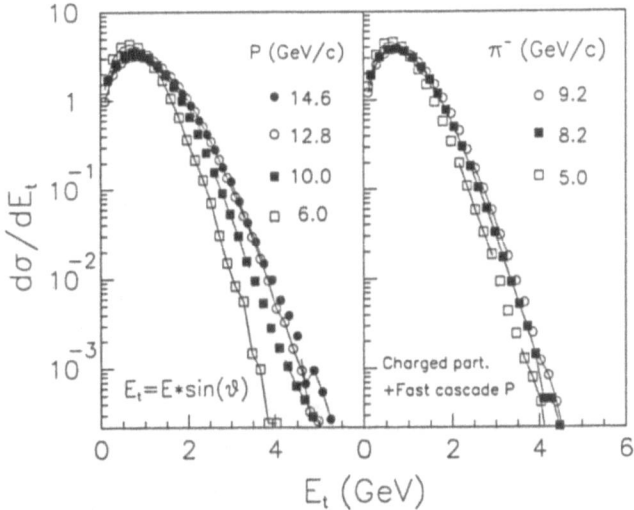

Figure 4. Total transverse energy distributions for: left frame: proton-induced reactions and right frame: π^- -induced reactions.

angular distributions are slightly forward peaked, consistent with source velocity estimates, and depend only weakly on collision violence.[1-3] Inclusive studies at higher energies, however, have indicated a peaking at angles in the vicinity of 60°.[3] In Fig. 5, the angular distributions for Li fragments are plotted for the 14.6 GeV/c p + ^{197}Au reaction as a function of Z_{obs}. For the inclusive and low Z_{obs} values, the data resemble the lower energy results. However, as Z_{obs} increases, a distinct peaking of the angular distribution develops near an angle of 50°.

The observed peaking, while not a dominant contributor to the total cross section, clearly indicates the presence of dynamical fragment-formation processes in these reactions. This phenomenon is suggestive of shock-wave production of fragments, previously hypothesized by Glassgold et al.[11] and several subsequent authors. This aspect of the data is currently being investigated in greater detail. Another aspect of these data that was originally stimulated by proton-induced reactions on complex nuclei is that of fragment charge distributions.[12,13] Under the assumption of thermalized emission, these studies have examined the charge distributions in terms of a power law, $\sigma(Z) \propto Z^{-\tau}$. In Fig. 6 representative charge distributions are shown as a function of Z_{obs}. As with the multiplicity distributions, the charge distributions are nearly identical for all seven systems. There is a strong dependence of the power-law exponent τ on Z_{obs}, again providing an underlying basis for understanding limiting fragmentation.

A power-law analysis has been performed on the data as a function of projectile type, incident energy and Z_{obs}. These are shown in Fig. 3. For low Z_{obs}, values of $\tau \sim 3.5$ are found, typical of dynamically-produced IMFs in preequilibrium or coalescence-like processes.[14] With increasing collision violence, the τ parameters systematically decrease to near $\tau \sim 1.3$ for large Z_{obs} values, becoming significantly smaller than values of well-documented equilibrated systems formed in both light- and heavy-ion reactions.[14] Fi-

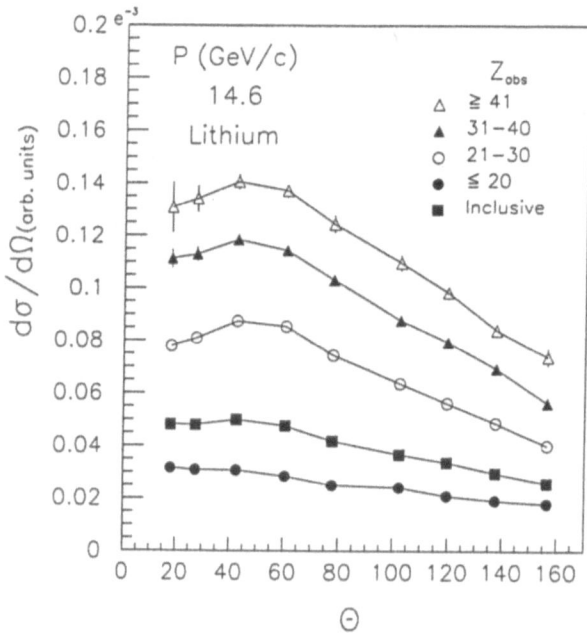

Figure 5. Angular distributions for 14.6 GeV/c p + ^{197}Au reaction for inclusive lithium spectra and as a function of Z_{obs}, as indicated on figure.

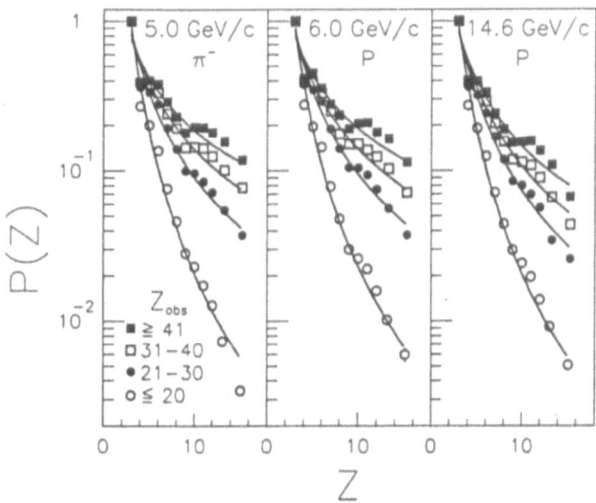

Figure 6. Total charge distributions for 5.0 GeV/c π^- (left), 6.0 GeV/c p (center) and 14.6 GeV/c p (right) bombardment of ^{197}Au. Results are shown as a function of Z_{obs} and are normalized to P(Z) = 1 for Li fragments in all cases.

nally, large-angle correlation studies suggest that the dominant equilibrium-like component of the most violent events is accompanied by dynamically-produced IMFs.[15] Since the charge distributions for such events fall off much more rapidly than for thermal-like processes, removal of this component from the power-law fits would lead to further reduced τ values.

CONCLUSIONS

Here we have described preliminary results from the first study of GeV-hadron-induced reactions on heavy nuclei in which both LCPs and IMFs have been Z-identified with large solid- angle acceptance and low energy thresholds. The observed LCP and IMF multiplicity and charge distributions show little dependence on either hadron type or beam momentum for beam momenta between 5 - 15 GeV/c. This suggests a saturation of deposition energy for hadron- induced reactions and a lack of sensitivity to hadron type in initiating the fast cascade.

Measured IMF multiplicities and charge distributions depend strongly on collision violence. The results are qualitatively consistent with a dynamic production mechanism for low deposition energies and equilibrium-like properties for the most violent events. However, the angular distributions are found to exhibit forward peaking near 50°, even for high deposition- energy events, providing evidence for some dynamical IMF production mechanism for these events as well.

Acknowledgments

This research was sponsored primarily by the U.S. Department of Energy with additional support from the U.S. National Science Foundation and the National Research Council of Canada.

REFERENCES

1. W.G. Lynch, Ann. Rev. Nucl. Sci. 37, 439 (1987).
2. K.B. Morley et al., Phys. Rev. C 53, 737 (1996); E. Renshaw Foxford et al., Phys. Rev. C 54, 737 (1996).
3. L.P. Remsberg and D.G. Perry, Phys. Rev. Lett. 35, 361 (1975).
4. K. Kwiatkowski et al., Nucl. Instrum. Meth. A 360, 571 (1995).
5. G. Rudstam, Z. Naturforsch. Teil A 21a, 1027 (1966); J. Benecke et al., Phys. Rev. 188, 2159 (1969).
6. G. Wang et al., Phys. Rev. C 53, 1811 (1996).
7. J. Pochodzalla et al., Phys. Rev. Lett. 75, 1040 (1995); J. Hubele et al., Z. Phys. A 340, 263 (1991).
8. G.F. Peaslee et al., Phys. Rev. C 49, R2271 (1994); M.B. Tsang, private communication.
9. N.A. Perfilov et al., Nuclear Reactions Induced by High-Energy Particles (Akademiya Nauk, SSSR, 1962; J. Hudis, in Nuclear Chemistry), Academic Press, ed. L. Yaffe, 1968, P.169.
10. V. Toneev et al., Nucl. Phys. A 519, 463c (1990).
11. A.E. Glassgold, W. Heckrotte and K.M. Watson, Ann. Phys. (New York) 6, 1 (1959)
12. R.E.L. Green et al., Phys. Rev. C 29, 1806 (1984).
13. N.T. Porile et al., Phys. Rev. C 39, 1914 (1989); J.E. Finn et al., Phys. Rev. Lett. 49, 1321 (1982).
14. K. Kwiatkowski et al., Phys. Lett. B 171, 41 (1986); J.L. Wile et al., Phys. Rev. C 45, 2300 (1992).
15. G. Wang et al., Phys. Lett. B 393,290 (1997).

SEARCH FOR A SHORT-LIVED DIBARYON IN Au + Au COLLISIONS: STATUS OF AGS EXPERIMENT E896

Morton Kaplan

Carnegie Mellon University
4400 Fifth Avenue
Pittsburgh, Pennsylvania 15213

for the E896 Collaboration

ABSTRACT

A new high-rate experiment has been initiated at the AGS to search for a short-lived H^0-dibaryon in 11.6 A GeV/c Au + Au collisions. The heart of the detection system is a large distributed-drift-chamber (DDC) in a 1.7 T magnetic field, which will provide tracking of charged particles from H^0 decay. Directly upstream of the DDC is a 6.8 T superconducting dipole sweeper magnet and a high-Z collimator system, which combine to remove most charged particles produced in the target and yield a low background in the critical tracking region of the DDC. The implementation of the system, including additional detectors, is described, and our experience from the recently ended first run is reported.

INTRODUCTION

The objective of experiment E896 at the Brookhaven Alternating Gradient Synchrotron is to search for a short-lived dibaryon and study hyperon production in 11.6 A GeV/c Au + Au collisions. To carry out the design, construction, and operation of the experiment, a new collaboration has been formed with membership representing the institutions listed in Fig. 1.

The H^0-dibaryon was first predicted theoretically almost twenty years ago. It consists of six quarks in the configuration $uuddss$, and is expected to have the lowest mass of all possible six-quark combinations. Many theoretical calculations of the H^0 mass have been made over the years, and appear to cluster around 2.15 GeV/c^2, which is below the $\Lambda\Lambda$ threshold of 2.23 GeV/c^2. Hence if the H^0 exists, it will most likely decay weakly rather than strongly, which places a lower limit of ~0.1 ns on the lifetime.

E896 Collaborating Institutes

Figure 1. Institutions (and logos) with membership in the E896 Collaboration.

However, the predicted mass is above the Λn threshold (2.06 GeV/c^2), so the decay must have $\Delta S=1$, which places an upper limit of $\sim$100 ns on the H^0 lifetime. Within this mass (lifetime) range, there are a number of possible decay channels for the H^0-dibaryon, and E896 has been designed for sensitivity in detecting several of the most probable of these. This broad capability of the experiment ensures that E896 will also be efficient in detailed studies of known strange baryons, such as Λ production systematics and correlations.

THE EXPERIMENT

Why do we think that E896 can improve significantly over previous H-dibaryon searches, none of which has yet produced a definitive positive result, or for that matter, a definitive negative result? There are several parts to the answer.

1) In 11.6 A GeV/c Au+Au collisions, the Λ production yield is greatly enhanced over p+A production. Hence the probability of Λ -Λ coalescence (as a possible H^0 production mechanism) is significantly increased.

2) As indicated above, mass estimates for the H^0 suggest weak decays with lifetimes in the 0.1-100 ns regime. Our experiment is designed for sensitivity in the range $c\tau \sim 4$ cm to $c\tau \sim 100$ cm, which spans the short-lifetime region very effectively.

3) With a strong sweeping field and a high-Z collimator, simulations indicate that the primary tracking detector will have a low background in the critical tracking region.

4) Multiple detectors in the experimental setup allow event selection and reconstruction, and provide redundancy for several possible H^0 decay channels.

5) The combined multiple detector system is designed for high-beam-rate running, to enable collection of a large data set of triggered central collision events.

We show in Fig. 2, as an example, a schematic diagram of the expected H^0 decay via the $\Sigma^- p$ channel, one of the more probable decay modes. Should an H^0-dibaryon

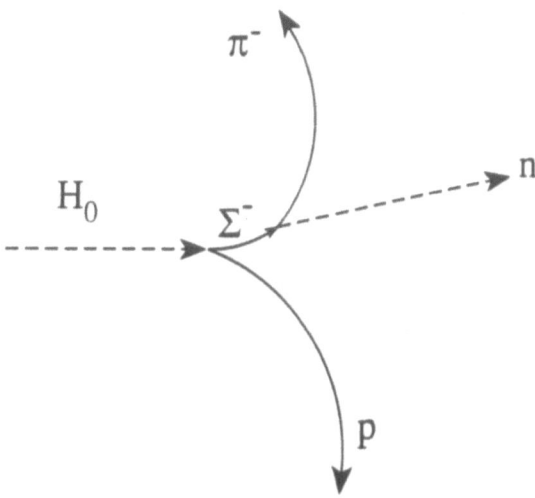

Figure 2. Schematic diagram indicating the decay of an H^0-dibaryon via one of its predicted probable decay modes. The curvatures of the charged-particle tracks in the magnetic field are exaggerated for illustrative purposes. The E896 detection system is expected to record the unique signature provided by the Σ^-, p, π^-, and n particles.

decay in the active region of the E896 primary tracking detector, a unique H^0 signature could be recorded by tracking the Σ^- and p originating from a neutral particle decay, and observing a kink in the Σ^- trajectory resulting from its decay to a π^- plus a neutron. A multifunctional neutron spectrometer (MUFFINS) placed downstream would detect the neutron in kinematic coincidence and measure its energy.

A detailed schematic view of the E896 experimental setup is presented in Fig. 3. Referring to this figure, we shall now briefly describe the major elements of the detector system.

The detection system centers around two large dipole magnets, a superconducting sweeper magnet with a central field of 6.8 Tesla and a "room temperature" analyzing

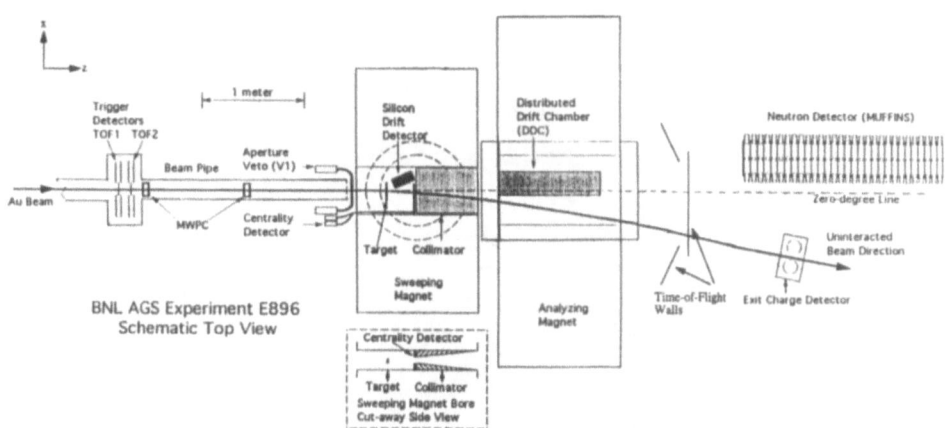

Figure 3. Experimental layout of the major components in the E896 multidetector system.

magnet with a 1.7 Tesla field. Within the pole gap of the sweeper magnet are housed the Au target (on a target wheel), a high-Z (W,Pb) reaction-product collimator, the centrality detector, and the silicon-drift detector. The adjacent downstream analyzer magnet contains a large distributed drift chamber (DDC), which serves as the primary tracking detector in the experiment. The magnets are oriented at an angle of approximately 3 deg relative to the incident Au beam, to optimize the acceptance of the DDC for H^0-dibaryon decays. Upstream of the dipole magnets are a series of detectors in the beamline which serve for beam-vectoring purposes and trigger detectors for the downstream time-of-flight (TOF) walls. In addition to the TOF scintillator arrays, the area downstream of the analyzing magnet contains an "exit charge" detector, positioned to collect surviving projectile-like fragments for use in event selection, and the MUFFINS neutron detector assembly. The latter consists of a cylinderical stack of 30 scintillator disks, each 40 cm dia., with multiple phototube readout on each disk. The determination of neutron hit positions throughout the stack permits derivation of the neutron momentum vector, which can be traced back and correlated with tracking information from the DDC.

The basic strategy in the experimental design is that essentially all charged particles produced by Au beam interactions in the Au target are strongly deflected by the sweeper magnet, leaving the DDC relatively free of primary charge tracks. The intense photon flux, however, produces many secondary charged particles which enter the DDC. Thus on the basis of detailed simulations we have designed and built the reaction-product collimator out of tungsten and lead, with a geometrical configuration and aperture which minimizes background in the DDC yet has high transmission for neutral particles originating from the target.

Figure 4 is a photograph of the superconducting dipole sweeper magnet, together with its cryogenic reservoir. The magnet was built to specifications for E896 by Oxford Instruments, and in addition to the maximum field of 6.8 T, must be able to operate continuously without quenching in close proximity to the 1.7 T analyzing magnet.

The collimator installed in the sweeper magnet is shown schematically in Fig. 5. The darker lines indicate sections made of tungsten, and the aperture can be seen to

Figure 4. The superconducting dipole sweeper magnet viewed from the upstream beam direction.

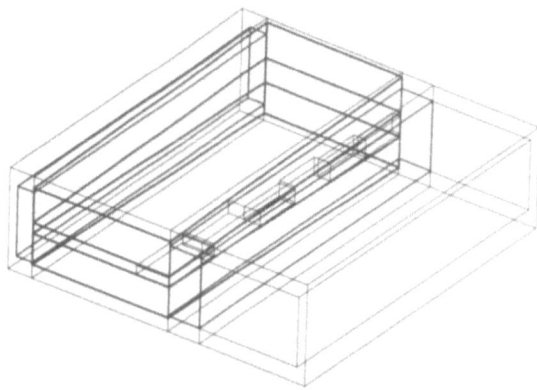

Figure 5. Diagram of the reaction-product collimator, which is mounted in the sweeper magnet gap.

increase in vertical dimension from the lower left to the upper center of the figure. The right hand section of the collimator assembly allows unreacted beam to pass through.

Also mounted in the sweeper magnet, in close proximity to the target, is the silicon-drift-detector array (SDDA). This device is a prototype of the silicon-vertex-tracker (SVT) being developed for the STAR detector at RHIC. The purpose of the SDDA in the present experiment is to provide tracking capability for possible H^0 particles in the very short lifetime region, $c\tau \sim 4$ cm.

As the production of H^0-dibaryons is most likely to occur in the more central Au+Au collisions, we have installed multiplicity detectors inside the sweeper magnet to allow event selection based on centrality criteria. Figure 6 shows the results of simulations which demonstrate the correlation between impact parameter (or centrality) and the number of hits recorded by the multiplicity detectors. These detectors consist of scintillator sheets covered with Pb converters and mounted on the upper and lower faces of the collimator assembly. They are read out via optical fibers coupled to photomultiplier tubes outside the sweeper magnet.

In addition to the multiplicity detectors just discussed, centrality criteria are also determined by the response of the exit-charge detector, which consists of scintillators

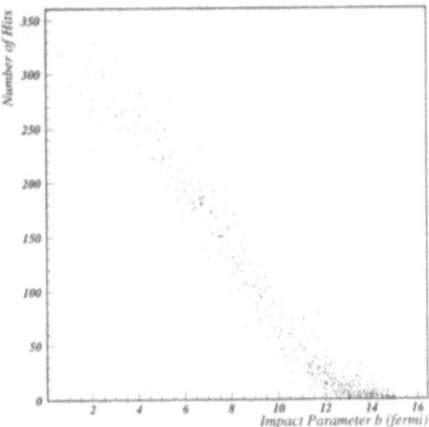

Figure 6. Simulation showing the correlation between impact parameter and the number of hits recorded by the multiplicity (centrality) detectors.

located downstream in the uninteracted beam direction, and hence is sensitive to projectile remnants.

The analyzing magnet is a standard BNL 48D48 copper coil magnet, with pole pieces modified to accomodate the DDC assembly and supporting structure. A view of this magnet from the downstream direction is shown in Fig. 7. Looking through the magnet coils and gap, the sliding rail support for the DDC can readily be seen, as well as the face and gap of the sweeper magnet immediately upstream.

Figure 7. The dipole analyzer magnet viewed from downstream.

During normal operation, of course, the DDC is mounted on the sliding rails and enclosed within the pole gap of the analyzing magnet. The assembled distributed drift chamber, minus most of its electronics and cabling (for clarity), is shown in Fig. 8.

Figure 8. The distributed drift chamber mounted on the sliding rail platform.

The DDC design consists of 12 nearly identical modules, each of which contains 12 planes. To date, however, 10 complete modules have been constructed. Half of the modules (the upstream half) have 3 mm spacing between the wires, and the other half (the downstream half) have 4 mm spacing. The individual modules are interconnected with O-ring seals and the entire assembly is made gas tight by O-ring sealed window frames on each end. The DDC is operated using a 50/50 Helium/Ethane gas mixture, a combination which has been found to provide adequate gain and drift velocity, while minimizing multiple scattering effects.

The expected DDC data acquisition rate is ~1000 events per AGS beam spill, a fact which when combined with the high track-finding efficiency provides a very sensitive search instrument for H^0-dibaryons which decay in the chamber. For very short-lived particles, the SDDA will have relatively favorable acceptance (since it is close to the target), and thus serves as a complementary tracking detector to the DDC.

Downstream of the DDC are the three time-of-flight walls, which are constructed from many long, narrow scintillator slats and serve for charged particle identification. The TOF system has high granularity in the horizontal dimension, excellent timing resolution of 80-100 ps, and is positioned to optimize detection of charged decay products from known strange baryons as well as predicted H^0-dibaryons. A photograph of the TOF system is given in Fig. 9. The center TOF wall (from AGS experiment E877) is being used by courtesy of McGill University, and the two outer TOF walls were built by Rice University.

Figure 9. The time-of-flight system consisting of three walls of long, narrow, scintillator slats.

CURRENT STATUS OF E896

The first Au beam run of experiment E896 took place in January, 1997, and data from the DDC were recorded in the period Jan. 21-31, 1997. At that time 10 (out of the planned 12) DDC modules were installed in the analyzer magnet, and 8 of these appeared to operate reliably. The DDC readout boards exhibited higher noise levels than expected, however the problem has been effectively diagnosed (but not yet fixed).

Due to a mechanical accident during delivery, the superconducting sweeper magnet suffered internal damage and could only be operated in a stable mode at a maximum field of 4.9 T, rather than its design value of 6.8 T. Even under these restrictive conditions, however, the effectiveness of the collimator and the favorable low track density in the DDC were established. The TOF walls became operational during the run and recorded good signals which should be useful in identifying fragments from known strange particle decays. The MUFFINS neutron detector ran continuously and well, yielding a large neutron data set. The SDDA development and assembly experienced unexpected delays, resulting in only two Si detector planes being installed in the sweeper magnet. However, particle signals were observed from these planes, and substantial progress was made in understanding the SDDA behavior and response.

During the 10-day data-taking period the experiment was run with an average beam rate of ~40K Au ions per spill (occasionally as high as 60K), and ~800 central triggered DDC events per spill were written to tape using a 10% Au target. Thus the high-rate performance of the DDC has been satisfactorily demonstrated. A total of ~50M central events were collected in this initial run, which is adequate to begin data analyses using software developed in parallel with the hardware construction.

In closing, we show in Figs. 10-12 a selected sampling of events taken from the DDC on-line display, illustrating the variety of "tracks" observed with one particular set of wire planes.

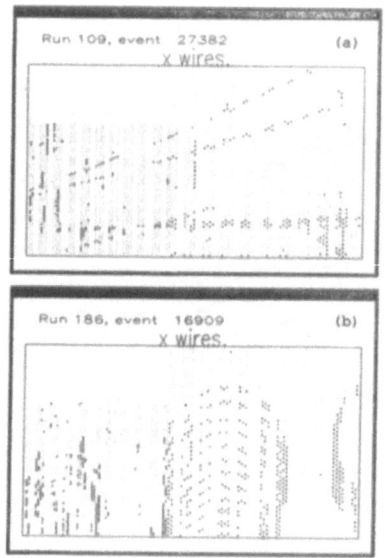

Figure 10. Display of two selected events recorded in the DDC. In (a) a pair of smooth tracks was produced by two negatively charged particles. In (b) the semicircular patterns result from low energy electrons spiraling in the magnetic field.

In looking at the interesting track patterns in Figs. 10-12, one should be particularly cautious about interpreting the significance of any individual track or group of tracks. Not only is the data very preliminary, but only a two-dimensional view is given of a three-dimensional trajectory, and no cuts or criteria (such as vectoring to the target) have been applied. Rather, the present objective is simply to demonstrate that tracks of various kinds are seen in the DDC, and that the overall track density and the behavior in the magnetic field are reasonable.

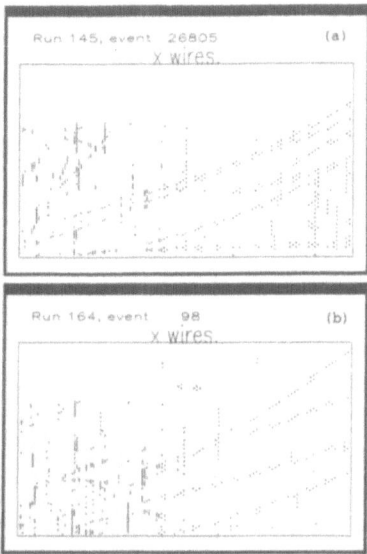

Figure 11. Display of two more selected events recorded in the DDC. In both (a) and (b) there are smooth tracks and some indications of possible decays of neutral particles.

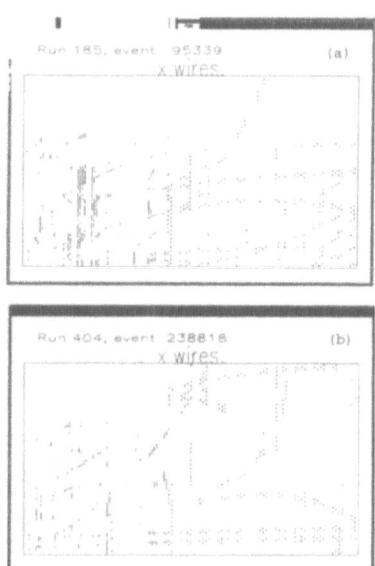

Figure 12. Yet another two selected events recorded in the DDC. In both (a) and (b) there appear to be tracks suggesting decay of a neutral particle into two oppositely charged fragments, and an apparent kink in the negative particle trajectory.

ACKNOWLEDGEMENTS

M.K. wishes to thank the Division of Nuclear Physics, U. S. Department of Energy, for support of this work at Carnegie Mellon University.

Directed Flow for p, d, t Emission at Target Rapidity in 10.2 A.GeV Au + Au Collisions

H. C. Britt,[1] M. N. Namboodiri,[2] and T. C. Sangster [2]
for Experiment E866 (the E802 Collaboration)

[1]University of Maryland and Lawrence Livermore National Laboratory
[2]Lawrence Livermore National Laboratory

INTRODUCTION

A major goal in the study of relativistic heavy ion collisions is the determination of the equation of state, EOS, of nuclear matter under conditions of temperature and density far removed from those normally encountered in the studies of nuclei near their ground states. An experimental tool sensitive to the EOS in hot dense systems is the study of nuclear flow[1]. For collisions at energies up to 1 A GeV, compressional bounce off of spectator nucleons and fragments in the reaction plane, squeeze out of participant nucleons and pions and collective radial expansion have been predicted[1, 2, 3] and observed[4]. At the higher energies, 10-200 A GeV, available at AGS and CERN similar flow effects have been predicted[5, 6, 7]. It is expected that in this energy regime it will be possible to study reactions in which initial compression can reach up to 3 times normal nuclear densities and in this case the resulting flow may become sensitive to new characteristics of the EOS such as the behavior of collision cross sections for nucleons and excited baryons and for mesons at high relative momentum[8, 9, 10]. At the AGS compressional bounce off has been reported from experiments E866[11], E877[12], and E875[13]. In addition, it has been calculated that radial flow may measurably effect the spectral shapes for composite fragments, (d, t, He), emitted at mid rapidities [7]. In this paper we report on measurements of directional flow, "bounce off", for protons, deuterons and tritons from the reaction Au+Au at a bombarding energy of 10.2 A GeV and on the relative production rates for these particles.

APPARATUS

The results reported are a part of the experiment E866 which represents modifications of the earlier E802 experiment[14] in order to study the Au+Au system. As part of experiment E866 a forward angle scintillator hodoscope was installed to determine on an event-by-event basis the orientation of the reaction plane from an observation of the mean position of the projectile spectator fragments. An improved array of PHOSWICH scintillator telescopes was also installed to measure protons, deuterons and tritons emitted from the target spectator. The projectile spectator hodoscope, HODO, was installed in front of the E802 zero degree calorimeter, ZCAL, at a distance of 11.5 meters from the target. It consists of a 40 element array of 1cm x 8mm x 40cm scintillator slats oriented in the X direction followed by an equivalent array oriented in the Y direction. The mean position of an undeflected beam particle in the array was determined by direct measurement by mixing a fraction of the beam events into the trigger. For each event the position of the spectator center of gravity was connected by a line to the estimated position of the undeflected beam to give an orientation for the reaction plane. The PHOS array was an upgrade of the array used previously [16] in experiment E859 to measure proton and deuteron spectra at target rapidities[17]. It consisted of 100 fast/slow scintillator modules arranged in the cylindrical geometry covering a pseudorapidity range, $-1.0 < \eta < 0.5$. All modules provided separation between p, d, and t and energy acceptance in the range 30 to 200 MeV.

METHOD

The trigger used in accumulating the flow data set was a minimum bias plus a requirement that at least one PHOS module had an event. For each identified p, d, or t the azimuthal angle of the appropriate module relative to the identified reaction plane was determined. Then distributions of the two dimensional momentum projections, P_x vs P_y, were developed for each module and particle. These distributions could also be developed as a function of the centrality of the collision as determined from the energy in ZCAL. There is a significant dispersion in the reaction plane determination due to several factors and this has the effect of decreasing the measured flow signal below the value characteristic of a precise reaction plane determination. The major sources of this dispersion are the fluctuations in the charged particle emissions relative to the reaction plane and the uncertainty in the position of the beam spot. This dispersion is minimum for "hard peripheral" collisions corresponding to 730 GeV$< E_{ZCAL} < $ 1580 GeV (impact parameter, b, approximately 5-10 fm). The dispersion in the measured reaction plane orientations was estimated using a GEANT Monte Carlo routine which reproduced conditions of the experimental geometry, target thickness and various absorbers. A gaussian fit to the estimated dispersion gives a standard deviation of approximately 40 degrees in the b=5-10 fm region.

RESULTS

Figure 1 shows the d/p and t/p ratios as a function of η for the limited energy acceptance of the PHOS array. Figure 2 shows d/p and t/p ratios as a function of ZCAL energy for the spectator region ($\eta < 0$). The relative decrease in d, t production at low values of E_{ZCAL} (i.e. more central collisions) is consistent with the expected trend based on a coalescence picture. In a simple coalescence model one expects a larger freezeout volume (and fewer complex particles) for the highly excited participant region as compared to the cooler spectator region.

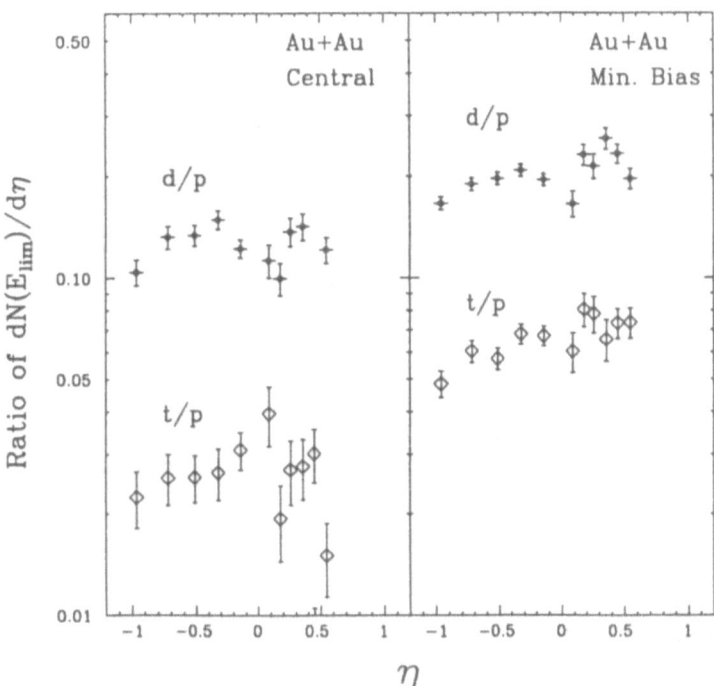

Figure 1. Ratios of deuteron to proton and triton to proton emission rates as a function of pseudo-rapidity for central and minimum bias Au+Au collisions.

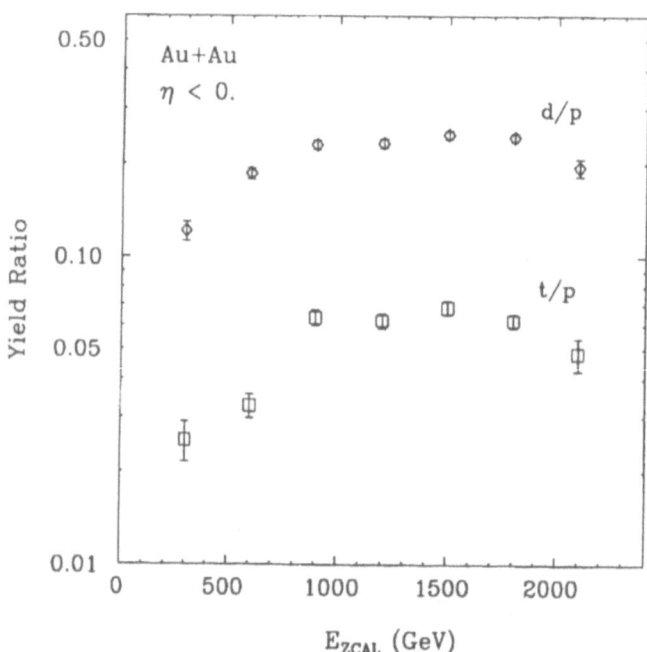

Figure 2. Ratios of deuteron to proton and triton to proton emission rates for pseudo-rapidity less than zero as a function of energy in the zero degree calorimeter, ZCAL.

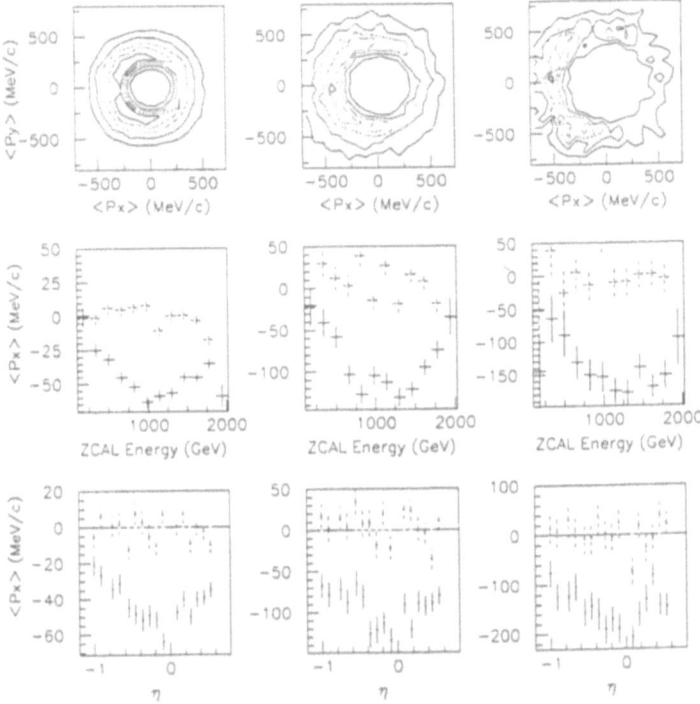

Figure 3. Mean transverse momentum, $\langle P_x \rangle$, as a function of $\langle P_y \rangle$, ZCAL energy, and pseudo-rapidity for protons (left), deuterons (center), and tritons (right). Points with dashed error bars are for $\langle P_y \rangle$.

A composite of flow results obtained for protons, deuterons and tritons is shown in Fig. 3. The 2D distribution P_x vs. P_y shows that the mean of the overall distribution of events is symmetric in P_y and shifted to negative values for P_x. The results can be seen more quantitatively as a function of ZCAL energy and pseudorapidity in other panels of Fig. 3. The results for $E_{ZCAL} > 1000$ GeV and target rapidities $-0.5 < \eta < 0.5$ show an "apparent" flow, equal within experimental uncertainties, of approximately 60 MeV/c/A for p, d and t.

Currently there are three theoretical models which have been able to quantitatively characterize hadron and produced particle inclusive distributions for AGS Au+Au experiments and have gone on to make predictions on flow quantities. These models (ARC[11], ART 1.0[12], and RQMD 1.07[13]) are all microscopic relativistic collision models including baryons, produced mesons and resonances. They all use experimental interaction cross sections where available. They, however, differ significantly in how they handle the particle transport during the collision process. To compare the experimental results with the theoretical calculations it is essential to take into account the dispersion and the limited acceptance present in the experimental data and to compare for a consistent range of impact parameters. In the comparisons below the reaction plane dispersion has been unfolded from the experimental data and the theoretical calculations have been performed for the energy and impact parameter acceptance of the experiment. Figure 4 shows a comparison of the dispersion corrected data for protons in the range b=0-10 fm compared to RQMD calculations (version 1.07) with and without the inclusion of nuclear potentials. Within statistics the data and the pure

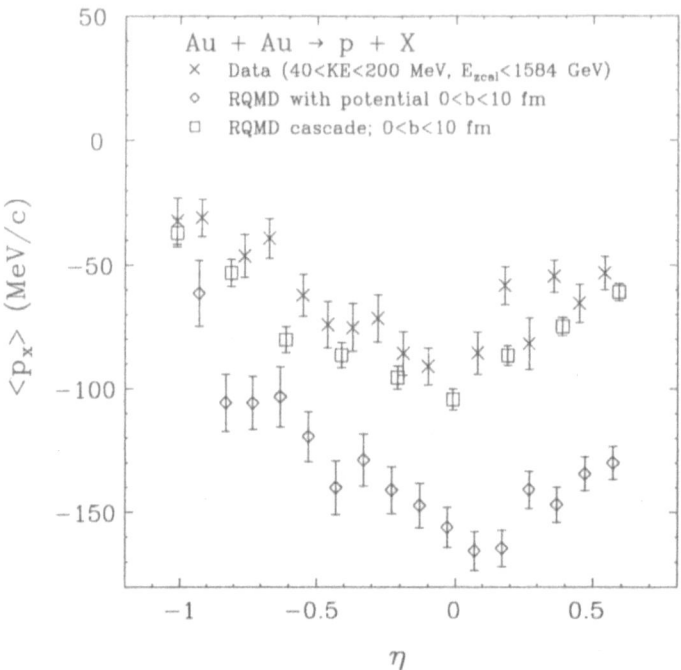

Figure 4. Comparison of mean transverse momenta, $\langle P_x \rangle$, for protons with predictions from RQMD (version 1.07) in cascade mode and with inclusion of nuclear potentials. Calculations and data are for the impact parameter range $0 < b < 10$ fm. Data have been corrected for estimated dispersion in the reaction plane determination.

cascade calculation agree. Thus, these results for the target spectator source do not indicate the need for the inclusion of nuclear potentials. This result can be compared to the case of radial flow in the participant region using RQMD (version 1.07) where the proton and deuteron inclusive spectra agree rather well with a calculation including potentials[7]. More recently comparison of the E877 flow data with RQMD (version 2.3) has indicated that in this version the inclusion of nuclear potentials is required to reproduce the experimental results[18]. Experimentally the E866 and E877 data appear equivalent (see below)

Figure 5 shows a comparison of dispersion corrected data for impact parameter ranges b=5-8 fm ($E_{ZCAL} = 730 - 1280$ GeV), b=8-10 fm ($E_{ZCAL} = 1280 - 1580$ GeV) and b=0-14 fm (min. bias) to calculations using the ART model. The results show that the data track the cascade calculation in all cases. The addition of a soft equation of state modifies the minimum $\langle P_x \rangle$ in an impact parameter sensitive manner which does not appear to be represented in the data set.

Additionally these results have been compared in detail to calculations from ARC[19]. Here it is again seen that the results are in good agreement with calculations from a cascade model without inclusion of nuclear potential effects. However, this cascade does have a restriction to repulsive scattering at low energies which is not included explicitly in the other models. A measurement[7] in the region of the projectile spectator in AGS experiment E877 shows a maximum, dispersion-corrected $\langle P_x \rangle$ of +100-120

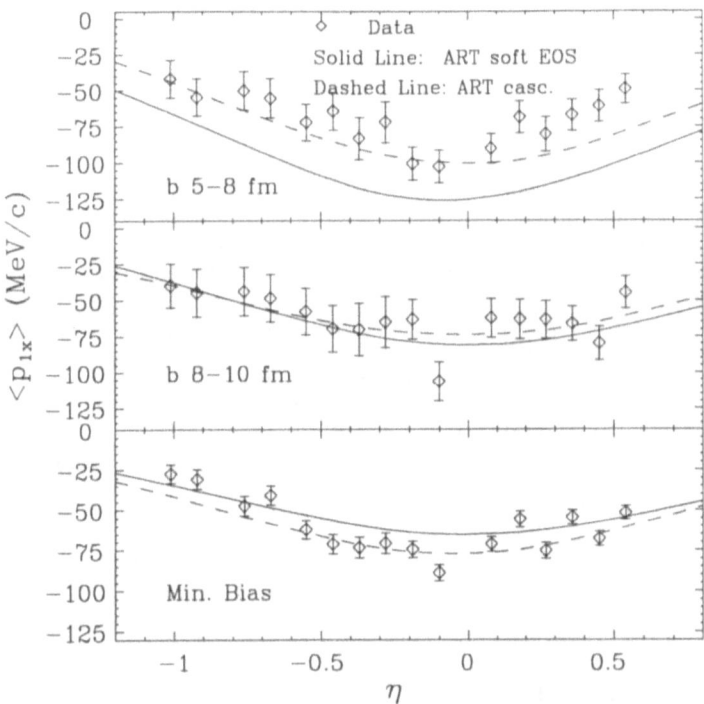

Figure 5. Comparison of mean transverse momenta, $\langle P_x \rangle$, for protons with predictions from ART in cascade mode and with inclusion of nuclear potentials. Calculations for impact parameter ranges b=5-8 fm and 8-10 fm are compared to data gated on appropriate ranges in ZCAL energy. Data have been corrected for estimated dispersion in the reaction plane determination.

MeV/c for protons at rapidities slightly less than beam rapidity. A comparison has been made where both the data presented here and the detailed flow tensors measured in experiment E877 were compared in detail to ARC calculations[19]. In both cases remarkably good agreement is found between data and predictions for ARC using realistic experimental acceptance filters. We conclude that within the stated statistical and systematic errors the two measurements are in agreement.

ACKNOWLEDGEMENTS

We are pleased to acknowledge the help of G. Peilert in performing some of the RQMD calculations. Experiment E866 is supported by the U.S. Department of Energy (ANL, BNL, UC-Berkeley, UC-Riverside, Columbia, LLNL, and MIT), by NASA (UC-Berkeley), and by the US-Japan High Energy Physics Collaboration Treaty.

REFERENCES

1. H. Stöcker, J. A. Maruhn, and W. Greiner, Phys. Rev. Lett 44, 725 (1982).
2. H. Stöcker, et al., Phys. Rev. C 25, 1873 (1982).
3. H. Stöcker and W. Greiner, Phys. Rep. 137, 277 (1986).
4. H. H. Gutbrod, A. M. Pozkznzer, and H. G. Ritter, Rep. Prog. Phys. 52,1267 (1989).
5. H. Sorge, H. Stöcker and W. Greiner, Ann. Phys. (NY) 192, 266 (1989); Nucl. Phys. A498, 567c (1989).
6. S. H. Kahana, Y. Pang and T. J. Schlagel, Nucl. Phys. A566, 465c (1993); Y. Pang, D. E. Kahana, S. H. Kahana and T. J. Schlagel, Nucl Phys. A590, 565c (1995); D. E. Kahana, D. Keane, Y. Pang, T. Schlagel and S. Wang, preprint 1995.
7. R. Mattiello, A. Jahns, H. Sorge, H. Stöcker, and W. Greiner, Phys. Rev. Lett. 74, 2180 (1995); R. Mattiello, H. Sorge, H. Stöcker, and W. Greiner, to be published.
8. J. Aichelin, A. Rosenhauer, G. Peilert, H. Stöcker and W. Greiner, Phys. Rev. Lett. 58, 1926 (1987).
9. F. deJong and R. Malfliet, Phys. Rev. C 46, 2567 (1992).
10. J. Ellis, J. Kapusta and K. Olive, Phys. Lett. B273, 122 (1991).
11. T. C. Sangster, Presentation at Quark Matter 95, Monterey, CA (1995).
12. J. Barrette, et al., Phys. Rev. Lett 73 2532 (1994), T. K. Hemmick for the E877 Collaboration in Proceedings of Quark Matter '96, Nuclear Physics A, to be published (1996).
13. G. Singh and P. L. Jain, Phys. Rev. C 49, 3320 (1994); P. L. Jain, G. Singh and A. Mukhodhyay, Phys. Rev. Lett. 74, 1534 (1995).
14. T. Abbott, et al., Nucl. Instr. Meth. A290, 41 (1990).
15. T. C. Sangster, et al., In preparation
16. J. B. Costales, et al., Nucl Instr. Meth. A330, 183 (1993).
17. L. Ahle et al. submitted to Phys. Rev. C.
18. T. K. Hemmick, in these proceedings.
19. D. E. Kahana in Proceedings of the HIPAGS '96 Conference, August 22-24, 1996, Wayne State University, Detroit, Michigan.

Projectile Like Fragments from ^{129}Xe + ^{nat}Cu reactions at E/A = 30, 40, 50 MeV

D.E. Russ,[1] A.C. Mignerey,[1] E.J. Garcia-Solis,[1] H. Madani,[1] J.Y Shea,[1] P.J. Stanskas,[1] O. Bjarki,[2] E.E. Gualtieri,[2] S.A. Hannuschke,[2] R. Pak,[2*] N.T.B. Stone,[2†] A.M. VanderMolen,[2] G.D. Westfall,[2] and J. Yee[2]

[1]Department of Chemistry and Biochemistry
University of Maryland College Park
College Park, MD 20742-2021
[2]National Superconducting Cyclotron Laboratory
and Department of Physics and Astronomy
Michigan State University, East Lansing MI 48824

INTRODUCTION

There has been a great deal of experimental and theoretical interest in the modes of dissassembly of highly excited nuclear matter. However, the mechanisms by which these hot nuclei are formed is also important to the study of the energy dependence of the influence of the nuclear mean field. One method to form an excited system is via a damped reaction. First observed at energies just above the Coulomb barrier, damped reactions were thought to occur only at low energies. The persistance of the damped reaction mechanism into the intermediate energy regime, between 20 and 100 MeV per nucleon, has recently been seen experimentally[1, 2]. But how high in energy do damped reactions occur? In order to measure an excitation function for damped reactions, an experiment was performed at the National Superconducting Cyclotron Laboratory (NSCL) on the campus of Michigan State University (MSU). The experiment consisted of a ^{129}Xe beam at energies of 30, 40, 50, and 60 MeV per nucleon incident on targets of ^{nat}Cu and ^{nat}Sc. Reaction products where detecting using the MSU 4π detector system[3] augmented by the Maryland Forward Array[4] (MFA), a detector that covers between 1.4° and 2.9° from the beam in the laboratory.

As signatures for the presence of damped reactions, velocity and charge distributions of projectile-like fragments (PLF) detected in the MFA are compared with the Tassan-Got stochastic nucleon exchange model[5]. The model codes Gemini[6]

---

*Present address: Dept. Of Physics and NSRL, University of Rochester, Rochester NY
†Present address: Nuclear Science Division, Lawerence Berkeley Laboratory, Berkeley, CA

and SMM[7] are used to simulate decay of the hot primary fragments formed by the Tassan-Got calculation. The Gemini calculation is a multistep sequential decay of a primary fragment, where the SMM is a one step simultaneous multifragmentation model. These two models represent two extremes in the decay of the hot system formed by the Tassan-Got calculation.

VELOCITY DISTRIBUTIONS

Velocity distributions scaled by the beam velocity are useful for comparing systems at different beam energies. The velocity distributions for ^{129}Xe + ^{nat}Cu at E/A = 30, 40, and 50 MeV are shown in Fig. 1. The circles are the data, and the lines represent the results of the Tassan-Got model calculations. The solid and the dashed lines are the results of using either Gemini or SMM, respectively, to generate secondary distributions from Tassan-Got primary distributions. In the 30-MeV ^{129}Xe system, the data fall near the Tassan-Got + Gemini calculation. As the energy increases to 40 and 50 MeV, the data look more like the Tassan-Got + SMM calculation. This is consistant with a damped reaction forming a hot primary system that decays sequentially for low beam energies, and simultaneously for higher energies.

The velocity distributions shown in Fig. 1 are integrated over all impact parameters. Since damped reactions occur in peripheral collisions, impact parameter selection should filter out other mechanisms occurring in the more central collisions. Charged particle multiplicity, m_c, as measured in the main ball of the MSU 4π detector, is used as an impact parameter selector. The multiplicity distributions shown in Fig. 2 are divided into 3 bins. Note that this is simply m_c, not the sum charge, since products with Z$\geq$2 are also detected in the ball. The first bin, representing the most peripherhal collisions, contains events with less than nine charged particles detected in the main ball of the MSU 4π; the second bin contains events with between 9 and 20 particles. The last bin contains events with over 20 charged particles, representing the most central collisions, and clearly contains charged particles from the projectile as well as the Cu target (Z=29).

The velocity distributions for each multiplicity bin are shown in Fig. 3. The circles with the solid lines are data from the most peripheral collisions. For this multiplicity bin, there appears to be a change in the shape of the velocity distributions as the incident energy is increased. The squares with the broken lines are the data in the middle multiplicity bin. For the 40-MeV data in this bin, there is an interesting plateau in the distribution. The diamonds with the dotted line are the data in the highest multiplicity bin. There is a similarity between the lowest multiplicity bin in the 50-MeV data and the mid peripherial bin in the 30-MeV data. A more central reaction is required to excite the 30-MeV system to similar energies as a peripheral reaction in the 50-MeV system.

CHARGE DISTRIBUTIONS

The centroids of the charge distributions as a function of the laboratory kinetic energy are shown in Fig. 4. The circles are the data. The lines are the result of Tassan-Got model calculation. In damped reactions for the ^{129}Xe + ^{nat}Cu system, the primary projectile-like charge distribution changes little with the laboratory kinetic energy of the

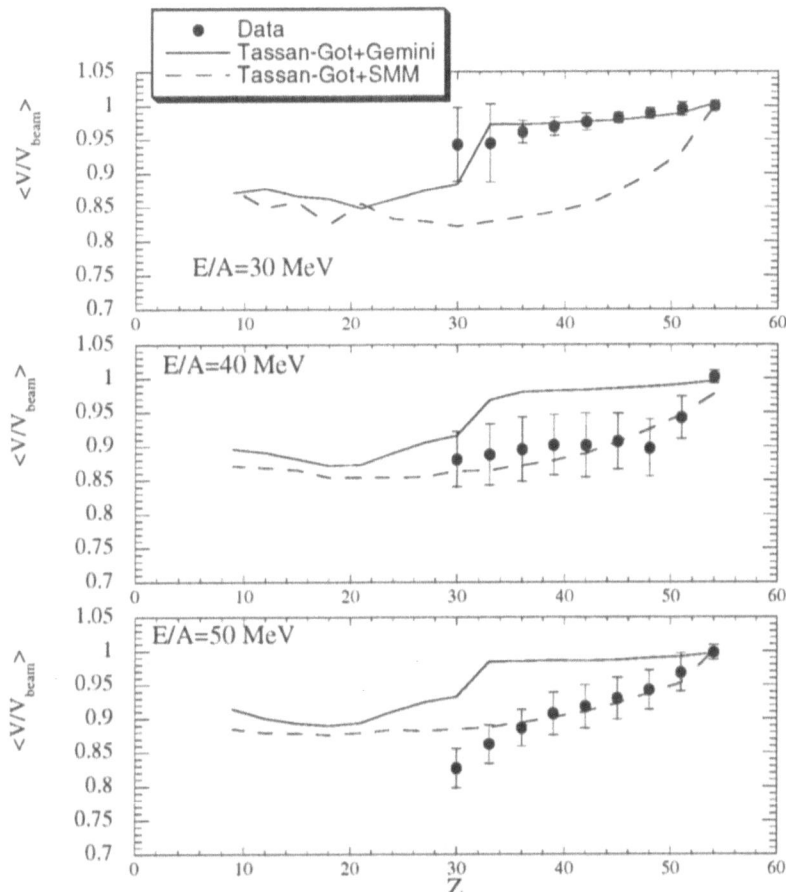

Figure 1. Velocity distributions for ^{129}Xe + ^{nat}Cu at E/A = 30, 40, and 50 MeV. The circles are the data. The lines are the results of Tassan-Got model calculations. The solid and the dashed lines are the results of using either Gemini or SMM, respectively, to generate secondary distributions from Tassan-Got primary distributions.

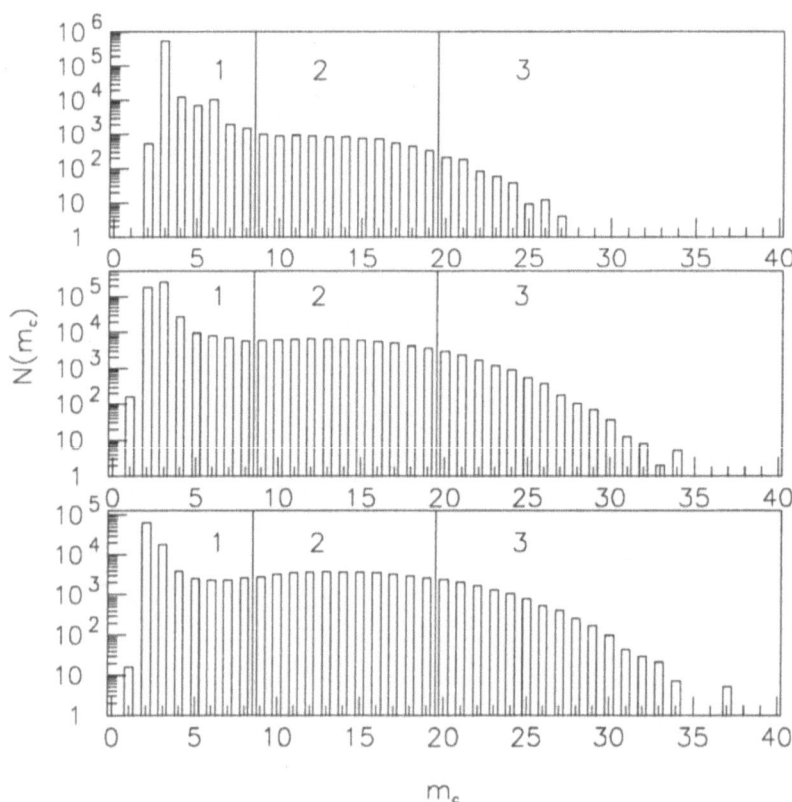

Figure 2. Charged particle multiplicity distributions for ^{129}Xe + ^{nat}Cu at E/A = 30, 40, and 50 MeV as measured by the main ball of the MSU 4π detector.

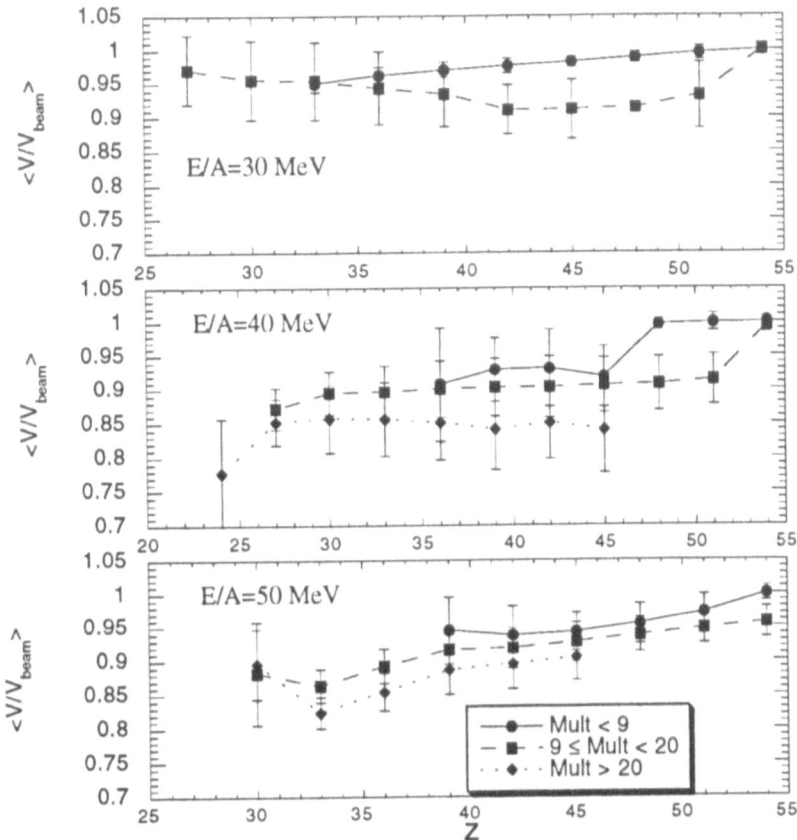

Figure 3. Velocity distributions for ^{129}Xe + ^{nat}Cu at E/A = 30, 40, and 50 MeV for different multiplicity bins as defined in Fig 2.

fragment[8]. The decrease in the observed $\langle Z \rangle$ with decreasing energy is therefore due to the secondary decay of the hot primary fragments. For the 30 MeV ^{129}Xe system, the data resembles the Tassan-Got + Gemini calculation over the entire fragment energy range. As the system energy increases, the more peripheral components (higher E) still agree with the Gemini calculation, but the intermediate energy products (with $\langle Z \rangle$ values significantly lower than the projectile Z=54) fall between the two limits. At the lowest E values (highest excitation), the calculations and data all converge. It appears as if the primary products have so much excitation that it is impossible to differentiate between the multiple sequential decay and simultaneous multifragmentation.

Conclusion

The charge and velocity distributions are in agreement with Tassan-Got's stochastic nucleon exchange model. It appears that over the energy range of study, the dominant reaction mechanism which produces forward focused products is not changing. The differences in the distributions may be accounted for in the disassembly of the hot system. This does not preclude other mechanisms which may give the same ob-

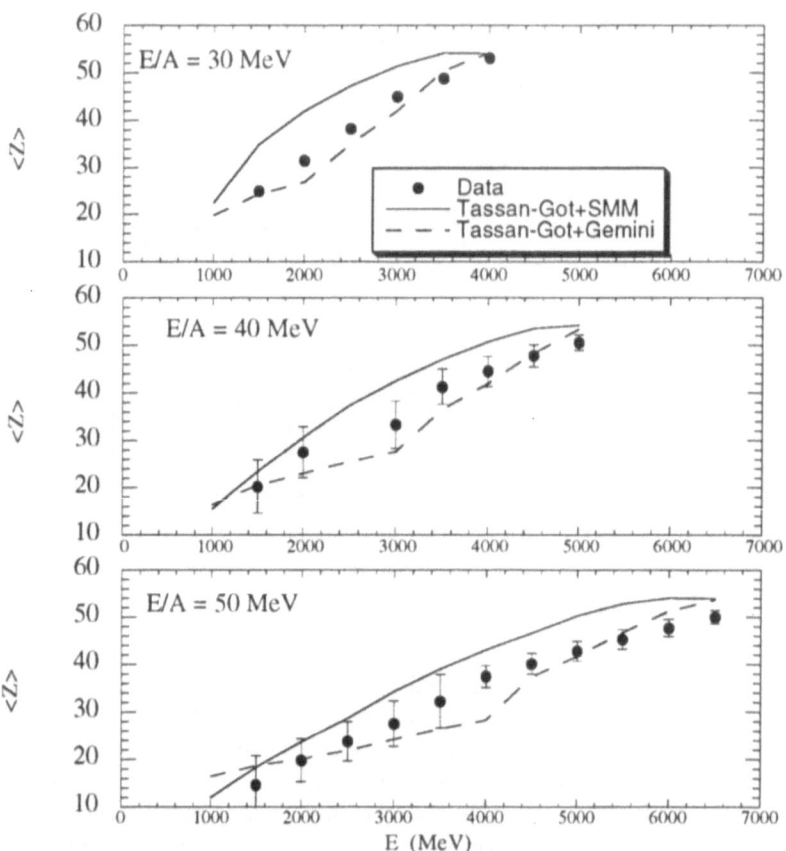

Figure 4. Charge distributions for ^{129}Xe + ^{nat}Cu at E/A = 30, 40, and 50 MeV. The circles are the data. The lines are the results of Tassan-Got model calculations. The solid and the dashed lines are the results of using either Gemini or SMM, respectively, to generate secondary distributions from Tassan-Got primary distributions.

servables or additional mechanisms for central collisions which produce particles over a much larger angular range. Continued work on the angular distributions of the PLF's will give additional evidence for the presence of damped reactions in the intermediate energy regime.

REFERENCES

1. B. Lott et al., Phys Rev. Lett. 68 3141 (1992).
2. S.P. Baldwin et al., Phys Rev. Lett. 74 1299 (1995).
3. G.D. Westfall et al., Nucl. Instr. and Meth. A238 347 (1985).
4. D.E. Russ et al., Nucl. Instr. and Meth., to be submitted, DOE/ER/40802-6 Appendix B.
5. L. Tassan-Got and C. Stéphan, Nucl. Phys. A 524 121(1991).
6. R.J. Charity, Computer Code GEMINI.
7. J.P. Bondorf et al., Phys. Rep. 257 135 (1995).
8. H. Madani et al., Advances in Nuclear Dyanmics, Edited by W. Bauer and A. Mignerey, Plenum 1996, p. 145.

RECENT RESULTS FROM EXPERIMENT E910

Vince Cianciolo[1], for the E910 Collaboration

Lawrence Livermore National Laboratory
Livermore, CA, 94551

[1] Current address:
Oak Ridge National Laboratory
MS 6372
Oak Ridge, TN 37831-6372

INTRODUCTION

BNL experiment E910 is the first high-statistics experiment with large-acceptance and particle identification to study proton-nucleus (pA) collisions at AGS energies. Our main goal is to survey these collisions for a variety of targets (Be, Cu, Au, U) over the range of energies available at the AGS (6–18 GeV/c). The results of these studies will increase our understanding of nucleus-nucleus (AA) collisions by:

1. Elucidating the mechanisms of energy deposition and particle production in excited nuclear matter.

2. Standing as baselines against which to judge "interesting" results from AA collisions.

3. Serving as vital input to cascade codes which attempt to model AA collisions.

E910 is also sensitive to H^0 dibaryons, if they are pro ' ed by a coalescence of two strange baryons and if their lifetime is short enough (c 10^{-8}sec).

EXPERIMENTAL APPARATUS

The E910 spectrometer, located in the AGS A1 l   is shown in figure 1. The heart of the experiment is the EOS TPC [1], which sits   ide the MPS dipole magnet and provides tracking and low-momentum particle iden   ation (PID, through dE/dx

measurements) over a large acceptance. Downstream of the TPC there are five drift chambers [2], two wire chambers, an atmospheric čerenkov counter and a time-of-flight scintillator hodoscope (TOF), which together provide high-momentum tracking and PID in the forward direction.

There is also a battery of beamline and trigger detectors upstream of the spectrometer. Čerenkov counters are used to ensure beam purity, wire chambers are used to measure beam momentum and locate the interaction vertex, and scintillator counters are used to provide a start-time measurement and to veto upstream interactions.

Three different "centrality" detectors were used to enrich the sample of violent collisions:

1. ST was a scintillator panel positioned immediately downstream of the target. A threshold set on the value of the total energy deposited roughly selected events with a minimum number of energetic charged particles.

2. During the run ST was replaced by SFIB, a silicon fiber hodoscope with one plane of fibers in each of two orthogonal directions mutually perpendicular to the beam. Trigger criteria for this detector were formulated in terms of a minimum number of fibers fired in each plane. Offline, SFIB also provides tracking information very close to the target.

3. The bull's eye detector (BE) was a four-element scintillator hodoscope (two slats in each of two orthogonal directions mutually perpendicular to the beam) used to select events in which the beam particle did not follow an unperturbed trajectory. This trigger was not as selective as the other two, but it allowed us to increase our low-rapidity acceptance by removing ST and SFIB and positioning the target closer to the TPC entrance. It may be possible to extend our acceptance even further (backwards of $y = 0$) by examining interactions in the TPC volume selected by the BE trigger.

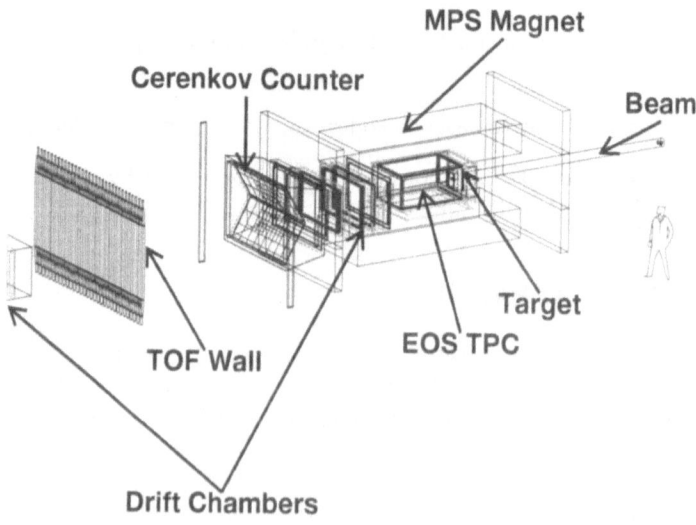

Figure 1. Schematic of the E910 experimental apparatus, see text for details.

RESULTS

All results reported here should be considered preliminary and it should be noted that they were obtained from a tiny fraction of the total data set. Nevertheless, the breadth of questions that will be addressed is clear.

Event Characterization

Figure 2 shows a cartoon of a pA collision. The projectile passes through the nucleus, undergoes a series of ν_c "primary collisions", and emerges from the nucleus as the fastest, or "leading" proton. At each primary collision there may be particle production, and the reaction products may be subject to rescattering.

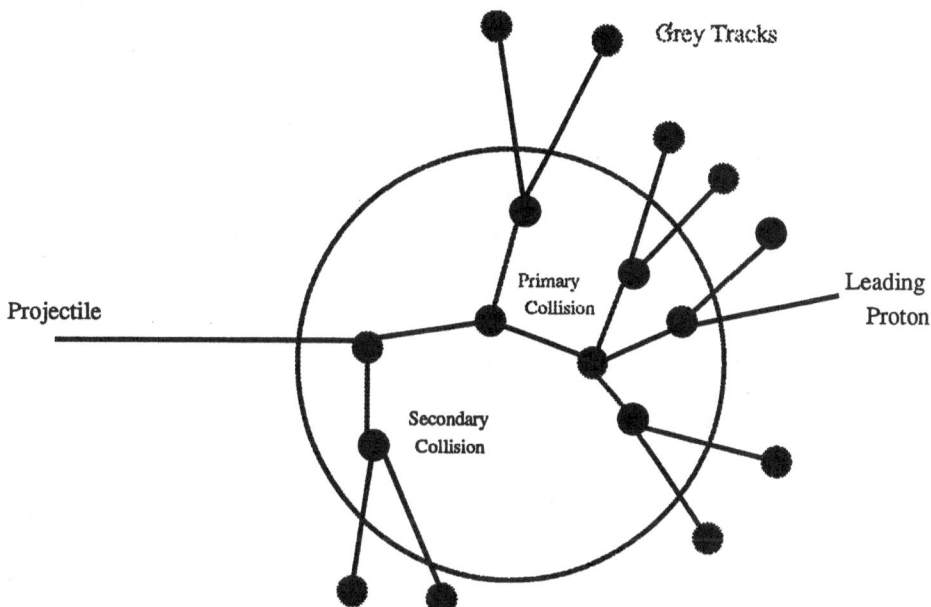

Figure 2. Cartoon of a proton-nucleus collision, see text for details. Courtesy, S. Mioduszewski.

Two of the measures of the violence of a pA collision that we are interested in are 1) the energy deposited by the projectile (usually quoted in units of rapidity loss, Δy), and 2) the number of primary collisions, ν_c. One of our major goals is to examine the relationship between these quantities and study strangeness production, anti-baryon production/absorption, etc., as functions of these quantities.

The projectile rapidity loss is determined by measuring the momentum of the leading proton, $\Delta y = y_{\text{beam}} - y_{\text{leading}}$. This measurement is complicated by the presence of high-momentum pions which can be confused for protons, and by charge-exchange reactions which turn the leading proton into a (undetectable) neutron. For this analysis, pion/proton discrimination relied on TPC particle identification, which is good at $18\,\text{GeV}/c$. For the final analysis, information from the TOF and čerenkov detectors will also be used. Charge exchange reactions are eliminated by cutting events in which more than x% of the total longitudinal momentum is lost (to neutral particles). This cut may also be refined in the final analysis.

The number of primary collisions has been related to the number of "slow protons", N_p, observed in the event, $\nu_c \propto \sqrt{N_p}$ [3, 4, 5]. To motivate this relationship, consider the number of primary collisions, which should scale like R/λ_{mfp}, and the number of secondary collisions, which should scale like $(R/\lambda_{\mathrm{mfp}})^2$, where R is the nuclear radius and λ_{mfp} is the nucleon-nucleon mean-free-path. The relationship is a simple result of identifying N_p as the number of secondary collisions. There are complicating factors dealing with slow protons resulting from tertiary collisions and failure of the naive scaling of the secondary collision probability for reaction products of primary collisions occurring at the edge of the nucleus, but these roughly cancel.

Previous studies examining this relationship [6] did not have high-momentum PID, and were therefore limited in their ability to define a "slow proton". We are free to optimize this definition, and for the results presented here a slow proton is defined to have $p < 1.4\,\mathrm{GeV}/c$. No acceptance or efficiency corrections have been applied yet.

Figure 3 shows the rapidity loss of the leading proton versus $\sqrt{N_p}$. If the relationship $\nu_c \propto \sqrt{N_p}$ holds for our energies and acceptance/efficiency, this figure shows that a projectile suffers a roughly constant amount of rapidity loss in each collision. Figure 4 shows the rapidity loss distributions for different targets. As expected, Δy increases with target mass (and thus N_p). A more detailed understanding awaits analysis of the full data set.

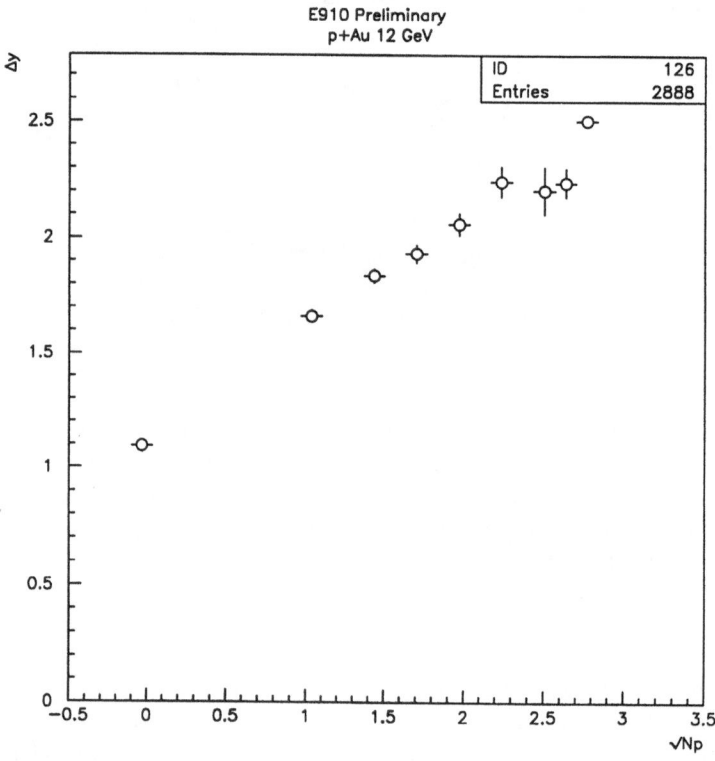

Figure 3. Leading proton rapidity loss versus the square root of the number of "slow protons", see text for details. Courtesy, S. Mioduszewski.

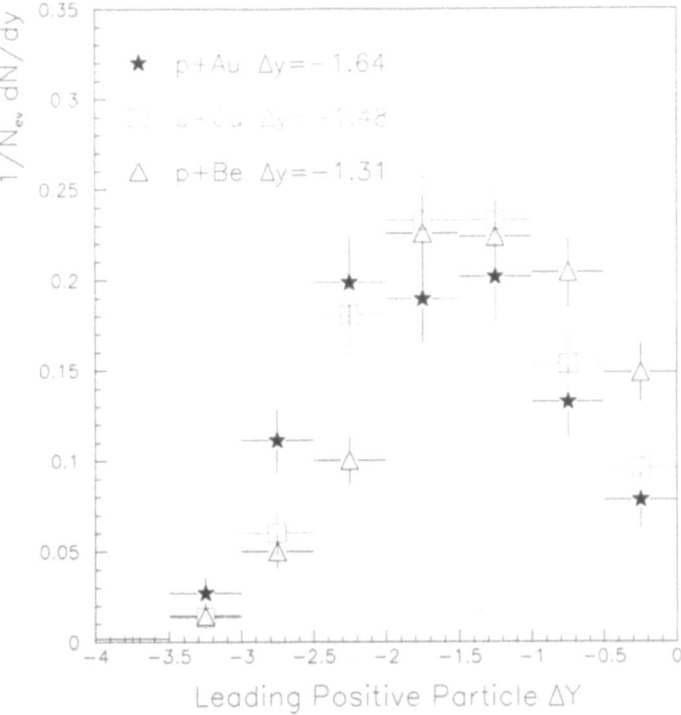

Figure 4. Leading proton rapidity loss for different target atomic mass. Courtesy, M. Rosati.

Pion Production

Low-rapidity pion production in pA collisions at these energies is unexplored territory. Not surprisingly, models differ in their prediction for the pion production cross-section $(\sigma_A^{tot}(\pi))$ in this region of phase space by more than a factor of two [7]. It is necessary to understand pion production in detail because:

- Pions are the baseline against which we measure strangeness enhancement.

- Pions are the most copious species in AA collisions making it important to correctly model their interactions in cascade codes.

Also, the value of $\sigma_A^{tot}(\pi)$ is a critical design parameter for a $\mu^+\mu^-$ collider which is being studied as a future high-energy accelerator [7, 8, 9]. The basic idea is to accelerate a proton beam into a heavy target and create pions, which would be cooled and allowed to decay into muons, which would then be accelerated to very high energy. Obviously, the achievable luminosity depends critically on $\sigma_A^{tot}(\pi)$. The contribution of E910 is made clear in figure 5, which shows the pion acceptance of experiment E802 (the experiment which came the closest to measuring the pions of interest) overlayed with a plot of reconstructed charged pions in the TPC.

In addition to charged pion spectra, not shown here for brevity, E910 can also measure neutral pions. Figure 6 is the invariant mass distribution for π^0's, reconstructed from the two e^+e^- pairs resulting from the $\pi^0 \to \gamma\gamma$ decay channel (the photons convert in the target). This new analysis shows our ability reconstruct resonances with photons in the final state.

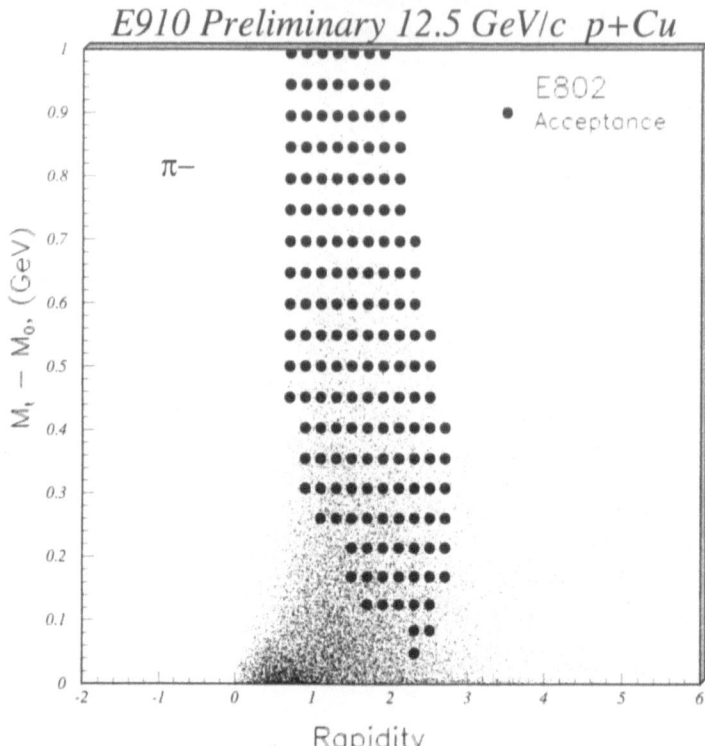

Figure 5. E910 charged-pion acceptance (transverse momentum versus rapidity). Small dots are reconstructed pions in the TPC, large spots indicate the acceptance for experiment E802. Courtesy, H. Kirk.

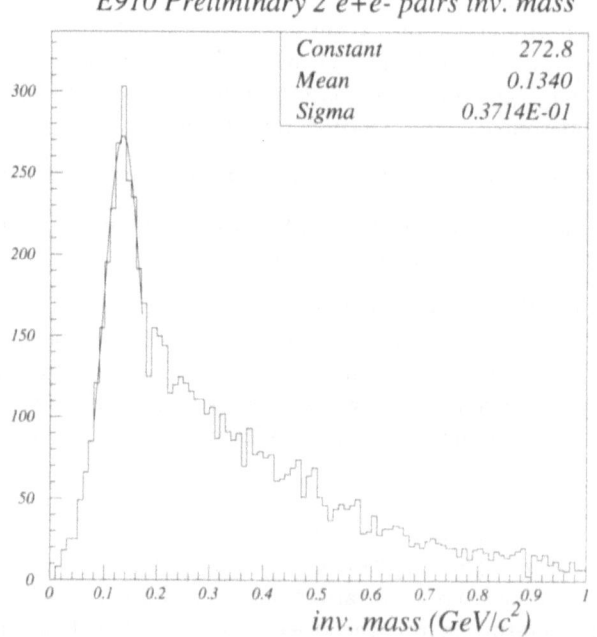

Figure 6. E910 invariant-mass distribution of π^0's reconstructed from four-electron events. Courtesy, H. Heijima.

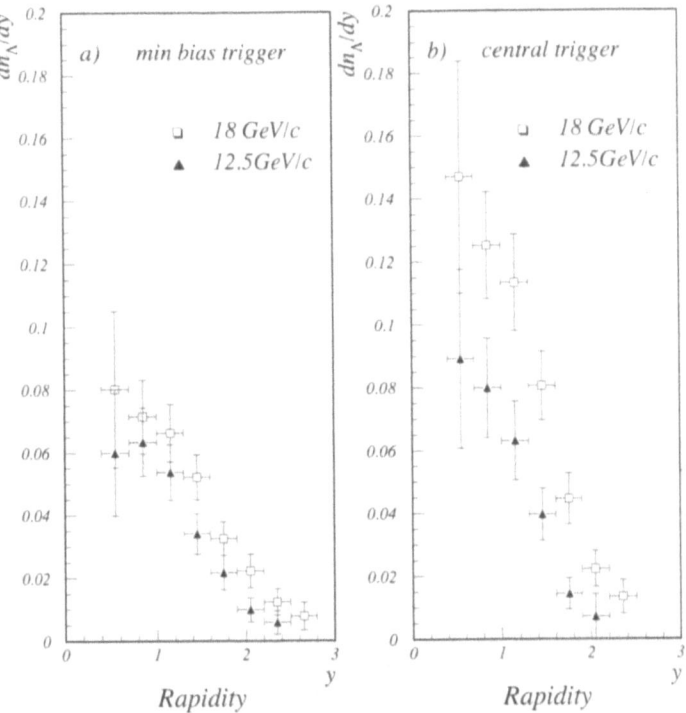

Figure 7. Centrality and beam-energy dependence of Λ Rapidity density distributions. Courtesy, X. Yang.

H⁰ Search

The H^0 dibaryon is a six-quark ($ssuudd$), $S = 0$ state, first predicted by Jaffe in 1977 [13]. Detection of the H^0 would provide the first evidence of a hadron that is neither a meson nor a baryon. Experiments to date have produced null [14, 15, 16, 17] or ambiguous [18] results. Dover first suggested coalescence of two strange baryons as a possible production mechanism [19]. pA collisions are an especially promising hunting ground for H production because of a large (relative to pp collisions) number of Λ pairs close in phase space, and a clean (relative to AA collisions) surrounding environment. Calculations using RQMD and the coalescence H^0-production mechanism predict 18 H^0's in our 18 GeV p+Au data set [20].

Strangeness Production

One of the initially intriguing results from AA collisions was the enhancement of strangeness production ($K^{\pm}/\pi^{\pm}$) relative to pp collisions [10]. Since these initial results, it has been shown that these same ratios are also enhanced in pA collisions [11] (in which we do not expect new physics, e.g., a phase transition). Furthermore, the observed enhancements may be explained by postulating multiple excitation of projectile nucleons into heavy resonances, thus storing energy for particle production. E910 will measure strangeness enhancement in great detail: charged kaon and strange resonance (K_0^s, Λ, Ξ) production, and strangeness correlations over much of phase space, and as functions of our event characterization measures. Initial studies of Λ's and K^0's by E910 are detailed in [12]. Λ rapidity density distributions are reproduced in figure 7.

CONCLUSIONS

Preliminary analysis from experiment E910 has been presented to show the breadth of questions that will be addressed by this data set. Answers to these questions are of basic interest, and are important to those studying AA collisions and to those interested in constructing a $\mu^+\mu^-$ collider.

ACKNOWLEDGMENTS

This work has been supported by the US Department of Energy under contracts with BNL (No. DE-ACO2-76-CH00016), Columbia University (No. DE-FG02-86-ER40281), Kent State University (No. DE-FG02-89ER40531), LBL (No. DE-AC03-76SF00098), LLNL (No. W-7405-ENG-48), and the University of Tennessee (No. DE-FG02-96ER40982); by the US National Science Foundation under contract with Florida State University (No. PHY-9523974); and by the Korean Ministry of Education under contract with Yonsei University (No. BSRI-96-2426).

REFERENCES

1. G. Rai, et al., A TPC detector for the study of high multiplicity heavy ion collisions, IEEE Trans. Nucl. Sci. 37:56 (1990).
2. S. Eisman, et al., The MPS II drift chamber system, IEEE Trans. Nucl. Sci. 30:149 (1983).
3. B. Andersson, I. Otterlund, and E. Stenlund, On the correlation between fast target protons and the number of hadron-nucleon collisions in high-energy hadron-nucleus collisions, . Phys. Lett. B73:343 (1978).
4. M.K. Hegab and J. Hufner, How often does a high-energy hadron collide inside a nucleus?, Phys. Lett. B105:103 (1981).
5. M.K. Hegab and J. Hufner, The distribution of grey particles in very high-energy hadron-nucleus collisions, Nucl. Phys. A384:353 (1982).
6. C. De Marzo, et al., Dependence of multiplicity and rapidity distributions on the number of projectile collisions in 200-GeV/c proton-nucleus interactions Phys. Rev. D29:2476 (1984).
7. The $\mu^+\mu^-$ Collider Collaboration, $\mu^+\mu^-$ collider: a feasibility study, BNL-52503:180 (1996).
8. D.B. Cline, Physics potential of a few hundred GeV $\mu^+\mu^-$ collider, Nucl. Instr. Meth. A350:24 (1994).
9. D.V. Neuffer, $\mu^+\mu^-$ colliders: possibilities and challenges, Nucl. Instr. Meth. A350:24 (1994).
10. T. Abbot, et al., Kaon and pion production in central Si + Au collisions at 14.6 A·GeV/c, Phys. Rev. Lett. 64:847 (1990).
11. T. Abbot, et al., Comparison of p + A and Si + Au collisions at 14.6 A·GeV/c, Phys. Rev. Lett. 66:1567 (1992).
12. X. Yang, Lambda production in p+Au collisions: preliminary results from E910 at the AGS, in Proceedings of the 1996 Workshop on Heavy-Ion Physics at the AGS (HIPAGS-96), 75 (1996).
13. R.L. Jaffe, Perhaps a stable dihyperon, Phy. Rev. Lett. 38:195 (1977).
14. A. Carroll, et al., The search for six-quark states, Phys. Rev. Lett. 41:777 (1978).
15. S. Aoki, et al., Direct observation of sequential weak decay of a double hypernucleus, Phys. Rev. Lett. 65:1729 (1990).
16. P.D. Barnes, et al., Search for the H particle: its production and weak decay, Nucl. Phys. A547:3c (1992).
17. P.H. Pile, Searches for strangelets and other exotics at the AGS, in Proceedings of the 1993 Workshop on Heavy-Ion Physics at the AGS (HIPAGS-93), 161 (1993).
18. B.A. Shabzian, et al., The observation of a stable dibaryon, Phys. Rev. Lett. 235:208 (1990).
19. C. Dover, P. Koch, and M. May, Production of the H dibaryon in relativistic heavy ion collisions, Phys. Rev. C40:115 (1989).
20. B.A. Cole, M. Moulson and W.A. Zajc, Coalescence production of H^0's in pA collisions, Phys. Lett. B350:147 (1995).

QUASI-PROJECTILE FORMATION AND DECAY COMPARISONS BETWEEN ^{58}Ni+C AND ^{58}Ni+Au REACTIONS AT 34.5A MeV

L. Gingras,[1] X. Bai,[1] G. C. Ball,[2] L. Beaulieu, [1*], D. R. Bowman,[2] B. Djerroud,[1†], D. Doré,[1‡], P. Gagné,[1] A. Galindo-Uribarri,[2] E. Hagberg,[2] D. Horn,[2] R. Laforest,[1 §], Y. Larochelle,[1] X. Qian,[1¶], R. Roy,[1] Z. Saddiki,[1] M. Samri,[1‖], C. St-Pierre,[1] and M. Vachon[1]

[1]Laboratoire de Physique Nucléaire, Département de Physique
Université Laval, Sainte-Foy, Québec, Canada G1K 7P4
[2]AECL, Chalk River Laboratories, Chalk River, Ontario
Canada K0J 1J0

1. INTRODUCTION

It is now well known that collisions between heavy ions in the Fermi energy domain produce mainly binary type events[1]-[5]. It seems that this binary character dominates even for the most violent reactions[1, 3, 4]. However, what is still not well understood is the deexcitation stage of the two principal emitters and the effects produced by the entrance channel dynamics. An important factor in this energy range is that many processes are possibly in competition and it is experimentaly difficult to isolate each of them. Processes such as the progressively vanishing fusion, binary deep inelastic collisions and the appearance of nucleon-nucleon scattering are all present in the Fermi energy range. Furthermore, detected particles could have been emitted on a large time scale from very different stages of the decay, ranging from pre-equilibrium process to evaporation. Within the statistical break-up hypothesis, where the two principal emitters are considered as thermalized nuclei, we expect only the excitation energy of each emitter, and not the way it is reached, to be a determinant quantity for the disintegration exit channels. On the other hand, typical violence of these collisions can also lead to important deformations of the two main products of the reaction. Such

---

*Present address: LBNL, Berkeley, CA 94720, USA.
†Present address: NSRL, University of Rochester, N. Y., USA.
‡Present address: IPN Orsay, BP 1, 91406 Orsay Cedex, France.
§Present address: Cyclotron Institute, Texas A&M, Texas 77843, USA.
¶Present address: GANIL, DSM/CEA-IN2P3/CNRS, BP 5027, 14021 Caen Cedex, France.
‖Present address: Département de Physique, Faculté des Sciences, Université Ibn Toufail, Kénitra, Maroc.

Advances in Nuclear Dynamics 3, edited by
Bauer and Mignerey, Plenum Press, New York, 1997

deformations were observed recently by the rupture of neck-like structures linking the reaction partners[6]-[9]. In an asymmetric collision, we could expect the biggest nucleus to sustain the largest deformation. By its subsequent disintegration toward a more stable state, it could be possible to observe the effects of such a deformation on the deexcitation mode of the nucleus.

In the present analysis, the effects of the different formation modes are investigated by comparing the decay of two nuclei of approximately the same mass, but formed in very different collisions. These nuclei are the quasi-projectiles (QP) in the following two heavy ions reactions: ^{58}Ni+C and ^{58}Ni+Au at 34.5A MeV.

2. EXPERIMENT

All the experiments presented here were performed at the Tandem Accelerator SuperConducting Cyclotron (TASCC) facility at Chalk River. A beam of ^{58}Ni at 34.5 MeV/nucleon bombarded a 2.7 mg/cm^2 gold target and a 2.4 mg/cm^2 carbon target. Charged particles issued from these reactions were detected in the CRL-Laval 4π array constituted by 144 detectors set in ten rings concentric to the beam axis and covering polar angles between 3.3° and 140°. The first four rings (3.3° to 24°) are each made of 16 plastic phoswich detectors with detection thresholds of 7.5 (27.5) MeV/nucleon for Z=1 (28). The next two rings (24° to 46°) are made of 16 CsI(Tl) crystals achieving isotopic resolution for Z=1 and 2 ions with a threshold of 2 MeV/nucleon and element identification for Z=3 and 4 ions with a threshold of 5 MeV/nucleon. The last four rings are also composed of CsI(Tl) crystals detectors set in groups of 12 per ring. Finally, three of these detectors were replaced in the third, fourth and fifth ring by Si-Si-CsI(Tl) telescope detectors achieving isotopic resolution for Z=1 to 5 particles with thresholds of 2.0 (3.1) MeV/nucleon for Z=1 (5). The main trigger for event recording was a charged particle multiplicity of at least 3 particles. More information on detectors and energy calibration can be found in Refs.[10]-[13]. Following the same systematic, two other systems are also presented in this study. They are ^{35}Cl+C and ^{35}Cl+Au reactions at 43 MeV/nucleon, where the experimental details can be found in Refs.[5, 9, 14, 15].

3. RESULTS AND ANALYSIS

3.1 Event Selection

Since the primary goal of this work is to study and compare the break-up of quasi-projectiles in different types of reactions, it is important to use a meticulous method that selects events where there is such a QP. To achieve this, the analysis must be focused on well characterized events, i.e. events where most of the charged particles are detected. For the ^{58}Ni+C system we have only kept events with total charge detected (ΣZ) of 33 or 34. With the reaction ^{58}Ni+Au, heavy fragments issued from the deexcitation of the quasi-target were not detected because of detection thresholds and $\Sigma Z=26$ to 30 was then used as a first analysis criterion. In the next step, an event shape analysis was performed to reject events composed of a single emitter. This was done only on the ^{58}Ni+C system where, within our experimental sample, the process seems possible but with a low probability. The method used consists in the construction event-by-event of

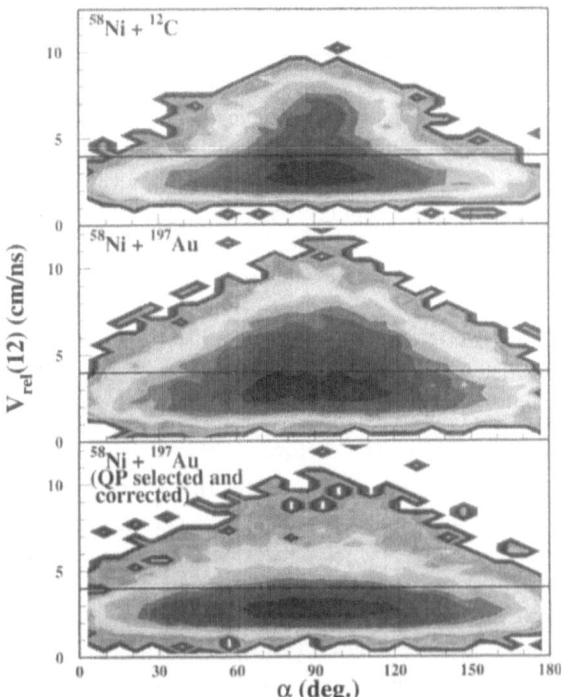

Figure 1. Logarithmic contour plot of $V_{rel}(12)$ as a function of α angle for events with $\Theta_{flow} < 50°$. The line at $V_{rel}(12) = 4$ cm/ns is the deduced limit of the QP emission.

the quadratic momentum tensor[16]:

$$P_{i,j}^2 = \Sigma_{n=1}^{N_{CP}} P_i^{(n)} P_j^{(n)}; i, j = 1, 2, 3 \tag{1}$$

where $P_i^{(n)}$ and $P_j^{(n)}$ are the ith and j^{th} Cartesian components of the nth charged particle momentum vector in the center of mass (c.m.) frame and where N_{cp} is the total number of charged particles in the event. Diagonalization of this tensor gives three eigenvalues and eigenvectors characterizing the event shape. The direction indicated by the eigenvector with the highest eigenvalue constitutes the reaction axis, where the quasi-projectile and the quasi-target, in binary events, should be aligned. The angle between this axis and the beam axis is called "flow angle" (Θ_{flow}). It was shown[5], for the ^{35}Cl+C system at 43 MeV/nucleon, that binary type events are mainly characterized by low flow angles, while single source events, although few, appear when Θ_{flow} approches 90°. This can be explained by the fact that an isotropic emission of particles coming from the c.m. velocity shows no privileged axis and that in these conditions the flow angle distribution is sinusoidal, with a maximum at 90°. However, binary events have a well defined reaction axis at low flow angles. This allows their selection, thus minimizing the quantity of one-source events, by using a second analysis condition that is: $\Theta_{flow} < 50°$.

From this step, events that are kept in the analysis sample are mostly well characterized binary type events, where a quasi-projectile was probably present in the first stages of the deexcitation. In order to describe the break-up of this quasi-projectile, and

241

especially the first stages of its break-up where dynamical effects should be stronger, for each event we have sorted particles by decreasing order of their charge and then focused on the two heaviest (particles 1 and 2) of each event. We have used here the hypothesis that the heaviest particles hold in average more information on the emission sources and the first stages of the decay. For the ^{58}Ni+C reaction, the detected particle with the highest charge (Z_1) is, in almost all cases, a projectile-like fragment. This was observed by its forward velocity ($<V(1)> \approx 1$cm/ns) in the c.m. reference frame and can be explained principally by the asymmetry level (projectile-target) of the system and by the high detection thresholds of the phoswich detectors. The same situation applies for the ^{58}Ni+Au reaction where total charge condition, centered around the projectile charge, and detection thresholds insure us that the heaviest fragment comes from the quasi-projectile. However, we must also be sure that the second particle (with charge Z_2) comes from the same emitter as the first one.

Figure 1 shows absolute relative velovity between the two heaviest particles as a function of α angle. This angle is formed by the relative velocity vector $\overrightarrow{V}_{rel}(12)$ (directed from 1 to 2) and a vector normal to the reaction plane that is defined by the reaction axis and the beam axis. Since the velocity vector of particle 1 (the heaviest particle) is in almost all cases very close to the reaction plane, α angle can tell if particle 2 is in this plane ($\alpha = 90°$) or out of it. Top panel of figure 1 shows, for the ^{58}Ni+C system, that particle 2 is in some cases a target-like fragment ($V_{rel}(12) \approx 6$ cm/ns and in-plane). Obviously, we must reject this kind of events and focus on the quasi-isotropic emission of the QP with $V_{rel}(12)$ lower than 4 cm/ns. This condition was also applied to ^{58}Ni+Au system (middle and bottom panels of fig. 1) where it has been possible to eliminate events with particle 2 coming from the target or the mid-rapidity zone.

Since we want to compare the break-up of quasi-projectiles formed differently, it is of interest to verify that the experimental set-up has the same effects on both systems. The thresholds and geometry effects are very important, because if they are not the same, they can contribute, along with the analysis conditions, to the selection of very different classes of events. However, they will be clearly similar if global character-istics of QP, such as angle and velocity, are the same. The mean velocity of heavy fragments has been evaluated in the laboratory reference frame for the two reactions studied, and both have been found very close to the same value (≈ 7.4 cm/ns). But the reaction kinematic permits the QP in peripheral events of the ^{58}Ni+Au reaction to be deflected at higher angles than in the ^{58}Ni+C reaction ($\Theta_{grazing}^{(Ni+Au)} \approx 7.7°$, $\Theta_{grazing}^{(Ni+C)} \approx 2.7°$). It is thus expected that 'many peripheral events in ^{58}Ni+C are lost near the beam axis whereas, in ^{58}Ni+Au, they can be detected. A method was then used to simulate on the ^{58}Ni+Au reaction the principal acceptance reduction effective on the ^{58}Ni+C reaction. The method used to make each system approximately equivalent for the detection effects, consists in creating an artificial hole of a few degrees around the QP velocity vector (in the lab. frame) and eliminating all particles going in this forbid-den velocity region (adapted from the correction method used in Ref.[17]). The most sensitive observable to this correction seems to be the charge of particle 1 (Z_1). The hole opening angle was then adjusted in a way that the maximum of the corrected Z_1 distribution of the ^{58}Ni+Au reaction matched the ^{58}Ni+C system. An opening angle of 4° (i.e. 2° with respect to the reconstructed QP velocity vector) was then found to be the best suited value. Lower panel of figure 1 presents effects of this correction combined with an iterative procedure that selects particles originating from the QP in

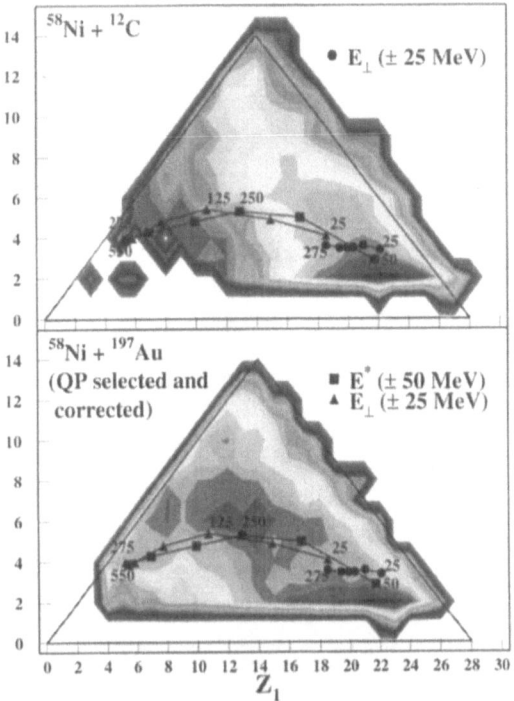

Figure 2. Logarithmic contour plot of the charge of the second heaviest particle (Z_2) versus the charge of the heaviest particle (Z_1) for events with $\Theta_{flow} < 50°$ and $V_{rel}(12) < 4$ cm/ns. Symbols are located at the average of the two quantities for different cuts in excitation observables: transverse energy in ^{58}Ni+C (circles) and in ^{58}Ni+Au reactions (triangles), and QP excitation energy in ^{58}Ni+Au reaction (squares).

peripheral events (details of this method can be found in Refs.[15, 18, 19]). As we can see on the figure, this procedure permits to eliminate most of the particles coming from the target and/or mid-rapidity zone, without modifications on the region of interest, that is: $V_{rel}(12) < 4$ cm/ns.

Finally, to verify that the conditions on flow angle and relative velocity can effectively select events where the two heaviest charges (Z_1 and Z_2) are issued from the quasi-projectile, we have summed for each event the charges of particles that are slower than the slowest between particle 1 and 2 for the ^{58}Ni+C reaction. For the complete range of transverse energy (which is a quantity related to the excitation level of the system[5]), there is always in average between five and eight such backward emitted slow particles, suggesting that the conditions have mostly eliminated events where the second heaviest particle was a target-like fragment.

3.2 Charge Correlations

To compare both systems we looked at correlations between Z_1 and Z_2. Figure 2 shows these correlations by means of logarithmic level curves. The triangular shape, emphasized by three line segments ($Z_2=0$, $Z_2=Z_1$, $Z_1+Z_2=Z_{proj}$), constitute limiting cases which should include all points if the event selection was perfect and if no net mass transfer between projectile and target had taken place. Evolution of the correlation

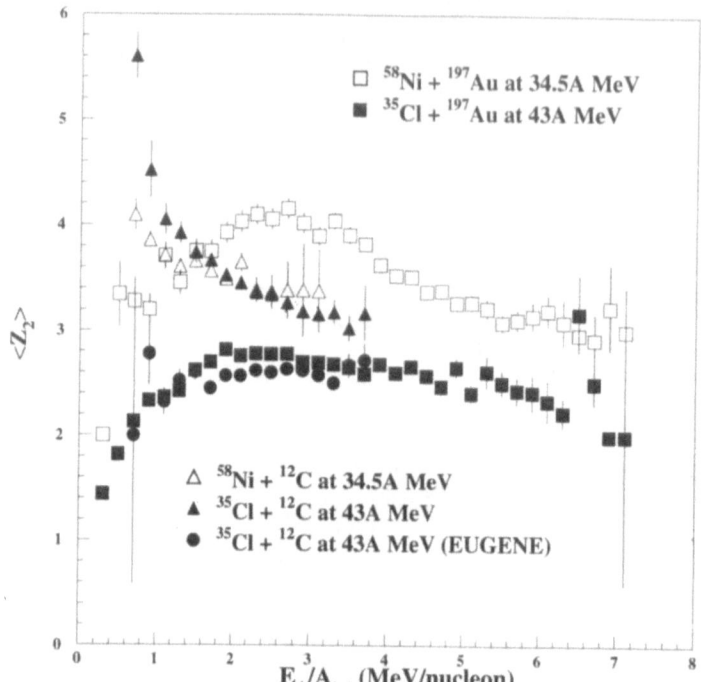

Figure 3. Average charge of the second heaviest particle (Z_2) versus average transverse energy per nucleon ($E_\perp/A_{tot}$). Analysis conditions are: $\Theta_{flow} < 50^o(60^o$ for ^{35}Cl) and $V_{rel}(12) < 4$ cm/ns.

between these quantities as a function of the system excitation has been evaluated with the total transverse energy, for both systems, as well as with the excitation energy of the quasi-projectile (calculated by calorimetry) in ^{58}Ni+Au peripheral events. Symbols on figure 2 are located at the mean coordinates in the Z_1-Z_2 plane for bands of these two types of energy distribution ($\Delta E_\perp = 50$ MeV, $\Delta E^* = 100$ MeV). On the bottom panel of fig. 2, it is clear that the correlation between Z_1 and Z_2 is quite the same whatever the quantity used to follow the system excitation evolution. For the ^{58}Ni+Au reaction, the correlation shows that the QP is mainly in a simple evaporation mode at low excitation energy (high Z_1, low Z_2). Then, the mean evaporated charge increases with a rise of excitation (and a fall of Z_1) up to a certain saturation value around 5 MeV/nucleon of excitation energy. For higher degrees of excitation, Z_1 and Z_2 decreases toward the vaporization region (low Z_1, low Z_2). For the ^{58}Ni+C reaction, the correlation (top panel of fig. 2) does not show such an evolution of the QP decay. Instead Z_2 stays almost constant with the decrease of Z_1 indicating a different decay mode.

Figure 3 presents the changes in the mean value of Z_2 as a function of the mean transverse energy per nucleon. On this figure, results of two other systems with a ^{35}Cl projectile and the same combination of targets (C and Au) were added. The two curves for both systems with the gold target (XX+Au) have the same thermal aspect (rise, saturation and fall of Z_2) except for a clear size effect between the two QP. Values of the $<Z_2>$ distribution appear to be directly proportional to the quasi-projectile charge (statistical scale invariance[14]). However, there is no noticeable size effect in the $<Z_2>$ distribution of the quasi-projectile decay in the XX+C reactions, where the two curves are very close to each other. Furthermore, they only present a decreasing function with the excitation of the system. This feature, very different in nature from XX+Au reactions, is even not reproduced by a filtered simulation of the ^{35}Cl+C system. This simulation was obtained with the code EUGENE[20], which performs an entrance

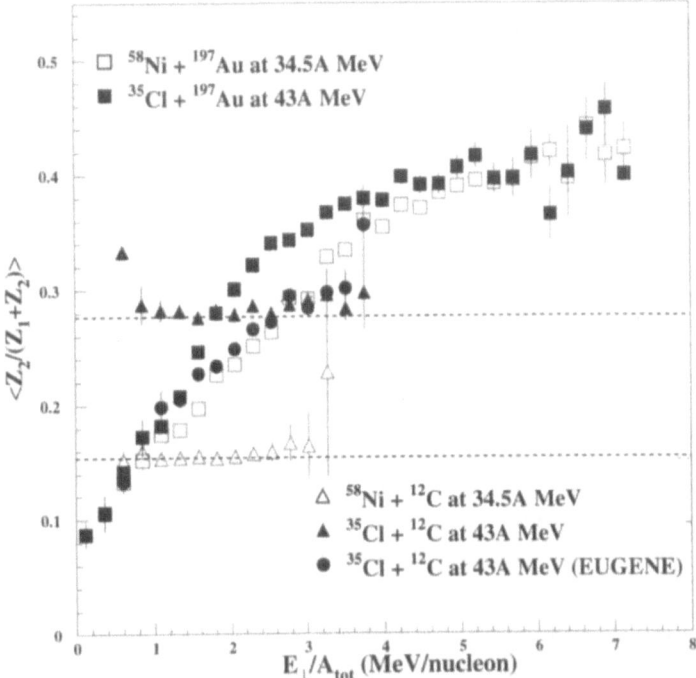

Figure 4. Average of the configuration ratio $(Z_2/(Z_1+Z_2))$ versus average transverse energy per nucleon $(E_\perp/A_{tot})$ with analysis conditions: $\Theta_{flow} < 50^o(60^o$ for ^{35}Cl$)$ and $V_{rel}(12) < 4$ cm/ns.

channel calculation and the statistical decay of two excited emitters. More details on the simulation parameters can be found in Refs.[5, 9], where the same calculation was used for the same system. It appears on the figure that results from this calculation are more compatible with data from the ^{35}Cl+Au reaction than from the ^{35}Cl+C reaction. This seems to indicate that the decay of the QP in ^{35}Cl+Au reaction can be statistically described and that a phenomenon not taken into account in the statistical model is responsible for the break-up of the QP in ^{35}Cl+C reaction, at least for the $<Z_2>$ distribution.

In order to characterize this phenomenon, the configuration ratio of particles 1 and 2 $(Z_2/(Z_1+Z_2))$ has been evaluated for all studied systems. Figure 4 shows the mean of this ratio as a function of the mean transverse energy per nucleon. For the two XX+Au systems, the ratio simply increases toward the limit value of 1/2 at the onset of the QP vaporization. However, the ratio appears to be strongly independant of system excitation for the XX+C reactions. This is probably an indication of the persistence of the non-statistical phenomenon on the whole range of excitation for accepted events in the analysis.

Within the hypothesis that the two heaviest particles of the XX+C reactions are the main remnants of one of the first stages of the QP decay (after fast nucleon-nucleon scattering processes), experimental evidences seem to point out that the relative exit configuration of this particular decay stage is on average always the same. This, combined with the fact that $<Z_2>$ is slightly smaller than Z_{target} (see fig. 3) and the observation of slower quasi-target particles, seems to point out that this special quasi-projectile break-up configuration is strongly driven by the target. Furthermore, the absence of size effects observed for the QP decay in XX+C reactions would be understood within this picture since the break-up is caused by the same target. This effect is maybe the result of a deformation of the projectile nucleus that is produced locally by the collision with a smaller nucleus (here the target). This dynamical deformation (probably beyond a saddle point configuration[21]) can then lead to an unavoidable break-up of the quasi-projectile that can be observed by a relatively recurrent decay configuration.

4. SUMMARY

In this work, we have investigated the deexcitation modes of differently formed quasi-projectiles. The first part of the analysis consisted in selecting well characterized binary events where the two heaviest particles were issued from the QP emitter. After a kinematic correction that allowed us to compare both types of reactions, the second part focused on the charge correlation evolution along the degree of excitation of the system. It was found, in variance with XX+Au reactions where we have noticed the thermal aspect of the QP decay, that the principal break-up of the quasi-projectile, in XX+C collisions, resulted in a relatively recurrent exit channel configuration. This effect, not reproduced by a statistical decay calculation, is suspected to be the result of a quasi-projectile deformation rupture induced by the local effect of the violent collision with a smaller nucleus.

ACKNOWLEDGMENTS

The authors would like to thank Dr. D. Durand (EUGENE) for providing his simulation code. This work was supported in part by the Natural Science and Engineering Research Council of Canada (NSERC) and by the Fonds pour la Formation de Chercheurs et l'Aide à la Recherche (FCAR, Québec).

REFERENCES

1.   B. Lott et al., Phys. Rev. Lett. **68**, 3141 (1992).
2.   J.F. Lecolley et al., Phys. Lett. **B 325**, 317 (1994).
3.   J. Péter et al., Nucl. Phys. **A593**, 95 (1995).
4.   Y. Larochelle et al., Phys. Lett. **B 352**, 8 (1995).
5.   L. Beaulieu et al., Phys. Rev. Lett. **77** , 462 (1996).
6.   C.P. Montoya et al., Phys. Rev. Lett. **73**, 3070 (1994).
7.   J.F. Lecolley et al., Phys. Lett. **B 354**, 202 (1995).
8.   J. Tõke et al., Phys. Rev. Lett. **75**, 2920 (1995).
9.   Y. Larochelle et al., Accepted in Phys. Rev. C **55**, (1997).
10. C. Pruneau et al., Nucl. Inst. and Meth. **A297**, 404 (1990).
11. D. Horn et al., Nucl. Instr. and Meth. in Phys. Res. **A320**, 273 (1992).
12. Y. Larochelle et al., Nucl. Instr. and Meth. in Phys. Res. **A348**, 167 (1994).
13. D. Fox, et al., Nucl. Instr. and Meth. in Phys. Res. **A374**, 93 (1996).
14. L. Beaulieu et al., Phys. Rev. C **54**, R973 (1996).
15. L. Beaulieu, Ph. D. Thesis, Université Laval, 1996 (unpublished).
16. J. Cugnon and D. L'Hote, Nucl. Phys. **A397**, 519 (1983).
17. Y. Larochelle et al., Phys. Rev. C **53**, 823 (1996).
18. P. Désesquelles et al., Phys. Rev. C **48**, 1828, (1993).
19. A. Lleres et al., Phys. Rev. C **48**, 2753 (1993).
20. D. Durand, Nucl. Phys. **A541**, 266 (1992).
21   D. Durand and B. Tamain Noyaux en Collision, Ecole Internationale Joliot-Curie de Physique Nucléaire, Maubuisson, France, 1996 11-16 Septembre, p.101, (1995).

Searching for the Quark-Gluon Plasma via Kaon Production at the AGS

Craig A. Ogilvie,[1] for the E866 experiment

[1]Physics Dept.
MIT
Cambridge, MA 02139

INTRODUCTION

The overarching goal of our current research is to determine whether anything unusual happens in an Au + Au collision at 10 A.GeV. Of the many possibilities, two are compelling: 1) Do we form a baryon-rich Quark-Gluon-Plasma (QGP)? 2) Are the properties of hadrons, their masses, cross-sections changed in a dense nuclear medium?

Before answering these questions, we need a solid understanding of the reaction dynamics, the roles played by multiple scattering, excited resonances and the balance between absorption and initial production of particles. This foundation is necessary because any signatures of new physics will be sitting in the dominant environment of hadronic rescattering.

As an overview of heavy-ion collisions at the AGS, we present the rapidity distribution of particles emitted in central collisions (Figure 1). The data are a compilation from E866[1] and E877[2] which between them cover most of the phase space of produced particles. The protons have a flat dN/dy at mid-rapidity (y_{nn}=1.6) hence much of the incoming kinetic energy has been lost, suggesting that a dense, baryon-rich system has been formed. A considerable part of this lost kinetic energy has been converted into particle production, dominated by large yields of pions (Figure 1) and also a substantial number of kaons. In this talk I will focus on kaon production, the details of which I will use to explore the reaction dynamics and search for new physics.

The kaon yields have been systematically measured as a function of the centrality and beam energy (2, 4, 6, 8 and 10.1, 10.7 A.GeV) of Au+Au collisions. By changing the impact parameter, we are controlling the fraction of matter that exists at the very highest density. By changing the beam energy we are changing the conditions, density and energy density, in the reaction zone. If new physics has a sharp onset with baryon density, then this could be observable as a discontinuity in kaon yield versus centrality and/or beam energy. More realistically, any new physics is likely to turn on smoothly and will be hard to observe just from the data alone. If we compare our data with hadronic models over a broad range of impact parameters and energy, then a smooth

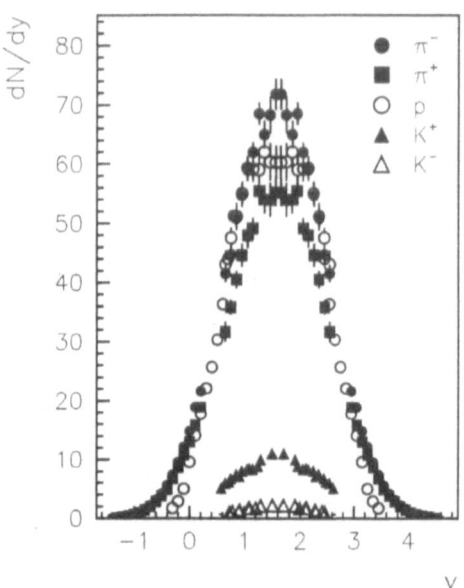

Figure 1. The measured dN/dy of protons, pions and kaons from central Au+Au at 11 AGeV/c. The data are preliminary and are from E866 and E877

onset in new physics might be observable as a discrepancy between model and data that grows with beam energy and centrality.

The third logical possibility is that the hadronic models reproduce all the data. Either there is no new physics or that information from early in the reaction, e.g. QGP formation, is lost because hadronic rescattering in the final expansion phase is too strong.

CHARACTERIZING THE EVENTS

The systematics of particle production as a function of centrality are key to the logic of this search. In E866 we have two complementary measures of the global characteristics of an event. Experimentally we measure the energy deposited in a zero-degree calorimeter that spans a cone of 1.5° around the beam axis. The energy in this acceptance (E_{ZCAL}) is dominated by nucleons from the projectile spectator. We can therefore form an approximate measure of the number of projectile participants

$$N_{pp} = 197 - E_{ZCAL}/(E_{kin}/nucl) \tag{1}$$

N_{pp} depends on the initial geometry of the collision. However, for a given initial geometry, there is still a wide range of events that have different energy densities in the participant region. Some selection within this range of events may be possible by gating on the overall multiplicity. We measure the multiplicity with a segmented, Cerenkov, lucite-array surrounding the target. For details see J. Chang's talk in this proceedings.

In Figure 2, we show the correlation between the event multiplicity and the energy deposited in the zero-degree calorimeter. Central collisions for Au + Au collisions at 11.6 AGeV/c fill the upper left corner with high multiplicity and little energy going forward into the calorimeter.

We will study kaon production in a 2-D grid of these event observables. Of particular interest is the event class that has the highest multiplicity and lowest value of

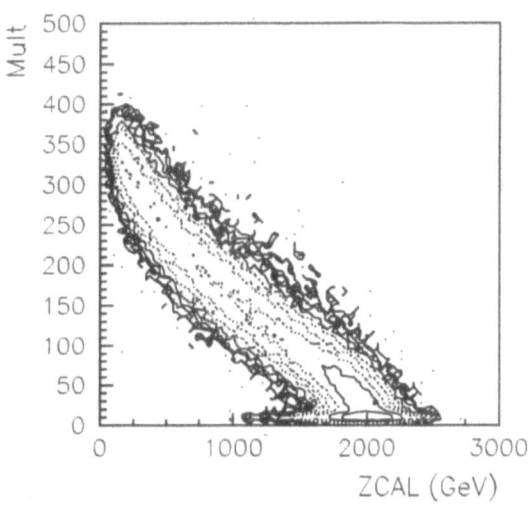

Figure 2. The measured correlation between the two global characteristic of an event, the total event multiplicity and the amount of forward-going energy E_{ZCAL}.

E_{ZCAL}. This event class contains the most violent, most geometrically central collisions with the highest chance of forming the QGP.

KAON PRODUCTION

The spectrometer we use to measure the spectra of identified particles has been described in many places[3]. Tracking in front and behind a dipole magnet provides the momentum (p) of each particle in the acceptance. A TOF-wall provides β. When combined with p, this uniquely identifies the particle. The spectrometer has a finite acceptance, but is rotatable from 14° to 50°; hence we build up coverage of phase space with four to five angle settings. The invariant cross-section of identified particles is approximately exponential in the transverse mass variable $(m_t^2 = p_t^2 + m^2)$. Such spectra can be integrated to give the average yield of particles dN/dy as a function of rapidity. For more details see Akiba et al.[1].

The dN/dy for kaons in geometrically-central events (as selected by E_{ZCAL}) is shown in Figure 3. In this data set, we are missing the coverage at back rapidities and near y_{nn}. Fitting the dN/dy with a Gaussian and integrating to get the total yield introduces an additional systematic error. Instead, we define the fiducial yield to be the sum of dN/dy over a limited range of y,

$$\text{fiducial yield} = \sum \frac{dN}{dy} \Delta y, \qquad 0.6 < y < 1.2. \qquad (2)$$

The fiducial yield of K^+ increases very smoothly with centrality as is seen in the left panel of Figure 4. There is no evidence for any change or onset of new kaon production in the most central collisions. Similar results occur for K^- production. The right panel of Figure 4 shows the same fiducial yields, but now plotted versus the inelastic cross-section corresponding to the upper range of the centrality cut.

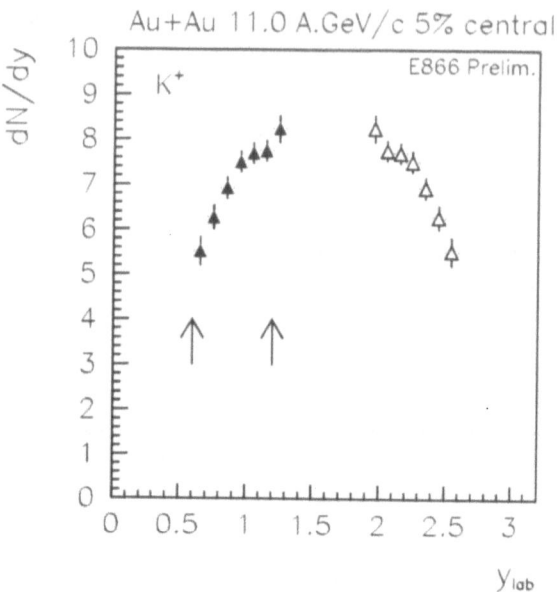

Figure 3. The measured dN/dy for K^+ from geometrically-selected central Au+Au at 11.0 AGeV/c. The data are preliminary and are from E866. The hollow symbols are reflected data points about y_{nn} and the arrows indicate the range used to calculate the fiducial yield.

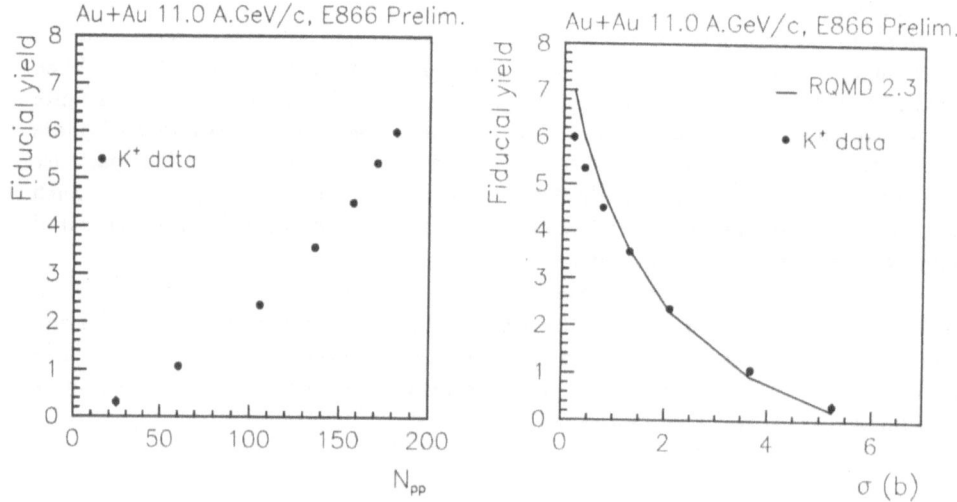

Figure 4. The measured fiducial yield (see text) versus the number of projectile participants from Au+Au at 11.0 AGeV/c. The data are preliminary and are from E866. The right panel shows the same fiducial yields, but now plotted versus the inelastic cross-section corresponding to the upper range of the centrality cut. The line is the prediction from RQMD 2.3.

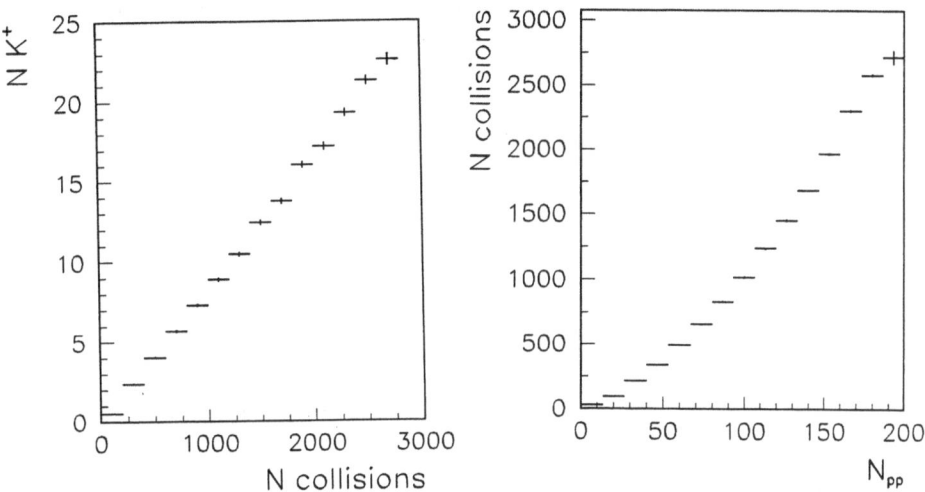

Figure 5. Left panel: the calculated correlation between the number of produced K^+ and the number of binary collisions in a minimum-bias distribution of RQMD Au+Au reactions at 11 AGeV/c. Right panel: the calculated number of binary collisions per participant nucleon. The participant nucleons are calculated via the energy deposited into a simulated ZCAL.

Two questions need to be addressed. How well do the hadronic models reproduce the magnitude and shape of the fiducial yields and what happens when we subdivide the most geometrically-central events into the different ranges of collision violence, i.e. multiplicity?

MODEL COMPARISON

RQMD (version 2.3)[4] is a hadronic cascade model that follows the trajectories of the individual hadrons, allowing them to scatter both elastically and inelastically. The inelastic collisions are modelled through the population of resonances, or at higher energies through strings and color ropes.

RQMD events have been selected on centrality by passing the event through the acceptance of the ZCAL and smearing the deposited energy with the device's resolution. The events are binned into classes based on this zero-degree energy, such that the event classes have the same part of the reaction cross-section as the experimental data.

Overall the agreement between model and data in Figure 4 is impressive, with the model fiducial yield slightly overpredicting the data for the most central collisions.

In addition to comparing the calculation to data, we would like to understand what aspect of the model drives the yield of kaons and potentially try to reveal if there is any physics in the current discrepancy between model and data.

The non-linear rise of the kaon yield with the number of projectile participants is equivalent to the observation that kaon production per participant is more effective in central reactions than in peripheral collisions. We can use the model to explore the reasons behind this. As a start to such studies, we plot the number of K^+ produced in RQMD versus the number of binary collisions in an event (Figure 5). The linear behavior implies that there is an average probability of producing a kaon in each binary collision and that this average probability is independent of the number of collisions. In

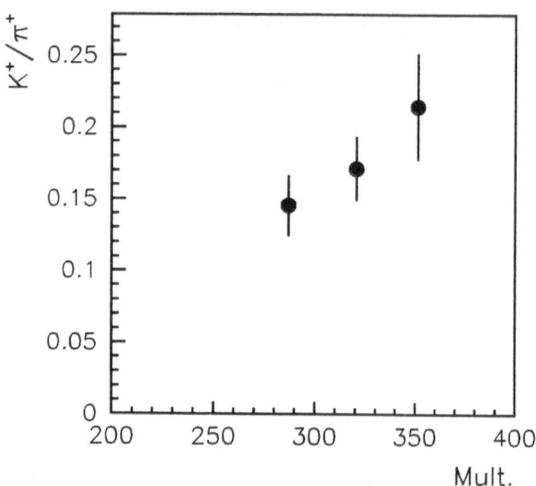

Figure 6. The ratio of fiducial yields of K^+/π^+ versus the multiplicity in geometrically-central Au+Au reactions at 11.6 AGeV/c. The data are preliminary and are from E866.

the right panel, the number of binary collisions increases non-linearly with the number of participants. Combined, this produces the non-linear rise in the model for K^+ versus participants and, most probably, the non-linear rise in the data. We plan to explore this further by dividing the collisions into meson-meson, meson-baryon etc., and by checking if the collision distribution of $\sqrt{s}$ is independent of the number of collisions. We would like to find an experimental check of this hypothesis.

KAON PRODUCTION IN MOST VIOLENT CENTRAL EVENTS

As discussed earlier, the geometrically-central events have a wide range of total multiplicity, suggesting that there is a range of energy deposited in reactions that have the same initial geometry. In Figure 6, the ratio of measured fiducial yields of K^+/π^+ increases with the event multiplicity. All the events in this sample are geometrically central, with $E_{ZCAL} < 250$ GeV. The analysis is still preliminary and is currently dominated by systematics. However, the trend suggests that the fraction of produced strange particles increases in the most violent collisions. This intriguing result will be explored with more statistics and better control of the systematics in the near future. It will also be repeated for other geometrical cuts on E_{ZCAL} and also for lower beam energies.

KAON PRODUCTION VERSUS BEAM ENERGY

As part of our systematic search for new physics, we have measured Au+Au collisions at 2, 4, 6 and 8 AGeV. The idea is to bridge the conceptual and data gap between AGS and SIS/Bevalac energies, and to search for the onset of new physics as the conditions in the participant region are changed.

In Figure 7 we plot the dN/dy at y_{nn} for K^+ as a function of beam energy. There is a smooth increase from 2 to 4 AGeV. The extrapolation to 11 AGeV is large. Definite

252

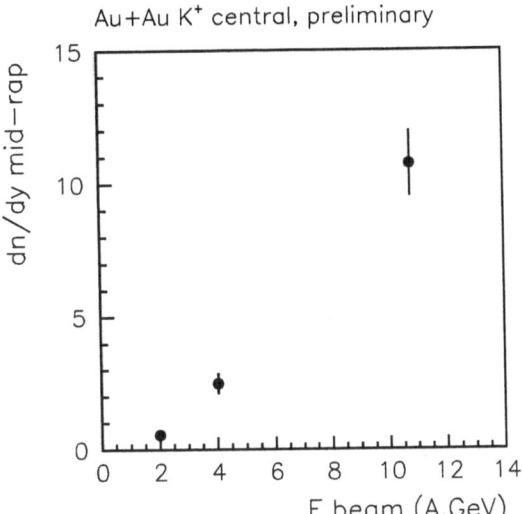

Au+Au K$^+$ central, preliminary

Figure 7. The dN/dy at y$_{nn}$ for K$^+$ as a function of beam energy. The data are preliminary and are from E866.

conclusions about the link to the highest beam energy at AGS will have to wait till we include the 6 and 8 AGeV data currently being analysed.

FLOW OF KAONS

To interpret the yield of kaons, it is important to know how strongly K$^-$ or K$^+$ are absorbed and whether their dynamics are altered by mean-field interactions with the dense medium. The mean-field interaction can be considered as a change to the effective mass of the kaons. These questions can be addressed by measuring the average component of the transverse momentum in the reaction plane, $< p^x > (y)$. For strongly absorbed particles, e.g. anti-protons or K$^-$, the asymmetric distribution of matter in coordinate space should produce an asymmetric momentum distribution. This is dubbed antiflow, because it is opposite the flow of the baryons. Anti-flow has yet to be observed for any particle species at any beam energy.

For K$^-$, an additional mean-field attraction to regions of high density is predicted[5]. This could counterbalance the absorption and change the sign of the flow till it is the same sign as the baryons. This is paired with a repulsive mean-field for K$^+$ away from the baryons leading to a prediction of reduced K$^+$ flow or even an anti-flow.

In E866 we measure the reaction plane by locating the azimuthal angle of the deflected projectile spectator. The reader is referred to H.C. Britt's talk (these proceedings) for more details.

In Figure 8, we plot the average $< p^x/p^t >(y)$ for K$^+$ in mid-peripheral collisions. The preliminary data are averaged over the transverse acceptance of the spectrometer for two angle settings, 14° and 19°. The plateau level at back rapidities between $0.6 < y < 1.4$ is one way of representing the average flow. This plateau level is plotted for K$^+$ and K$^-$ as a function of the event multiplicity in Figure 8.

For the mid-peripheral collisions (mult~200) the K$^+$ flow is smaller in magnitude than the K$^-$ flow, and both are in the same direction as the protons. The difference

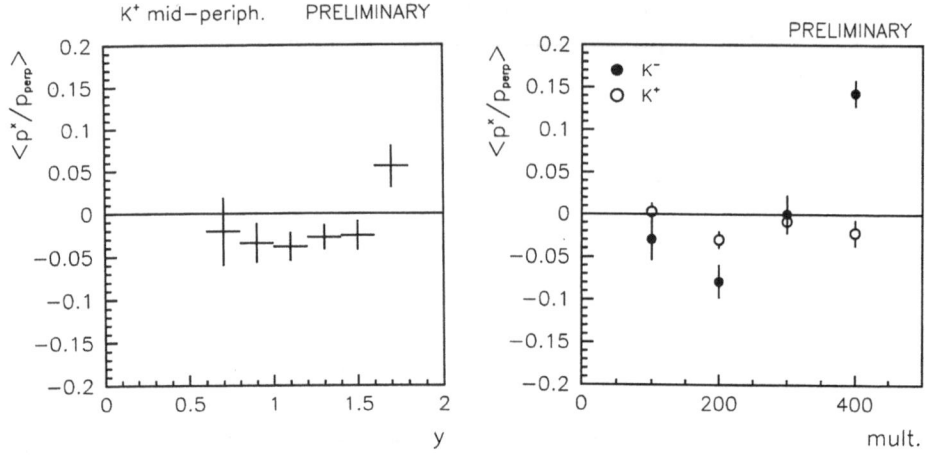

Figure 8. Left panel:The average $< p^x/p^t >$(y) for K^+ in mid-peripheral Au+Au collisions at 11.6 AGeV/c. Right panel: The back rapidity average of $< p^x/p^t >$(y) for K^+ and K^- as a function of the event-multiplicity. The data are preliminary and are from E866.

is consistent with the predicted mean-field interaction of kaons being attractive for K^- and repulsive for K^+. For more central collisions (upper 10% of the inelastic cross-section), the K^- show a distinct anti-flow, which is consistent with the absorption of K^- dominating the flow signal. It is not understood why the interplay between absorption and possible mean-field effects should change with impact parameter. We will further check our results by calculating the triple-differential cross-section of kaons, and using this to calculate the average $< p^x >$(y).

DISCUSSION

The overall question covered by this talk is whether anything unusual occurs in a Au+Au collision at AGS energies. To answer this we simultaneously needed a firm understanding of the dynamics of a HI collision and broad experimental systematics. Having a wide lever arm permits a data-driven search for the onset of new phenomena.

The yield of K^+ and K^- depend very smoothly on the initial geometry of the reaction. They rise non-linearly with the number of participants, with kaon production being more effective in the central collisions. There is no evidence for any new physics from the yields versus geometry. RQMD, a hadronic cascade, reasonably reproduces the kaon yields.

It is possible that the events that have a greater chance of forming the QGP are rarer and can be filtered by gating on the total multiplicity or violence of the event. Indeed the K^+/π^+ ratio shows that in the high multiplicity events, kaons are a greater fraction of the produced particles. This result will be compared to hadronic models and repeated for other geometrical regions and also for lower beam energies.

The flow of kaons has an intriguing centrality dependence. One interpretation is that the kaon mean-field interaction dominates the flow signal for peripheral collisions, while for the central events, the absorption of K^- on asymmetric matter dominates.

This interpretation needs to be checked with triple-differential cross-sections that do not integrate over the acceptance of the spectrometer.

The path for the next few years is very clear. There is currently no smoking gun indicating new physics, but we have just started to critically examine the rare, most-violent collisions for the QGP signatures. Sensitive probes such as flow, production of strange anti-baryons, and pion interferometry are also being actively pursued. In parallel we are steadily improving our quantitative understanding the role of reaction dynamics, multiple-scattering, and absorption in heavy-ion collisions. Combined, the extensive systematics, the use of sensitive probes, and the knowledge of the dynamics, make a very powerful search for new physics in heavy-ion collisions at the AGS.

This work is supported by the US Department of Energy under contracts with BNL,Columbia University, LLNL, MIT, UC Riverside, by NASA under contract with University of California, by the Ministry of Education and KOSEF in Korea, and by the Ministry of Education Science and Culture of Japan under Japan-US agreement on cooperation in High Energy Physics.

REFERENCES

1. Y. Akiba et al., Nucl. Phys. A610:139c (1996).
2. R. Lacasse et al., Nucl. Phys. A610:153c (1996).
3. T. Abbott et al., Nucl. Inst. Meth A290:41 (1990).
4. H. Sorge, Phys. Rev. C52:3291 (1995).
5. G.Q. Li, C.M. Ko, B.A. Li, Phys. Rev. Lett. 74:235 (1995).

INTERPRETING MULTIPLICITY-GATED FRAGMENT DISTRIBUTIONS FROM HEAVY-ION COLLISIONS

L. Phair, L.G. Moretto, Th. Rubehn, G.J. Wozniak,
L. Beaulieu, N. Colonna*, R. Ghetti†, and K. Tso,

Nuclear Science Division
Lawrence Berkeley National Laboratory, Berkeley, CA

INTRODUCTION

In recent years, multifragmentation of nuclear systems has been extensively studied, and many efforts have been made to clarify the underlying physics[1]. However, no clear consensus exists on the mechanism for multifragmentation. Is the emission of intermediate mass fragments (IMF: $3 \leq Z \leq 20$) a dynamical process (brought on by the occurrence of instabilities of one form or another) or a statistical process (i.e. the decay probabilities are proportional to a suitably defined exit channel phase space)?

Historically the charge (mass) distribution has played and still plays a very important role in characterizing multifragmentation. Since this subject's inception, the near power-law shape of the charge and mass distributions was considered an indication of criticality for the hot nuclear fluid produced in light ion and heavy ion collisions [2, 3]. More modern studies still infer critical behavior from the moments of the charge distribution [4, 5, 6, 7, 8, 9]. Furthermore, a charge distribution is readily predicted by most models and easily compared with data.

In what follows, we point out what the charge distributions might reveal regarding the mechanism of multifragmentation.

STATISTICAL SIGNATURES

Recently, it has been experimentally observed in many heavy-ion reactions that for any value of the transverse energy E_t ($E_t = \sum E_i \sin^2 \theta_i$ where E_i and θ_i are the kinetic energy and polar angle of charged particle i in an event), the n-fragment emis-

---

*INFN-Sez. di Bari, 70126 Bari, Italy.
†Department of Physics, Lund University, Sweden

Advances in Nuclear Dynamics 3, edited by
Bauer and Mignerey, Plenum Press, New York, 1997

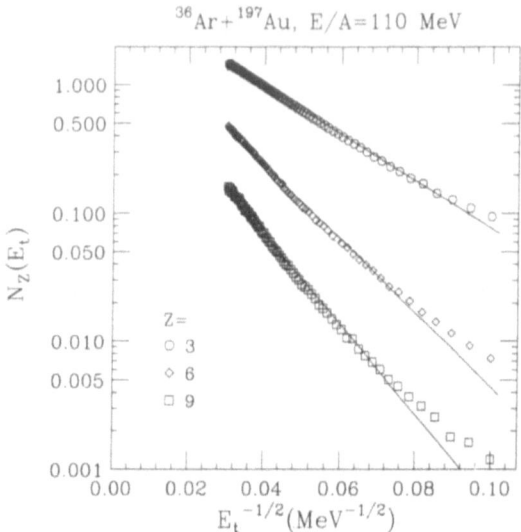

^{36}Ar + ^{197}Au, E/A=110 MeV

Figure 1. The average yield per event of lithium (circles), carbon (diamonds), and fluorine (squares) as a function of $1/\sqrt{E_t}$. The lines are linear fits to the data.

sion probability P_n is reducible to the one-fragment emission probability p through a binomial distribution [10, 11, 12]

$$P_n^m = \frac{m!}{n!(m-n)!}p^n(1-p)^{m-n}. \tag{1}$$

This empirical evidence indicates that multifragmentation can be thought of as a special combination of nearly independent fragment emissions. The binomial combination of the elementary probabilities points to a combinatorial structure associated with a time-like or space-like one-dimensional sequence [12]. It was also found that the log of such one-fragment emission probabilities ($\log p$) plotted vs $1/\sqrt{E_t}$ (Arrhenius plot) gives a remarkably straight line. This linear dependence is strongly suggestive of a thermal nature for p,

$$p = e^{-B/T} \tag{2}$$

under the assumption that the temperature $T \propto \sqrt{E^*} \propto \sqrt{E_t}$ where E^* is the excitation energy. These observations were made with data integrated over a broad range of fragment atomic numbers ($3 \leq Z \leq 20$). The difficulty of a thermal interpretation of the probability p averaged over Z was tentatively resolved by observing that if B is weakly (polynomial) dependent on Z then

$$p = \int e^{-\frac{1}{T}(B_0 + aZ^s)}dZ = \left(\frac{T}{a}\right)^{1/s}e^{-\frac{B_0}{T}} \tag{3}$$

and therefore p retains the form of Eq. (2).

These aspects of reducibility and thermal scaling in the integrated fragment emission probabilities lead naturally to the question: Is the charge distribution itself reducible and scalable?

By limiting our study to a single value of Z, the emission probabilities become small. The binomial distribution reduces to a Poisson distribution. The observed average multiplicity is now experimentally equal to the variance. Thus we are in the Poisson reducibility regime and we can check the thermal scaling directly on $\langle n \rangle$. If the charge distributions show "thermal" scaling then the average yield of a given charge should have a Boltzmann form,

$$N_Z \propto e^{-B_Z/T}, \tag{4}$$

where N_Z is the average yield of a given charge Z and B_Z is the Z dependent barrier. For a Poisson distribution $\log N_Z$ should scale linearly then with $1/\sqrt{E_t}$. This can be seen experimentally for the average yield of individual elements of a given charge (see Fig. 1) for the reaction $^{36}\text{Ar}+^{197}\text{Au}$ at E/A=110 MeV. In this case of isolating a single species, the reducibility is Poisson, and the thermal (linear) scaling with $1/\sqrt{E_t}$ is readily apparent.

The thermal nature of the charge distributions of this particular data set has been addressed before [13]. In its broadest form, reducibility demands that the probability $p(Z)$, from which an event of n fragments is generated by m trials, is the same at every step of extraction. The consequence of this extreme reducibility is straightforward: the probability distribution for IMF charges emitted from the one-fold events is the same as that for the n-fold events and equal to the singles distributions, i.e.,

$$P_{(1)}(Z) = P_{(n)}(Z) = P_{singles}(Z) = p(Z). \tag{5}$$

If the one-fold = n-fold = singles distributions is thermal, then

$$P(Z) \propto e^{-\frac{B(Z)}{T}} \tag{6}$$

or $T \ln P(Z) \propto -B(Z)$. This suggests that, under the assumption $E_t \propto E^*$ [10, 11, 12, 13, 14], the function

$$\sqrt{E_t} \ln P(Z) = D(Z) \tag{7}$$

should be independent of E_t.

In particular, since the charge distributions (i.e. the probability $P_n(Z)$ to emit an IMF of a given charge Z and a given IMF multiplicity) are exponential,

$$P_n(Z) \propto e^{-\alpha_n Z}, \tag{8}$$

we would expect for α_n the following simple dependence

$$\alpha_n \propto \frac{1}{T} \propto \frac{1}{\sqrt{E_t}} \tag{9}$$

for all folds n. Thus a plot of α_n vs $1/\sqrt{E_t}$ should give nearly straight lines. This is shown in Fig. 2 for $^{36}\text{Ar}+^{197}\text{Au}$ at E/A=110 MeV.

The expectation of thermal scaling appears to be met quite satisfactorily. For each value of n the exponent α_n shows the linear dependence on $1/\sqrt{E_t}$ anticipated in Eq. (9). On the other hand, the extreme reducibility condition demanded by Eq. (5), namely that $\alpha_1 = \alpha_2 = ... = \alpha_n = \alpha$, is not met. Rather than collapsing on a single straight line, the values of α_n for the different fragment multiplicities are offset one with respect to another by what appears to be a constant quantity.

In fact, one can fit all of the data remarkably well, assuming for α_n the form

$$\alpha_n = \frac{K'}{\sqrt{E_t}} + nc \tag{10}$$

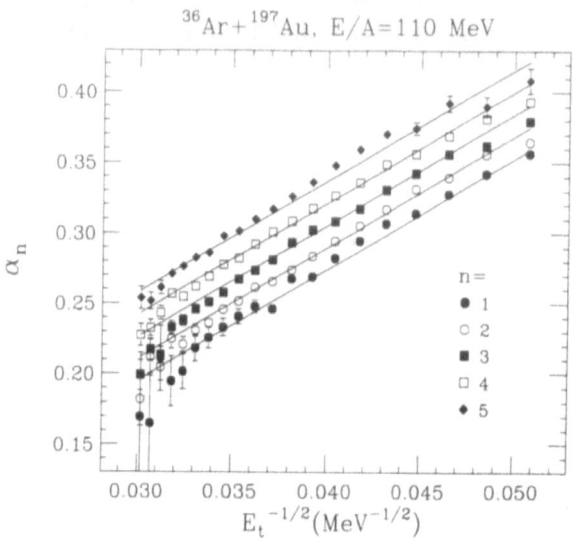

Figure 2. The exponential fit parameter α_n (from fits to the charge distributions, see Eq. (9)) is plotted as a function of $1/\sqrt{E_t}$. The solid lines are a fit to the values of α_n using Eq. (11).

which implies

$$\alpha_n = \frac{K}{T} + nc \qquad (11)$$

or more generally, for the Z distribution

$$P_n(Z) \propto e^{-\frac{B(Z)}{T} - ncZ}. \qquad (12)$$

Thus, we expect a more general reducibility expression for the charge distribution of any form to be

$$[\ln P_n(Z) + ncZ] \sqrt{E_t} = F(Z) \qquad (13)$$

for all values of n and E_t. This equation indicates that it is possible to reduce the charge distributions associated with any intermediate mass fragment multiplicity to the charge distribution of the singles [13].

We stress that the reduced quantity in Eq. (13) is *independent* of the functional form of the charge distribution. However, we have used the fact that the charge distributions are well described by exponential fits in the ^{36}Ar+^{197}Au reaction to summarize the reducibility of an enormous amount of data. Nearly one hundred different charge distributions are represented in Fig. 2.

The origin of the regular offset c has been addressed elsewhere [12, 13, 15, 16].

DYNAMICAL SIGNATURES

It has also been suggested that fragment production might be related to the occurrence of instabilities in the intermediate system produced by heavy ion collisions [17, 18, 19, 20, 21, 22, 23, 24, 25, 26, 27, 28]. In particular, two kinds of instabilities are extensively discussed in the literature: volume instabilities of a spinodal type (see e.g. Ref. [27]) and surface instabilities [18]. Spinodal instabilities are associated with the transit of a homogeneous fluid across a domain of negative pressure, where the homogeneous fluid becomes unstable and breaks up into droplets of denser liquid. Surface instabilities can be subdivided into Rayleigh or cylinder instabilities which are responsible for the decay of shapes like long necks or toroids [17], and sheet instabilities which cause the decay of bubbles or disklike structures [18]. Many models predict the formation of these exotic geometries which may develop after the initial compression of nuclei in the early stage of the collision for both symmetric and asymmetric systems [18, 19, 21, 22, 25, 26, 27, 28]. Although the scenarios and the models vary, breakup into several *nearly equal-sized* fragments has been discussed for both kinds of instabilities [29]. We have examined model independent signatures that would indicate decay into a number of nearly equal-sized fragments by investigating charge correlations from both experimental data and simulations [30].

We have experimentally studied the reactions Xe+Cu at E/A=50 MeV. The measurements were performed at the National Superconducting Cyclotron Laboratory of Michigan State University using the Miniball [31] and a Si-Si(Li)-plastic forward array [32]. Detailed information on the experiment can be found in Ref. [33].

For comparison, and to determine the sensitivity of our analysis, Monte Carlo calculations have been performed. The created events obey two conditions: the sum charge of all fragments is conserved within an adjustable accuracy, and a fragment is produced according to the probability resulting from the experimental finding, that the charge distributions for n intermediate mass fragments are nearly exponential functions [13],

$$P_n(Z) \propto \exp(-\alpha_n Z). \tag{14}$$

In our simulations, we have chosen $\alpha_n = 0.3$ (different values between 0.2 and 0.4 do not change our findings). The size of the decaying source ($Z_{source} = 83$) was chosen to be equal to the sum charge of Xe and Cu. Events with N_{IMF} equal sized fragments of charge Z_{art} were randomly added with probability P to simulate a dynamical breakup of the system into nearly equal-sized pieces. Furthermore, the charge distributions of the individual fragments from such events were smeared according to a Gaussian distribution. This smearing of the charge distribution roughly accounts not only for the width of the distribution due to the formation process, but also for the subsequent sequential decay of the primary fragments (i.e. the evaporation of light charged particles). In the following, the full width at half-maximum of this distribution is denoted by ω. We have demanded that at least 75% of the total available charge is emitted according to Eq. (14); i.e. the production of particles was stopped in the simulation once this percentage had been reached. We note that in this simple approach the transverse energy E_t has not been simulated. Furthermore, we restrict our analysis to IMFs only.

We have investigated two-particle correlations. Both the experimental and the simulated events have been analyzed according to the following method. The two-particle charge correlations are defined by the expression

$$\left. \frac{Y(Z_1, Z_2)}{Y'(Z_1, Z_2)} \right|_{E_t, N_{IMF}} = C[1 + R(Z_1, Z_2)]|_{E_t, N_{IMF}}. \tag{15}$$

Here, $Y(Z_1, Z_2)$ is the coincidence yield of two particles of atomic number Z_1 and Z_2 in an event with N_{IMF} intermediate mass fragments and a transverse energy E_t. The background yield $Y'(Z_1, Z_2)$ is constructed by mixing particle yields from different coincidence events selected by the same cuts on N_{IMF} and E_t. The normalization constant C ensures equal integrated yields of Y and CY'.

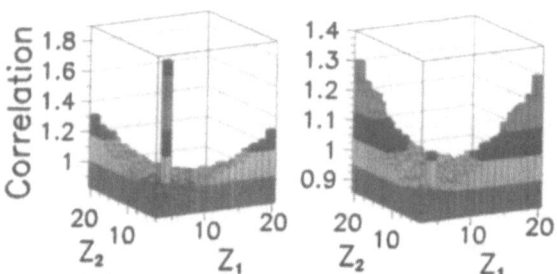

Figure 3. Two-particle charge correlations of IMFs from simulations investigating events with $N_{IMF} = 6$ and a source size of 83. Randomly, 1% (left panel) and 0.1% (right panel) of the events were chosen to have equal sized fragments ($Z_{art} = 6$).

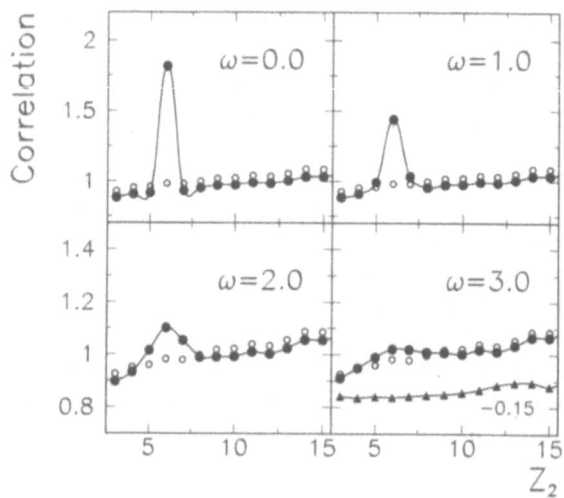

Figure 4. Two-particle charge correlations resulting from simulations for $Z_1 = 6$ as a function of the fragment charge Z_2 for different values of the width of the charge distribution ω. Randomly, 1% of the events were chosen to have nearly equal sized fragments (full circles). For comparison, we have also plotted a calculation where no additional events with equal-sized fragments have been added (open circles). Experimental results for the case $N_{IMF} = 6$ are shown in the right lower panel (full triangles). For clarity, these values are vertically shifted by a value of -0.15. The error bars are smaller than the size of the symbols.

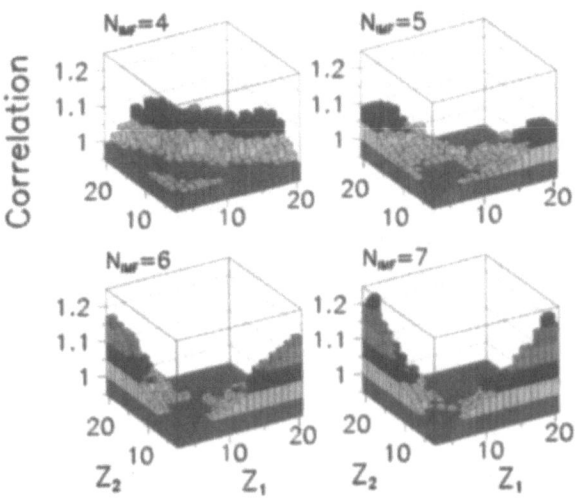

Figure 5. Experimental two-particle charge correlations for the reaction Xe+Cu at E/A=50 MeV. The different figures correspond to N_{IMF} cuts between 4 and 7.

To demonstrate the sensitivity of our method to breakup configurations producing equal-sized fragments, we show in Fig. 3 the results of simulations for the case $N_{IMF} = 6$. Here, a "contamination" of 1% of the events consisting of fragments which all have the size $Z_{art} = 6$ has been added to the data set. The peak produced by these fragments is clearly visible, even if we decrease the yield of equal-sized fragments to only 0.1%. A different choice of Z_{art} does not change our results; higher values would produce an even larger signal since the denominator value is smaller.

The magnitude of the peak shown in Fig. 3 depends not only on the yield, but also on the width of the charge distribution of the nearly equal-sized fragments. In Fig. 4, we show the correlation functions (solid circles) for different widths ω and for $Z_1 = 6$. For comparison, we have plotted the results of a calculation (open circles) where no additional events with equal-sized fragments have been added: As expected, a dependence of the size of the peak on the smearing can be observed which limits the sensitivity of the two-particle correlation functions to an enhancement of events where the charge distribution is relatively narrow. Thus, for possible secondary decay resulting in large values of ω, our analysis might not be adequate. The same analysis used for the simulation has been applied to experimental data. In Fig. 5, we show the results for central collisions (top 5% of events sorted by E_t) of Xe+Cu at E/A=50 MeV for different N_{IMF} cuts. With higher fragment multiplicity the distribution peaked along the line $Z_1 + Z_2 \approx 30$ changes into a distribution peaked at values where one fragment is heavy and its partner is light. However, an enhanced signal for breakup into nearly equal-sized fragments (a signal appearing along the diagonal) was not observed in *any* of the N_{IMF} bins. As an example, we show in Fig. 4 the experimental two-particle correlation function vs. Z_2 for $N_{IMF} = 6$ and $Z_1 = 6$ (triangles).

Furthermore, we have investigated the correlation functions obtained by our simulations without enhanced breakup for several IMF multiplicities. The evolution of the shape of the distribution with increasing values of N_{IMF} is very similar to that observed in the experimental data of Fig. 5. Simulations with different system sizes show that the charge correlations decrease as Z_{source} increases; this can be attributed to the definition of an IMF $(3 \leq Z_{IMF} \leq 20)$ relative to Z_{source}. We have also performed calculations using a percolation code and have observed a dependence similar to that presented in Fig. 5. To further study the evolution of the distributions' shape with multiplicity, we have investigated the breakup of an integer number Z_0 (chain) into n pieces. The calculated two-particle correlation functions for different multiplicities n have an evolution with n similar to that shown in Fig. 5. These findings suggest that the observed experimental evolution of the shape of the two-particle charge correlation distribution with fragment multiplicity may be due to the limited number of possibilities to create fragments if charge is conserved and the number of fragments is fixed. A signal of enhanced emission will sit on top of such a background.

CONCLUSIONS

In conclusion, we have studied different aspects of the charge distributions. The implications of the experimental evidence presented above are potentially far reaching.

On the one hand, the thermal features observed in the n-fragment emission probabilities for the ^{36}Ar$+^{197}$Au reaction [10] extend consistently to the charge distributions and strengthen the hypothesis of the important role of phase space in describing multifragmentation.

On the other hand, we have investigated charge correlation functions of multifragment decays to search for the enhanced production of nearly equal-sized fragments predicted in several theoretical works. The analysis of experimental data for the reactions Xe+Cu and Xe+Au at E/A=50 MeV and Ar+Au at E/A=50 and 110 MeV, however, shows no evidence for a preferred breakup into nearly equal-sized fragments. Recently, two groups have reported experimental signatures of possible formations of non-compact geometries in the reactions ^{86}Kr on ^{93}Nb at E/A=65 MeV and Pb+Au at E/A=29 MeV, respectively [34, 35]. It would be interesting to analyze these data using the method outlined above.

ACKNOWLEDGEMENTS

This work was supported by the Director, Office of Energy Research, Office of High Energy and Nuclear Physics, Nuclear Physics Division of the US Department of Energy, under contract DE-AC03-76SF00098 and by the National Science Foundation under Grant Nos. PHY-8913815, PHY-90117077, and PHY-9214992.

REFERENCES

1. L.G. Moretto and G.J. Wozniak, Ann. Rev. of Nucl. & Part. Sci., **43**, 379 (1993).
2. J.E. Finn *et al.*, Phys. Rev. Lett. **49**, 1321 (1982).
3. A.D. Panagiotou *et al.*, Phys. Rev. Lett. **52**, 496 (1984).
4. X. Campi, Phys. Lett. **B208**, 351 (1988).
5. W. Bauer, Phys. Rev. **C38**, 1297 (1988).
6. W. Trautmann, U. Milkau, U. Lynen, and J. Pochodzalla, Z. Phys. **A344**, 447 (1993) and refs. therein.
7. T. Li et al., Phys. Rev. Lett. **70**, 1924 (1993).
8. P. Kreutz *et al.*, Nucl. Phys. A **556**, 672 (1993).

9.   M. L. Gilkes *et al.*, Phys. Rev. Lett. **73**, 1590 (1994).

10. L.G. Moretto et al., Phys. Rev. Lett. **74**, 1530 (1995).

11. K. Tso *et al.*, Phys. Lett. B **361**, 25 (1995).

12. L.G. Moretto *et al.*, Phys. Rep., in press.

13. L. Phair *et al.*, Phys. Rev. Lett. **75**, 213 (1995).

14. L. Phair *et al.*, Phys. Rev. Lett. **77**, 822 (1996).

15. A. Ferrero *et al.*, Phys. Rev. C **53**, R5 (1996).

16. L.G. Moretto *et al.*, Phys. Rev. Lett. **76**, 372 (1996).

17. U. Brosa, S. Grossmann, A. Müller, and E. Becker, Nucl. Phys. **A 502**, 423c (1989); Phys. Rep. **197**, 167 (1990).

18. L.G. Moretto, K. Tso, N. Colonna, and G.J. Wozniak, Nucl. Phys. **A 545**, 237c (1992); Phys. Rev. Lett. **69**, 1884 (1992).

19. W. Bauer, G.F. Bertsch, and H. Schulz, Phys. Rev. Lett. **69**, 1888 (1992).

20. D.H.E. Gross, B.A. Li, and A.R. DeAngelis, Ann. Physik **1**, 467 (1992).

21. S.R. Souza and C. Ngô, Phys. Rev. C **48**, R2555 (1993).

22. H.M. Xu *et al.*, Phys. Rev. C **48**, 933 (1993).

23. L. Phair, W. Bauer, and C.K. Gelbke, Phys. Lett. **B 314**, 271 (1993).

24. T. Glasmacher, C.K. Gelbke, and S. Pratt, Phys. Lett. **B 314**, 275 (1993).

25. B. Borderie, B. Remaud, M.F. Rivet, and F. Sebille, Phys. Lett. **B 302**, 15 (1993).

26. S. Pal, S.K. Samaddar, A. Das, and J.N. De, Phys. Lett. **B 337**, 14 (1994).

27. Ph. Chomaz, M. Colonna, A. Guanera, and B. Jacquot, Nucl. Phys. **A 583**, 305c (1995).

28. D.O. Handzy *et al.*, Phys. Rev. C **51**, 2237 (1995).

29. M. Bruno *et al.*, Phys. Lett. **B 292**, 251 (1992); Nucl. Phys. **A 576**, 138 (1994).

30. L.G. Moretto *et al.*, Phys. Rev. Lett. **77**, 2634 (1996).

31. R.T. de Souza *et al.*, Nucl. Inst. Meth. **A 311**, 109 (1992).

32. W.C. Kehoe *et al.*, Nucl. Inst. Meth. **A 311**, 258 (1992).

33. D.R. Bowman *et al.*, Phys. Rev. C **46**, 1834 (1992).

34. N.T.B. Stone *et al.*, Phys. Rev. Lett. **78**, 2084 (1997).

35. D. Durand *et al.*, submitted to Phys. Lett. B; J.F. Lecolley *et al.*, submitted to Phys. Lett. B.

INDEX

Hot nuclei, 189
Hyperon production, 205

Interferometry, 35; see also Hanbury-
        Brown Twiss
Intermediate mass fragments, 75;
        see also Fragments
Instability, 257
Isospin
        dependence, 67, 129, 171
        equilibration, 67
Isotope ratio, 53

Kaon, 123, 247

Lambda, 123
Limiting fragmentation, 197

Molecular dynamics, 83; see also
        Fermionic Molecular
        Dynamics
Monte-Carlo , 9
Multifragmentation, 75, 83, 91, 257
        hadron-induced, 197
        heavy-ion induced, 223
        limiting 197
Multiplicity, 67, 145
        charged-particle, 75
        gated, 257
        neutron, 75

n/p ratio, 67
NA44, 99
NA49, 61
Non-equilibrium, 1
Nucleon
        emission, 67
Neutron
        multiplicity, 75
Neutron-rich nuclei, 171

Pairs, 137
Particle production, 17, 61
        exclusive, 17
Phase transition, 1, 91
PHOBOS, 179
Positron, 137
Pre-equilibrium, 67; see also Equilibrium
Pressure, 1
Projectile-like fragments, 223
Proton
        emission, 205
        spectra, 61
        stopping, 61

transverse mass spectrum, 99
Pseudorapidity, 145

QCD: see Quantum chromodynamics
Quantum chromodynamics, 9
Quantum electrodynamics, 137
Quark-gluon-plasma, 161, 179, 231, 247
Quark matter, 1, 161
Quasi-projectile, 239

Reaction
        mechanism, 239
        plane, 35, 153
Relativistic Heavy Ion Collider, 161, 179
Relativistic Quantum Molecular
        Dynamics, 1
Resonance matter, 1
RHIC: see Relativistic Heavy Ion
        Collider
RQMD: see Relativistic Quantum
        Molecular Dynamics

Scaling,
        global, 45
        thermal, 257
Scattering,
        secondary, 9
Semihard process, 115
Sequential, 67
Silicon detector, 179
Solenoidal Tracker at RHIC, 161
Spectrometer, 179
Spin physics, 161
SPS, 9, 35, 61, 123; see also CERN
Squeeze-out, 153
STAR: see Solenoidal Tracker at RHIC
Statistical, 91
Strangeness, 107, 123
Sweeper magnet, 205

Temperature
        apparent, 91
        measurement, 91
        nuclear 53, 91
Thermal scaling, 257
Time Projection Chamber, 17, 61, 231
TPC: see Time Projection Chamber
Transport model, 107
Transverse flow, 129; see also Flow
Triton emission, 215

Ultra-relativistic: see Collisions

WA98, 35